D. Readey

Springer
Proceedings in Physics 5

Springer Proceedings in Physics

Springer Proceedings in Physics is a new series dedicated to the publication of conference proceedings. Each volume is produced on the basis of camera-ready manuscripts prepared by conference contributors. In this way, publication can be achieved very soon after the conference and costs are kept low; the quality of visual presentation is, nevertheless, very high. We believe that such a series is preferable to the method of publishing conference proceedings in journals, where the typesetting requires time and considerable expense, and results in a longer publication period. Springer Proceedings in Physics can be considered as a journal in every other way: it should be cited in publications of research papers as *Springer Proc. Phys.*, followed by the respective volume number, page and year.

Physics of Finely Divided Matter

Proceedings of the Winter School
Les Houches, France, March 25 – April 5, 1985

Editors: N. Boccara and M. Daoud

With 168 Figures

Springer-Verlag
Berlin Heidelberg New York Tokyo

Professor Nino Boccara
Université Scientifique et Médicale de Grenoble, Centre de Physique, Côte des Chavants
F-74310 Les Houches, France

Dr. Mohamed Daoud
Commissariat à l'Energie Atomique Service de Spectrometrie et Diffraction Neutroniques
C.E.N. Saclay, F-91191 Gif-sur-Yvette Cedex, France

ISBN 3-540-15885-5 Springer-Verlag Berlin Heidelberg New York Tokyo
ISBN 0-387-15885-5 Springer-Verlag New York Heidelberg Berlin Tokyo

Library of Congress Cataloging-in-Publication Data. Main entry under title: Physics of finely divided matter. (Springer proceedings in physics; v. 5). Includes bibliographies and index. 1. Matter - Properties - Congresses. 2. Colloids - Congresses. 3. Porous materials - Congresses. 4. Percolation - Congresses. 5. Chemistry, Physical and theoretical - Congresses. I. Boccara, Nino. II. Daoud, M (Mohamed), 1947- . III. Centre de physique des Houches. IV. Series.
QC173.28.P47 1985 541.3′45 85-22254

Offset printing: Weihert-Druck GmbH, 6100 Darmstadt
Bookbinding: J. Schäffer OHG, 6718 Grünstadt
2153/3130-543210

Preface

The Second Winter School on the "Physics of Finely Divided Matter" was held at the Centre de Physique des Houches from 25 March to 5 April 1985. This meeting brought together experts from the areas of gels and porous media. People with different backgrounds - chemists, physicists - from university as well as industrial laboratories, had the opportunity to compare their most recent experimental and theoretical results. Although the experimental situations and techniques may seem at first sight unrelated, the theoretical interpretations are very similar and may be divided roughly into two categories: percolation and aggregation. These are present for the description of the synthesis of some gels as well as for a description of the structure of packings. They are also a precious help for understanding flows in porous media and hydrodynamic instabilities such as viscous fingering.

A different aspect, still in its early stages, deals with the influence of a random medium on a phase transition. This leads to metastable states and is interpreted in terms of random fields.

The following topics were covered:

- introduction to physical and chemical gels
- structure of packings and porous media
- microemulsions
- percolation
- aggregation
- elastic and dielectric properties of ill-connected media
- properties of gels near and far from the gelation threshold
- flow, diffusion and dispersion in porous media
- transitions in porous media.

Most of these are rapidly growing subjects, and we hope that these proceedings will serve as a reference for those entering this fascinating area.

This Conference was supported by the French Ministry of Research and Technology (Department of Mechanics), the French Ministry of Defense - D.R.E.T., The Centre National de la Recherche Scientifique - PIRMAT, M.P.B., chimie, S.P.I., Commissariat à l'Energie Atomique - DPh.G -, and by Exxon Research and Engineering Co. In these difficult budgetary times, we wish to express our deep gratitude to them for their efforts.

The program of this Winter School was put together by a scientific committee whose members were: H. BENOIT, S. CANDAU, M. DAOUD, P.G. de GENNES, E. GUYON and T.A. WITTEN. We wish to thank them for their wonderful work. It was proposed by the director of the of the Centre, N. BOCCARA, to the members of the council: M. TANCHE, J.C. LACOUME, P. AVERBUCH, R. BALIAN, N. BOCCARA, R. CARIE, C. DEWITT, J.P. HANSEN, S. HAROCHE, M. JACOB, R. MAYNARD, R. PAPON, Y. ROCARD, R. ROMESTAIN, R. STORA, D. THOULOUZE, N. VINH MAU, and G. WEILL, who accepted the project. We wish to thank them for the confidence they had in us.

Les Houches, France
July 1985

N. Boccara
M. Daoud

Contents

Part I Introduction to Chemical and Physical Gels

Covalent Macromolecular Gels

Henri Benoit
Institut Charles Sadron (CRM-EAHP)-(CNRS-ULP Strasbourg), 6, rue Boussingault, F-67083 Strasbourg Cedex, Fance

In this paper I would like to present a summary of the work which has been made on gels in Strasbourg during the last decade. Since this is a meeting of physicists, and since the discussion of the properties of gel is strongly related to their preparation methods, the first part of this paper will be devoted to the chemistry of the preparation of gels ; in the second part, some of the thermodynamical properties as well as neutron scattering results will be discussed.

Network preparation

We define a network which in solution gives a gel as an ensemble of macromolecules chemically linked one to the other. In order to prepare gels one has two groups of methods : the first one consists in the simultaneous preparation of the chain and the crosslinks, the second in the preparation of the chains, followed by the crosslinking reaction. Since they all involve polymerization, it is therefore necessary to begin with a summary of the different techniques of polymerizations [1][2].

A. Radical polymerization[3]

a) Propagation reaction

Its principle is the following

$$R\,M^{*} + M \rightarrow R\,M\,M^{*}$$

The star indicates that the monomer M has been activated into a radical. A classical example of this reaction is the polymerization of polystyrene. The monomer styrene $CH_2 = CH\,\Phi$ where Φ is a phenyl group reacts like :

$$CH_2\text{-}\underset{\Phi}{CH}^{*} + CH_2 = \underset{\Phi}{CH} \quad CH_2\text{-}\underset{\Phi}{CH}\text{-}CH_2\text{-}\underset{\Phi}{\overset{*}{CH}}$$

This chain reaction is usually very fast, and in classical reaction conditions the life time of a reactive chain is of the order of seconds or less.

A few remarks have to be made :

i) There is a possibility to have what is called "head to head" addition giving for the chain the structure

$$- CH_2-\underset{\Phi}{CH}-\underset{\Phi}{CH}-CH_2-CH_2-\underset{\Phi}{CH}-$$

contrarily to the classical "head to tail" addition. For polystyrene one has only "head to tail" addition, for PVC it is possible to obtain 15 % "head to head".

ii) The chain is not stereregular. The spatial placement of the substituents can be either :

$$- CH_2-\underset{\Phi}{CH}-CH_2-\overset{\Phi}{CH} \qquad \text{or :}$$

$$- CH_2-\underset{\Phi}{CH}-CH_2-\underset{\Phi}{CH}$$

the first one being called syndiotactic, the second isotactic. In radical polymerization these dyads have the same probability and are randomly distributed, leading to what is called an atactic polymer which cannot crystallize.

iii) The case of dienes. Butadiene $CH_2 = CH-CH=CH_2$ or isoprene $CH_2=\underset{CH_3}{C}H-CH=CH_2$ can polymerize by radical polymerization. One obtain simultaneously 1-2 addition giving

$$\left(-CH_2-\underset{CH=CH_2}{CH}-C- \right)_n$$

and 1-4 addition giving either the trans unit

$$\left(C \diagdown C - C \diagdown C \right)_n$$

or the cis unit

$$\left(C \diagup C = C \diagdown C \right)_n$$

In classical radical polymerization one obtains 10 % 1-2, 65 % 1-4 trans, 25 % 1-4 cis. Only the 1-4 cis units contribute to the rubber-like properties of the material.

b) Initiation reactions

1) Thermal initiation :

Spontaneous polymerization can be initiated by temperature, specially in the case of polystyrene and polymethylmethacrylate. In the first case polymerization is slow at 30°C and provides very high molecular weights (of the order of 30 millions). At higher temperatures, this kind of polymerization can be very fast, therefore usually inhibitors are added to monomers.

2) Chemical initiation

The most common method to start polymerization is to use a molecule which decomposes easily into free radicals. One of the most common is dibenzoyle peroxide

$$\phi - \underset{O}{\overset{\|}{C}} - O - O - \underset{O}{\overset{\|}{C}} - \phi$$

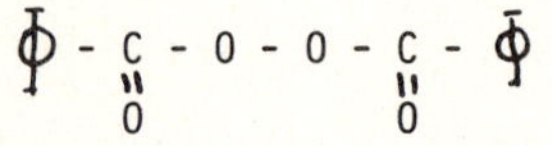

which gives

$$2 \phi \overset{O}{\overset{\|}{C}} - O^* \rightarrow 2 \phi^* + 2\ CO_2$$

The radical ϕ^* initiates the polymerization by its action on a monomeric unit. Many other peroxides can be used. They should exhibit an adequate decomposition temperature.

Another possibility of making radicals, which is used mainly in the case of emulsion polymerizations, is to use a redox system like

$$H_2O_2 + Fe^{++} \rightarrow OH^- + OH^* + Fe^{+++}$$

3) Photochemical initiation

The principle is the following : the substance m absorbs the energy h ν reaches an excited M· and decomposes into two radicals R^* and R'^* which initiate the polymerization. This can be done for instance with azobis-iso by tyronitrile

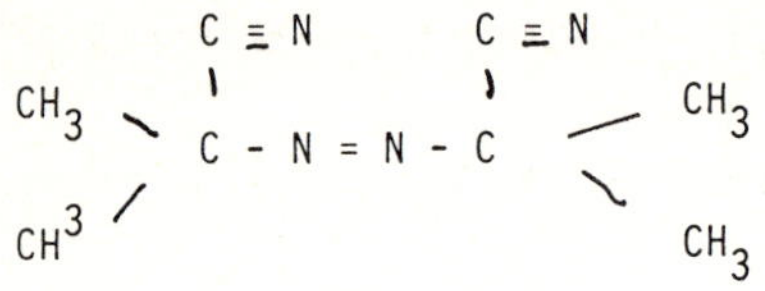

$$\begin{array}{c} C \equiv N C \equiv N \\ CH_3 \diagdown | | \diagup CH_3 \\ C - N = N - C \\ CH_3 \diagup \diagdown CH_3 \end{array}$$

which gives

$$N_2 + 2\ (CH_3)_2\ C^* - C \equiv N$$

The radical $(CH_3)_2 C^*$ initiating the polymerization.

Many radiations other than light can be used to start a polymerization α, β and γ rays as well as X rays. In these cases, an initiator is not needed, the effect of the radiations on solvent breaks it into radicals which initiate the polymerization. The mechanism is complex, but the advantage of this technique is that it works on a mixture of solvent and monomers without addition of any initiator.

All these reactions give polymers which are polydisperse (all molecules do not have the same molecular weight). This polydispersity is usually characterized by a polydispersity index, which is the ratio of the weight average molecular weight M_w to the number average molecular weight M_n. Values of M_w/M_n between 1,5 and 2 are usually obtained - provided the conversion into polymer is kept low.

B. Anionic polymerization

a) Propagation

$$- CH_2 - \underset{\phi}{CH^-} (N_a^+) + CH_2 + \underset{\phi}{CH}$$

$$CH_2 - \underset{\phi}{CH} - CH_2 - \underset{\phi}{CH^-} N_a^+$$

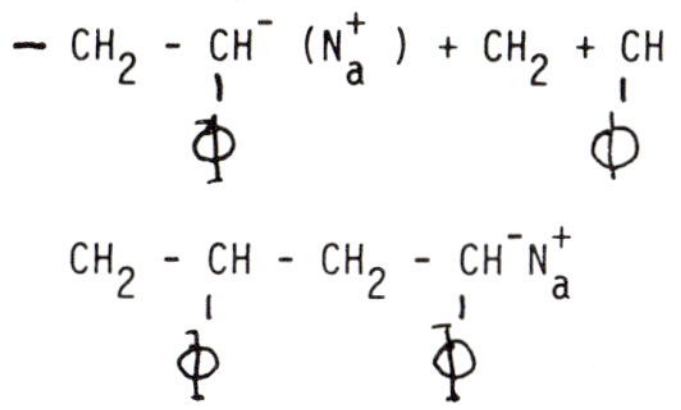

This reaction is only possible in the absence of oxygen (in vacuo or in the presence of an inert gas). All protonic impurities have to be destroyed, and one has to work in an aprotic solvent such as benzene or tetrahydrofuranne.

i) This reaction is very interesting, because there is no termination reaction. The life of the macroion is very long. One can successively add different monomers, making this way block copolymers.

ii) Another great advantage of this technique is to give almost monodisperse system. If one works carefully one can obtain without too much difficulty values of M_w/M_n less than 1,1 and even 1.01.

b) Initiation

Many initiators are used for starting the polymerization, each of them having its specific properties, the choice being made following the requirement of the planned reactions.

I will quote only one of them : the potassium ethylbenzene which is used in Strasbourg and added to the monomer :

$$\phi - \overset{-}{CH} - CH_3 \overset{K^+}{+} H_2 = CH\bar{\phi} \rightarrow CH_3 - \underset{\phi}{CH} - \underset{\phi}{C_2} - \overset{-}{CH} \quad K^+$$

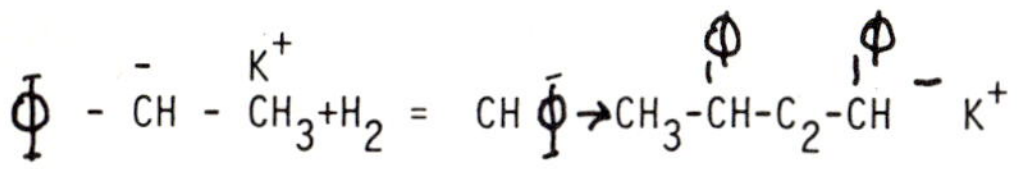

The initiator is attached to the growing chain and, since it has the same structure as a monomeric unit, the nature of the chain ends and the backbone is the same.

c) Difunctional initiation

If one uses a difunctional initiator, two chains start from it. At a given time the chain has two living ends. After completion of the polymerization, the initiator is in the midle of the chain which has two active ends. The most common of these initiators are naphtalene sodium and the tetramer of α methyl-styrene

$$Na^+ \overset{-}{C}-C-C-C-C-C-C-\overset{-}{C}\ Na^+$$
$$\quad\ \ \phi\ \ \phi\quad\quad\ \phi\ \ \phi$$

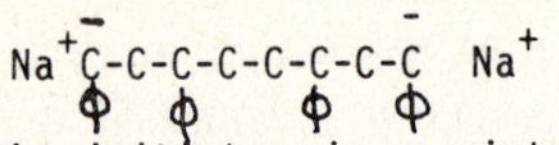

In fact, this initiator is a mixture of different oligomers, but this is not a problem when one uses polymers of large degree of polymerization.

Preparation of network (4)(5)(6)

a) The most classical way to prepare a network is to start from a mixture of mono and difunctional monomers and let the polymerization proceed. In order to make, for instance, a polystyrene gel, one adds to the styrene divinylbenzene (D.V.B)

$$H_2C = CH-\phi- CH=CH_2$$

The first double bond polymerizes in one polymer chain leading to the structure

$$\begin{array}{c}
\text{————} \underset{|}{C} \text{————} \underset{|}{C} \text{————} \\
\phi \qquad\qquad \phi \\
| \qquad\qquad | \\
CH+CH_2 \qquad CH=CH_2
\end{array}$$

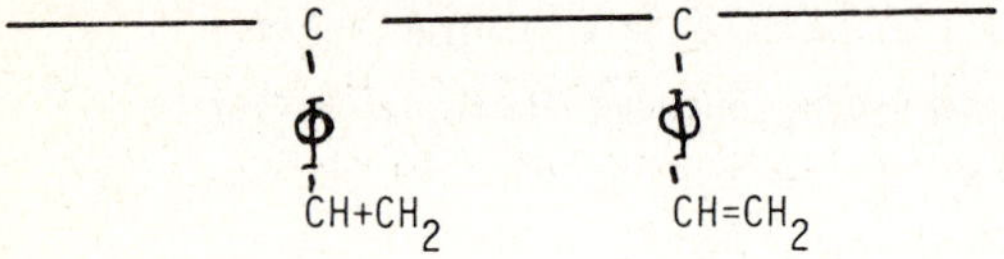

The dangling double bond can enter in a new chain giving the following structure

$$\begin{array}{c}
\text{————————} \\
| \\
\phi \\
| \\
\text{————————}
\end{array}$$

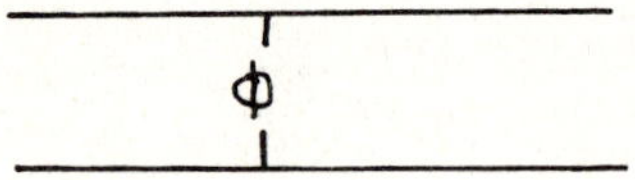

where two chains are linked together. One speaks of a crosslink of functionality $f = 4$ (four chains start from the crosslink); f can have other values but must be at least equal to three. If the number of D.V.B. molecules is sufficient to provide a minimum of two connections per chain, one obtains a system with infinite molecular weight i.e. a network or a gel, depending on the presence or the absence of solvent. It is easy to evaluate the length of the elastic chains, which is the length of the chain between crosslinks. If N is the number of monomers and Nf the number of monomers with functionality f, the total number of chain will be fNf/2 and their average length in term of monomeric units

$$\frac{2N}{fNf} = p$$

In the preceeding case (Polystyrene and DVB) if Nf/N = 0,04, p = 12.
This calculation gives an average length but no information on the distribution of the chain length. Until now no theory has been able to predict this distribution and no experience to measure it.

b) Utilization of a difunctional polymer crosslinked after polymerization

The following technique introduced by Rempp has been extensively used in Strasbourg. After preparation of a living polystyrene with two active ends by using a difunctional initiator, one adds into the medium a few per cent of DVB.

The last monomer adds to the preexisting living polymers and polymerizes, making small noduli to which are attached the elastic chains. The reaction is terminated by adding methanol, for instance. This gives a crosslinked system of infinite molecular weight in which we know that all elastic chains have the same length. These gels which could be called "ideal" are not an absolutely perfect model.

i) The functionality is not well known (from experimental results and their interpretation f should be of the order of 6)

ii) There can be chains having only one end attached to the network. By careful washing, one always extracts a few free chains. If there are free chains, there are also chains attached only by one end.

iii) One can also have loops i.e. chains attached by both ends to the same D.V.B. modulis and

iiii) One can also have trapped entanglements i.e; permanent entanglements which act more or less as extra crosslinks (cf. figure).

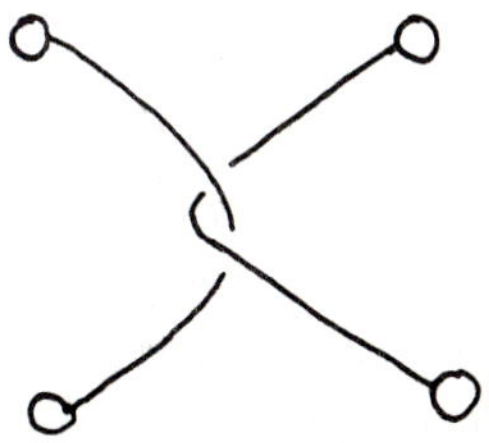

Preparation of networks with a given functionality

If, instead of using D.V.B. we add to a difunctional living polymer an f functional compound which reacts by addition, we obtain an f functional network. One of the best reagents is the tri-allyl-oxy 135 triazine

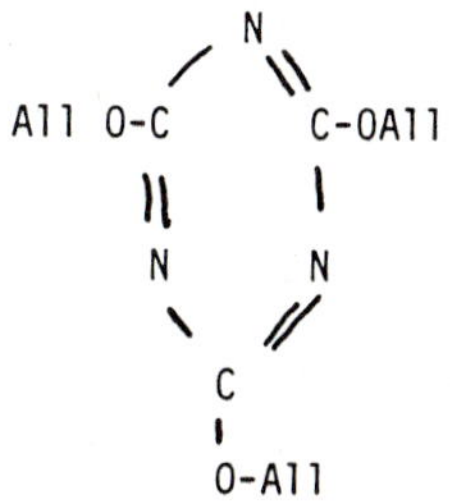

with All = $-CH_2-CH=CH_2$

it adds to three living polystyrene chains

$$+ 3PS^-M^+$$

$$PS-\underset{\underset{N}{\overset{N}{\|}}}{C}\underset{\underset{C-PS}{\overset{N}{\diagup}}}{C}-PS \quad + \ 3\ MO-CH=CH_2$$

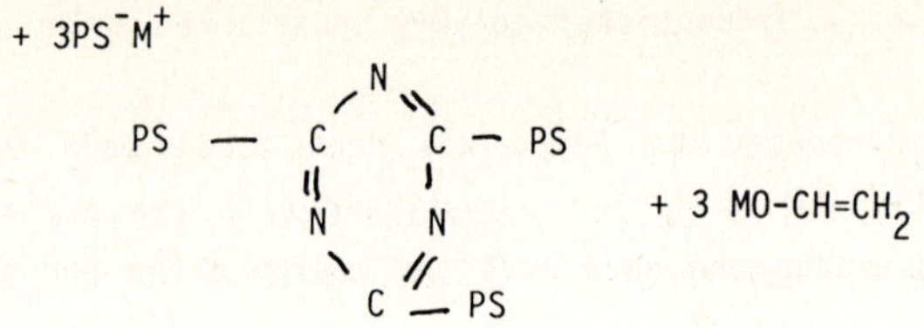

This is an improvement compared to the D.V.B., but the existence of dangling chains loops and entanglements cannot be avoided.

By using more complex reagent f can be increased. Values of four and six have been used in Strasbourg. This kind of reaction can also be used on polydimethylsiloxane having an H at both ends

$$H-(\underset{\underset{CH_3}{|}}{\overset{\overset{CH_3}{|}}{Si}}-O)_n-H$$

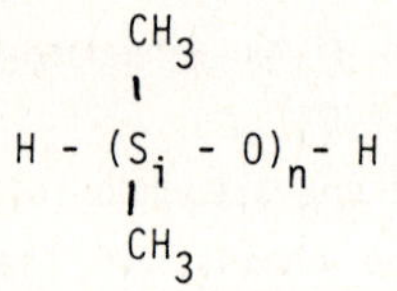

The interesting fact is that the network can be prepared in bulk and concentrated solutions, contrarily to the anionic polymerization, which has to be made in dilute solution.

Crosslinking of preexisting chains

This is the oldest technique, and is still used extensively in rubber technology. It consists of adding to the polymer a reactant (sulfur very often) which crosslinks the material. Many reagents can be used. In order to have a very clean reaction, we used gamma ray irradiation on polystyrene in cyclopentane. If one works in the absence of oxygen, the crosslinking reaction is much more effective than the chain scission, and this last can be neglected.

Preparation of labelled networks

As we shall see in the applications of neutron small angle scattering to the characterization of the structure of the networks, it is useful to prepare networks having some well-defined parts labelled with deuterium. Three methods have been used which we shall now describe briefly.

1) Labelling of elastic chains

This is easily done if one mixes before the crosslinking reaction two polymers, one deuterated, the other not. After having made the mixture with the desired relative concentration of both species, one adds the divinylbenzene or the polyfunctional reagent.

2) One can also label the crosslinks

As an adaptation of the proceding method we have used the following procedure : After the preparation of the living polymer with two functional ends, one adds a few percent of deuterated monomer, obtaining a block polymer of the structure D-H----H-D. This reaction being completed, one crosslinks as usual. By this technique one obtains a network with labelled end chains attached to the crosslink.

3) If one uses a mixture of ordinary and deuterated

polymer in a classical radical crosslinking reaction, one obtains labelled chains which are much longer than the elastic chains. They allow to examine at a larger scale the variation of the distances at the molecular level with swelling or under a stress.

Thermodynamic of networks

Before describing experiments which allow to obtain information on molecular conformations in gels, we shall briefly write some thermodynamical equations which are useful for a better understanding of networks' and gels' properties (7)(8)(9).

A) The network without solvent (10)

In the dry state, one assumes that the chains obey gaussian statistics (this will be shown experimentally later). Therefore the probability for the ends of a chain to be at a distance r is given by the formula :

$$w(r) = \left(\frac{3}{2\pi r_0^2}\right)^{3/2} \exp - \left(\frac{3\, r^2}{2\, r_0^2}\right)$$

where r_0^2 is the average for the square of the end to end distance. From this equation one obtains immediately the entropy of a chain

$$S_1 = K \ln w = S_1 - (3/2)\, K\, (r^2/r_0^2)$$

If we have N elastic chain $S = NS_1$. Now let us assume that during an elastic deformation the volume of the rubber does not change, and that the energy of the chain does not vary (this has been proved experimentally to be exact with a good precision). The elastic free energy for a given state is :

$$F_{el} = (3/2)\, K\, T \sum_{i=1}^{N} (r_i^2/r_{oi}^2)$$

where r_i is the end to end distrance for a deformed chain having r_{oi} as end to end distance in the reference state. In order to evaluate the sum, one uses the hypothesis of affine deformation, which is characterized by the three coefficient λ_x, λ_y, λ_z. Any macroscopic vector which at rest has for components x, y, z becomes $x\lambda_x$, $y\lambda_y$, $z\lambda_z$.

This affinity is not justified at the monomer level, but if one assumes that it does exist at the crosslinks points, one obtains immediately

$$F_{el} = \frac{N K T}{2} (\lambda_x^2 + \lambda_y^2 + \lambda_z^2 - 3)$$

where N is number for elastic chains.

This theory is a good first approximation, but many authors have developed more sophisticated approaches. The crucial point is the hypothesis of affinity, but we cannot enter into the details of these discussions. The more recent results keep the same functional relation, but the coefficient is no more 1/2 but depends on the functionality of the network. It is difficult to verify these theories because the number of elastic chains is difficult to evaluate precisely.

Knowing the free energy, it is easy to obtain an equation of state (assuming incompressibility) and from it the tensile and the shear moduli.

B. Swollen networks

As in the case of single chain swollen in a good solvent when a network is immersed in a solvent, there are two competitive forces acting on it. One is due to the osmotic pressure of the gel or to the free energy of dilution ; it tends to swell the network, the second one is due to the elasticity which tends to retract the network.

As usual in this kind of problem, one writes that the total free energy is the sum of these terms, and one minimizes it to obtain the equilibrium swelling.

In the frame of Flory's theory one obtains

$$Q^{5/3} = (\frac{1}{2} - \chi) \; v_1 \; \bar{v} \; Mc$$

where v_1 is the molar volume of the solvent, $\bar{v}$ the specific volume of the polymer and M the molecular weight on an elastic chain. Q is the swelling ratio i.e. the ratio of the volume of swollen gel to the volume of the dry gel. This depends strongly on the interaction parameter through the quantity $(1/2 - \chi)$ which, in a mean field theory, is the excluded volume parameter. This theory has been used extensively in the rubber industry to evaluate M the molecular weight of the elastic chain, but it has never been possible to check it because of the impossibility of making networks of rigorously knwon structure.

C. The c* theorem

It is well known that if you take a polymer solution and increase the concentration, there is a cross-over concentration called c*. Below c* the molecules are independent and free to move, above c* they interpenetrate, because the volume occupied by one molecule is larger than the volume it has at its disposal in the solution.

Therefore, the picture of a solution above c* looks like the figure on the next page. Between two crosslinks, the chain is more or less a free chain and obeys the statistics of a free chain in a good solvent $R^2 \sim N^{2\nu}$. At large distance the interactions are screened and the statistics are gaussian. One says that the chains are forming "blobs". The solution behaves like a bulk of blobs. By scaling arguments it is possible to obtain the size of blobs and the value of c*.

Now let us try to describe what happens when the chains are crosslinked. They tend to separate one from another, but they are linked by crosslinks and can expand until each chain forms a blob. Therefore, the equilibrium concentration is of the order of the overlap concentration for chains having the length of the elastic chains.

De Gennes gives for the swelling factor Q at equilibrium

$$Q^{5/3} \approx v\, M^{4/3}$$

where v is the excluded volume parameter. This differs from Flory's formula by the exponent 4/3 instead of one for the influence of the molecular weight of the elastic chain.

This c* theorem is very simple. It says that for equilibrium properties a swollen gel behaves like a concentrated solution in which the blobs have the size of the elastic chains. It has been verified by quasielastic light scattering and by neutron scattering (11).

I will just give results of M. Beltzung obtained by neutron small angle scattering with tetrafunctional gels of polydimethylsiloxane having a few percent of deuterated chains. It has been possible to measure the molecular weight of the elastic chains and to compare the radius of gyration of the free chain in dilute solution and the radius of gyration of the network swollen in the same solvent. These results are given in the following table.

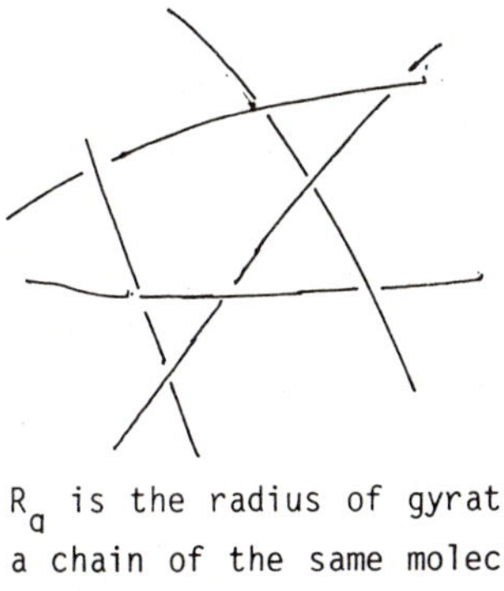

M_n	R_g	R_F	f	v_c
3050		27	0	
3100	27		4	0,71
6100		39		
6200	41		4	1.00
6200	40		4	0,71

R_g is the radius of gyration in the swollen gels, R_F the radius of gyration of a chain of the same molecular weight in dilute solution. The solvent used was cyclohexane, f is the functionality of the gel and v_c the volume fraction of the elastic chains used for making the gel, 1 means crosslinking reaction made in bulk, 0,71 % means that a solution with 21 % of solvent (in volume) was used.

The results are very convincing. They show an exact correspondence between the gel and the solution, which is more the c* theorem predicts.

D. Improvements of the thermodynamics of gels (12)(13)

The theory of solutions has been improved by des Cloizeaux, who showed that in semi-dilute regime one should replace the excluded term $v\varphi^2$ by $v\varphi^{9/4}$ calling φ the volume fraction of the polymer. Leibler, Bastide and Candau introduced a free energy F written as

$$F = v^{9/4} + F_{el}$$

using for F_{el} the scaling law $F_{el} = K\,T\,N\,f(\frac{\varphi}{\varphi^*})\,(\lambda_x^2 + \lambda_y^2 + \lambda_z^2)$ where N is the number of elastic chain. If one uses for f a power law with exponent ℓ one obtains for the free energy per site in the swollen state

$$F = K\,T\left\{v\,\varphi^{9/4} + B\,\varphi^{\ell}\,\varphi^{*\,\frac{9}{4}-\ell}\right\}$$

Using this expression for ΔF one recovers the c* theorem and one can evaluate the elasticity modulus and the equilibrium properties of a gel immersed in a polymer solution of the same monomer (partition coefficient and osmotic deswelling) as well as the inelastic scattering diffusion coefficient.

Characterization of gels by small angle neutron scattering

Studying swollen gels by elastic light scattering, X rays or neutron small angle scattering gives a signal I(q) where $q = (4\pi/\lambda)\sin(\theta/2)$ (λ , being the wavelength of the incident beam and θ the scattering angle) which is not very informative. It is always weak : for a gel swollen in a good solvent it should decrease like $q^{-5/3}$ for $q\,R_F \gg 1$ and like q^{-2} for $Q\,R_F \ll 1$ but this is difficult to test considering the dimensions of the elastic chains having R_F as radius of gyration. Moreover, one observes at small angle a sharp increase of the intensity when q goes to zero, increase which cannot be explained without assuming the presence of inhomogeneities. Considerable efforts have been made to prepare gels which are clear at small angle, but until now without success. It is not yet clear if these inhomogeneities are the result of chemical defects or if they are inherent to the structure of the networks.

The only way to obtain information is to use labelled network as described in the chemical part. This enhances the contrast, giving a stronger scattering and allowing to obtain structural information. In the following part we shall summarize experiments made in Strasbourg on different types of networks.

A. Networks labelled at the crosslinks

R. Duplessix (14) in his thesis used networks made of polystyrene elastic chains D — H — D with two labelled ends crosslinked with divinyl benzene. The results show a remarkable peak, in bulk as well as in the swollen state. The position of the peak varies like the radius of gyration of the elastic chains. When the amount of solvent increases the value of q maximum decreases. In order to check if this maximum was due to some order in the solution we prepared a solution of stars having exactly the same functionality and branches with half of the molecular weight of the elastic chains. The scattering diagrams are very similar. There is no qualitative difference between a solution of star at the same concentration as the swollen networks.
This can be explained at least qualitatively invoking the random phase approximation technique as developed by de Gennes (8) and Leibler (15). It has been shown that the intensity scattered by a copolymer monodisperse in composition starts from zero at $q = 0$ goes to a maximum around $qR_g = 1$ and decreases afterwards. This maximum persists in solution even in very dilute solutions.

Since the stars labelled in the middle are copolymers, one can explain easily the diagram obtained for stars. The fact that there is no difference between the stars and the network tells us that from scattering point of view they behave in a very similar manner, and that the junction between stars does not have a strong influence on the scattering function. This can be shown also for a different approach. Let us assume that we have in bulk of two block copolymers A B with equal length of A and B blocks. Now keeping constant the number of segments, we evaluate the scattering function of the multiblock copolymer $(AB)_n$ and let n go to infinity. The scattering function is only slightly modified in shape and intensity (16).

B. Networks with labelled elastic chains

It is well known that in the case of randomly mixed polymers identical but with a fraction $x = N_D/N$ of labelled molecules (N is the total number of molecules) the scattered intensity satisfy the relation

$$I \approx N(a-b)^2 \ x(1-x) \ P(q) \qquad (1)$$

where P(q) is the structure factor of one polymer molecule, a the coherent scattering length of one species and b the same quantity for the other species. What is interesting is that the intermolecular interferences do not appear in the result and that, regardless of the composition x one obtains the structure factor of one chain. This expression has been generalized to solution by C. Williams and al., Z. Akcasu and al. (17), they arrive at the expression

$$I(q) \quad N(a-b)^2 \ x(1-x)P(q) + (\bar{a}-s)^2 \ NP(q) + A^2Q(q) \qquad (2)$$

In this expression $\bar{a}$ is the average scattering length of the sample $\bar{a} = x a + (1-x)b$, s the scattering length of the solvent and Q the intermolecular interference term.

If, by using for the solvent a mixture of deuterated and ordinary molecules to fulfil the condition a=s equation 2 reduces to 1 and one obtains directly the structure factor of the molecules.

As a verification,this has been applied to gels. Evidently we cannot speak of molecules since all chains are tied together, but the argument can be extended to this case giving the same result. It has been checked by making gels from a mixture of deuterated and hydrogenated chains otherwise identical. The main result is that, taking into account the precision of the determination of I(q) there is no difference in the scattering envelope of a free chain and a chain engaged in a network. This is a good verification of what has been assumed by theoreticians.

The memory term introduced mainly by Dusek and Prins (9) does not play an important role,since this result is obtained regardless of the conditions of preparation.

One interesting point is to characterize the distances below which the "affine" deformation is no more valid. This will be explained in detail in this book by J. Bastide, therefore I just quote this problem. In the first order the classical theory of elasticity as reformulated by Flory (18) seems to be approximately correct,but much remains to be done in order to have a detailed understanding on the deformation at the molecular level of the chains of a network or a gel.

References

1 Groupe Français des Polymères : Initiation à la chimie et la physicochimie Macromoléculaire (1980)

2 G. Champetier : Chimie macromoléculaire (Herrmann, Paris 1970)

3 C.H. Bamford : Free Radical polymerization (Elsevier 1976)

4 P. Rempp, p. 265 : Reactions on polymers (J. Moor, D. Reidel publishing company, Boston 1973)

5 P. Weiss, G. Hild, J. Herz, P. Rempp : Macromolecular Chemistry, 135, 249 (1970)

6 M. Beltzung, C. Picot, P. Rempp, J. Herz : Macromol. 15, 1594 (1982)

7 P.J. Flory : Principles of Polymer Chemistry (Cornell Univ. Press 1953)

8 P.G. de Gennes : Scaling Concepts in Polymer Physics (Cornell Univ. Press 1979)

9 K. Dusek, W. Prince : Advances in Polym. Sci. 6, 1 (1969)

10 J.J. Aklonis, W.J. MacKnight : Introduction to Polymer Viscoelasticity (John Wiley and Sons, second ed. 1983)

11 M. Beltzung : thèse Strasbourg (1982)
J.P. Munch, S. Candau, G. Hild, J. Herz : J. Physique (Paris) 38, 971 (1977)

12 F. Brochard : J. de Physique (Paris) 42, 505 (1981)

13 J. Bastide, S. Candau, L. Leibler : Macromol. 14, 719 (1981)

14 R. Duplessix, thèse Strasbourg (1975)
see also P. Rempp, J. Herz, G. Hild, C. Picot : Pure and Applied Chemistry 43, 76 (1975)

15 L. Leibler : Macromol. 15, 1283 (1982)
16 Unpublished result
17 C.E. Williams et al. : J. Polym. Sci. LeHers Ed. 17, 379 (1979)
A.Z. Akcasu et al. : J. Polym. Sci. Phys. Ed. 18, 8633 (1980)
H. Benoît, J. Koberstein, L. Leibler : Macromol. Chem. Suppl. 4, 85 (1981)
18 B. Erman, P.J. Flory : Macromol. 15, 800-806 (1982).

Physical Gels and Biopolymers

M. Rinaudo
CERMAV, CNRS - BP 68
F-38042 Saint Martin d'Heres, France

We review some recent experimental results on physical gels made of biopolymers which are either vegetal (pectin, agarose) or animal (gelatin). These are either neutral (agarose) or electrically charged (pectin, gelatin). Two types of gels are considered :

- Thermoreversible gels, stabilized by secondary links and/or pseudo crystallization of multiple helices. There is usually hysteresis when cooling the system down or heating it up. For both gelatin and K-carraghenan, it is shown that gelation occurs in two successive stages corresponding to two helicoidal distributions whose relative importance depends on time.

- Ionic gels that are formed by selective fixing of bivalent ions such as calcium with the "egg-box" mechanism (pectin). The importance of the microstructure (a random sequential distribution of carboxylic groups) is analyzed. With light scattering and viscosity measurements we show the influence of charge density on gel formation.

Introduction

Nous souhaitons présenter un certain nombre de résultats expérimentaux récents sur des gels physiques formés par quelques biopolymères.

Les biopolymères sont certainement les premiers polymères connus pour former des gels physiques, c'est à dire des réseaux tridimensionnels non covalents avec, par exemple :

- les pectines ou l'agarose qui sont des polysaccharides gélifiants d'origine végétale
- ou la gélatine, protéine d'origine animale

Ces gélifiants sont neutres du point de vue ionique (l'agarose) ou de nature polyélectrolyte (pectine, gélatine).

Ces polymères naturels sont d'abord stéréoréguliers ce qui va permettre d'obtenir la stabilisation d'une structure tridimensionnelle par un système coopératif de liaisons secondaires (liaisons hydrogène ou de Van der Waals) ou de liaisons ioniques.

Très généralement la transition de phase sol → gel des solutions aqueuses sera associée à la transition pelote → hélice du biopolymère.

Nous nous proposons d'examiner deux types de gels :

- les gels thermoréversibles stabilisés par liaisons secondaires et/ou une pseudo-cristallisation d'hélices multiples ou de segments d'hélices associée au changement de phase. Le rôle de la qualité thermodynamique du solvant doit être primordial.

- les gels ioniques pour lesquels la réticulation procède par fixation sélective d'ions divalents tels que le calcium.

1 - Les gels thermoréversibles

Nous prendrons comme exemple les gels de gélatine (collagène dénaturé) et de K-carraghénane (polysaccharides d'algues marines) en essayant de mettre en évidence un certain nombre d'analogies dans le mécanisme de la transition sol → gel à partir des travaux récents de M. DJABOUROV et coll [1-3], J.Y. CHATELLIER [4-5], D. DURAND et coll [6-7] et de C. ROCHAS, M. RINAUDO [8-11] respectivement.

Les conditions opératoires sont cependant très différentes : 5 à 20 % en poids pour les travaux sur la gélatine en solution dans 0,1 M NaCl ou dans l'eau ; 0,1 à 1 % en poids pour le K-carraghénane à concentration variable en électrolyte.

a) La gélatine

Le collagène est une protéine constituée par trois chaînes hélicoïdales associées en triple hélice ; dans les solutions aqueuses chauffées au dessus de 40° C, il se produit un changement conformationnel de type hélice-pelote bien mis en évidence par la variation du pouvoir rotatoire $[\alpha](T)$. La position de la courbe dépend des vitesses de variation en température et les courbes obtenues en montée et descente de température présentent toujours une hystérésis.

Lorsque cette solution est ramenée à une température inférieure à la température de fusion conformationnelle du polymère T_M (de l'ordre de 35° C), on observe une cinétique de renaturation à partir de la fonction $[\alpha]$ (temps) qui a été étudiée par DJABOUROV et coll [3], CHATELIER [4] et GODARD et coll [12]. Les résultats expérimentaux sont analysés en proposant deux régimes cinétiques au moins :

- une cinétique d'évolution rapide de $[\alpha]$ donc du taux d'hélicité qui correspond à la nucléation avec croissance fibrillaire (l'application de la théorie d'Avrami conduit à un indice $n = 1$).
- une cinétique plus lente associée à un accroissement de viscosité précédant le point de gel (4).

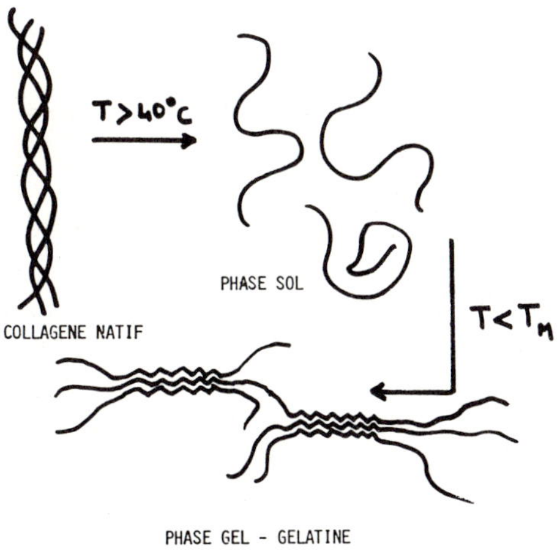

Figure 1 : Schéma de la gélification de la gélatine.

D. DURAND et coll [6] montrent que le taux d'hélicité (h) au point de gel ne dépend que de la concentration en polymère soit : 14 % (h) et 7 % (h) pour les concentration en polymère de 6 % et de 11 % respectivement ; ils montrent de plus que la concentration en hélice au point de gel est pratiquement constante. M. DJABOUROV [1] montre par ailleurs que ces gels sont hors d'équilibre ; un polymère placé pendant 100 heures aux températures 22° C et 26,5° C, présente à la fusion deux populations de domaines en hélices caractérisées par des températures de fusion de 29° C, 36° C et 32° C, 36° C respectivement. Le pic haute température a la même caractéristique que le collagène natif et plus la température de vieillissement du gel est proche de T_M plus la stabilité des domaines ordonnés augmente. Tous ces résultats semblent suggérer une réticulation par au moins des segments de triples hélices agrégées.

La formation du gel est contrôlée par la cinétique de renaturation des protéines ; elle dépend de la concentration en gélatine et de la température adoptée. Le rôle du solvant n'est cependant jamais examiné.

b) K-carraghénanes

Avec ce polysaccharide, la cinétique de formation du gel n'est jamais dissociée de celle de la transition hélice-pelote beaucoup plus rapide que celle de la gélatine.

A faible concentration ionique en présence du cation K^+ ($C<C_T$ 7.10^{-3} équiv/l), on observe une transition thermique réversible hélice-pelote en phase sol ; à partir de mesures thermodynamiques, on démontre que la conformation ordonnée est un dimère d'hélices [8]. Au dessus de cette concentration C_T, c'est à dire en réduisant la qualité thermodynamique du solvant, on obtient la formation d'un gel et une hystérésis entre montée et descente en température. Au voisinage de C_T, nous démontrons un processus de gélification en deux étapes correspondant à deux distributions de domaines hélicoïdaux dont les proportions relatives varient avec le temps de vieillissement (Fig. 2) [11]. Ce comportement est très analogue à celui de la gélatine.

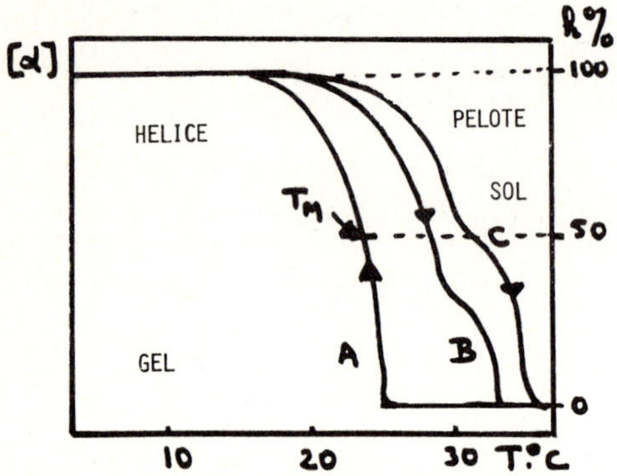

Figure 2 : Influence de la température et du viellissement sur le pouvoir rotatoire et le taux d'hélicité (h %) d'une solution de K-carraghénane. Courbes A et B : cycle thermique avec hystérésis (18° C/heure). Courbe C : viellissement 15 h à 21° C.

Pour les carraghénanes nous excluons la nécessité de défauts de structure pour former un gel ainsi qu'un mécanisme impliquant un pont potassium tel que le suggèrent MORRIS [13] et SMIDSROD [14].

Nous proposons un mécanisme de réticulation par agrégation d'hélices (Fig. 3a) plutôt que par pontages par des segments de double hélice (Fig. 3b).

Parmi les gels thermoréversibles il faudrait enfin citer les gels mixtes obtenus en mélangeant des polysaccharides dont les structures moléculaires sont compatibles. C'est le cas du xanthane sous forme hélicoïdale et des zones lisses des galactomannanes. Il apparaît que l'ensemble de ces systèmes implique des macromolécules de grandes anisotropies ayant un comportement de polymères

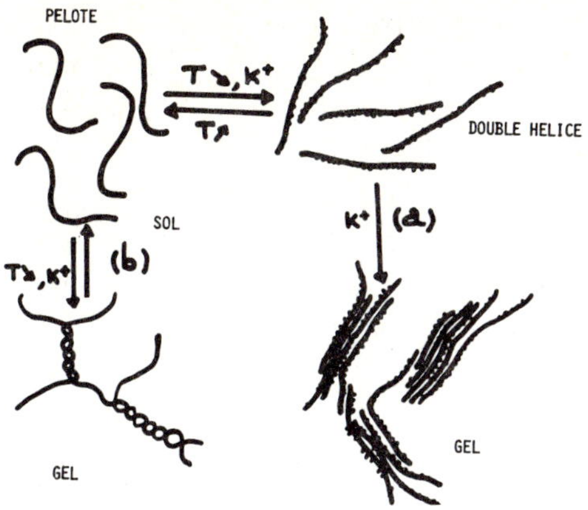

Figure 3 : Schéma de la gélification du K-carraghénane.
a) par agrégats de doubles hélices,
b) par zones hélicoïdales.

rigides. Leur comportement est cependant difficile à étudier en raison de la relation étroite entre les propriétés thermodynamiques et la transition conformationnelle qui se superpose au changement de phase. Cependant les comportements sont très proches de ceux obtenus sur le polyglutamate de benzyle, modèle de batonnet rigide dont le comportement est bien rendu par le traitement de Flory sur les cristaux liquides polymères [15]. Pour ce système, il vient d'être proposé un modèle de réticulation donné à la figure 4.

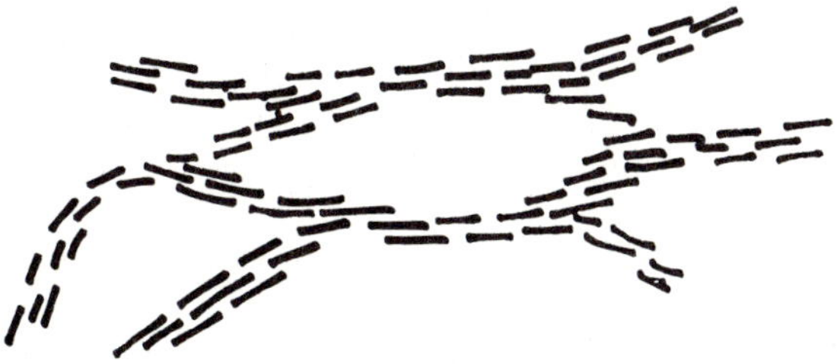

Figure 4 : Schéma de gélification d'un polymère rigide.

2 - Les gels ioniques

Les pectines faiblement méthylées (acide polygalacturonique à taux d'estérification variable) et les acides alginiques (copolymères d'acides mannuronique et guluronique) présentent de grandes analogies de structure moléculaire et forment des gels en présence d'ions calcium [16-17].

Il a été montré que dès que le degré de polymérisation était supérieur à 10 ou 15, il se formait des dimères d'hélices 2_1 en présence de calcium dans le mécanisme dit de "boite à oeufs" (Fig. 5).

La fixation de calcium est spécifique et conduit avec le polymère à la formation de gel stable assimilable à un gel covalent ; en revanche, l'ion magnésium ne forme ni dimères ni gels.

Nous avons également montré :

1) le rôle de la microstructure du polymère en comparant à même degré d'estérification un polymère partiellement hydrolysé par la soude (les groupes carboxyliques libres sont distribués au hasard sur la chaîne) et par voie enzymatique (les sites carboxyliques sont en séquence) [17-18].

2) L'influence de la densité de charge sur l'aptitude à former des gels. Les méthodes utilisées pour mettre en évidence l'agrégation sont des mesures de l'intensité de lumière diffusée et de viscosité.

Le comportement de ces gels ressemble à celui des gels covalents compte tenu de la durée de vie grande des noeuds du réseau.

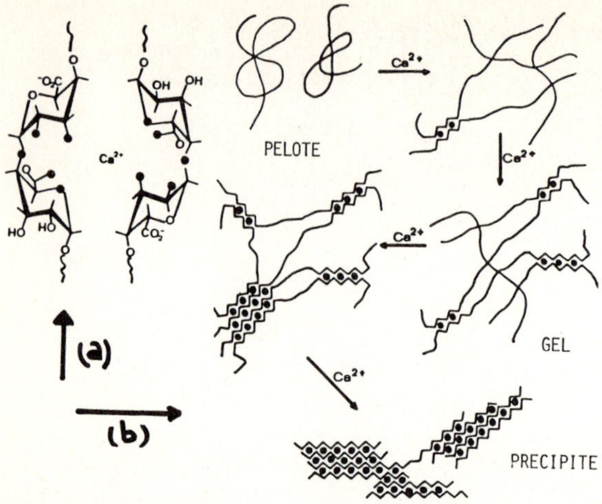

Figure 5 : Schéma de la réticulation par pont calcium.
a) liaison spécifique avec Ca^{+2}
b) gélification par addition progressive de calcium.

CONCLUSION

Il existe actuellement quelques biopolymères bien caractérisés dont le comportement est bien maîtrisé et sur lesquels il serait intéressant de développer des études physiques.

De toutes façons, avec ces polymères il faudra être prudent dans le choix de l'échantillon et sa qualité ou son degré de pureté si on veut établir les lois de comportement. Il est en effet clair que des différences de microstructure des échantillons modifieront leur comportement physique. C'est ce que nous avons bien montré sur des pectines dont on avait contrôlé la distribution des sites ioniques.

Références

1 - M. DJABOUROV, P. PAPON : Polymer 24, 537-542 (1983).
2 - J. MAQUET, H. THEVENEAU, M. DJABOUROV, P. PAPON : Int. J. Biol. Macromol. 6 162-163 (1984).
3 - M. DJABOUROV, J. MAQUET, H. THEVENEAU, J. LEBLOND, P. PAPON : British Polymer J. (à paraître).
4 - J.Y. CHATELLIER : Thèse, Le Mans (France) (1983).
5 - J.Y. CHATELLIER, D. DURAND, J.R. EMERY : Int. J. Biol. Macromol. (sous presse).
6 - D. DURAND, J.R. EMERY, J.Y. CHATELLIER : Int. J. Biol. Macromol. (sous presse).
7 - J.R. EMERY, J.Y. CHATELLIER, D. DURAND : J. Physique (à paraître).
8 - C. ROCHAS : Thèse, Grenoble (France) (1982).
9 - C. ROCHAS, M. RINAUDO, Carbohydr. Res. 105 227-236 (1982).
10 - M. RINAUDO, C. ROCHAS, B. MICHELS : J. Chim. Phys. 80 305-308 (1983).
11 - C. ROCHAS, M. RINAUDO : Biopolymers 23 735-745 (1984).
12 - P. GODARD, J.J. BIEBUYCK, M. DAUMERIE, H. NAVEAU, J-P. MERCIER : J. Polym. Sci. Polym. Phys. Ed. 16 1817 (1978).
13 - E.R. MORRIS, D.A. REES, C. ROBINSON : J. Mol. Biol. 138 349 (1980).
14 - O. SMIDSROD : I.U.P.A.C. 27th International Congress of Pure and Applied Chemistry on charged polysaccharides. Pergamon Press (1980).
15 - P.J. FLORY, I. UEMATSU, Y. UEMATSU : in Liquid Cristal Polymers I (1-73). Advances in Polymer Science. Volume 59, Springer Verlag (1984).
16 - G. RAVANAT : Thèse, Grenoble (France) (1979).
17 - J. F. THIBAULT, M. RINAUDO : British Polymer J. (sous presse).
18 - J.F. THIBAULT, M. RINAUDO : Biopolymers (sous presse).

Gelation of Physical Gels: The Gelatin Gels

Madeleine Djabourov
ESPCI, 10 rue Vauquelin, F-75231 Paris Cedex 05, France

Besides porous media, microemulsions, colloïdal aggregates, physical gels are also an example of finely divided matter. The physical gels are binary solutions made of a polymer + a solvent which acquire in some particular conditions, remarkable mechanical properties, due to an aggregation process of the polymer chains. This can be induced by a modification of the thermodynamical parameters of the medium, such as the ionic strength, the temperature... The gel state is reversible : the system recovers its fluidity when these parameters are changed back. Numerous examples of such systems are known, either of natural (agarose, pectin, ...) or synthetic (polyethylene, polyacrylonitrile, ...) polymers.

We shall deal in this paper with the gelation process of gelatin gels. Gelatin is a protein (denatured collagen). The gels are thermoreversible : they are formed by lowering the temperature of an aqueous solution (initially heated at T = 50°C), below 30°C and they "melt" when the temperature is raised back.

The gels are used for many industrial purposes, but the fundamental aspects implied in the gelation process are still not well understood.

It is well established [1] that gelation is related to the conformational coil→helix transition which arises when the temperature is lowered. The chains tend to recover their native conformation, which is that of collagen. The collagen rod (300 nm long) is made of three strands, wrapped together in a triple right-handed helix (pitch of 9 nm) and each strand itself is a left-handed helix (pitch of 0.29 nm). This structure is stabilized by hydrogen bonds between CO and NH groups of *adjacent* chains. The collagen rods are packed in fibrils, in the tissues, before extraction and denaturation. The way that the helices are renaturated, when a gelatin solution is cooled, is controverted.

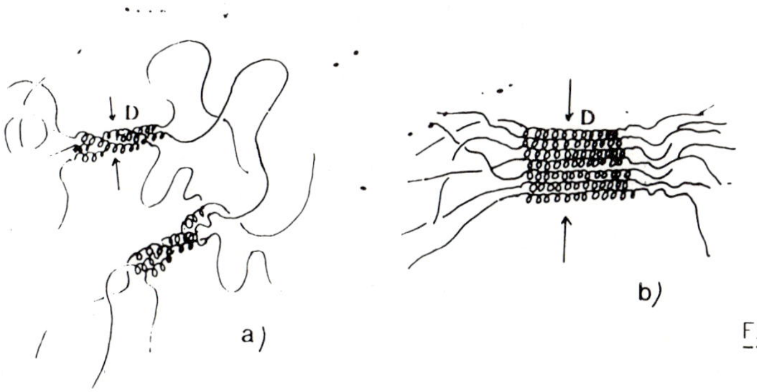

Figure 1

Basically, two schemes have been proposed (see Figure 1) :

a) A conformational transition [2] : coil ⇄ (left-handed) helices which is not an all-or-none transformation, each chain being partly helical and partly random coil. The chains are locally associated by groups of three. The diameter D is of the order $D \sim 1.2$ nm.

b) A crystallization of the helical parts of the chains, equivalent to the fringed micelle model, which means the nucleation and growth of fibers of a diameter of the order of $D \simeq 5 - 15$ nm according to GODARD et al. [3].

In both cases, the formation of the helices implies an aggregation of the chains, via the hydrogen bonds which are needed to stabilize the helices. So the network is created. However, its structure will mainly depend on the mechanism of aggregation.

We have studied the influence of different temperature treatments on the amount of left-handed helices renaturated which can be measured by polarimetry [4]. We have shown that during cooling and heating of the solutions at constant rates, hysteresis curves were observed. Also, the amount of helices varies with the rate of temperature change.

In quenching and isothermal annealing experiments, at different temperatures, $10 < T < 30°C$; the increase of the helix amount versus time depends strongly on the temperature; the lower the temperature, the higher the helix amount (Figure 2). The kinetics of helix formation can be described by means of a two-step process : the first one is exponential in time, starts immediately after quenching and lasts for a short time (ten minutes) ; it is followed or accompanied by a second process, logarithmic in time, which is responsible for a slow increase of the helix amount between one hour and one thousand hours, or so. Using curve-fitting methods [5], we could not decide whether these two mechanisms are independent or not (Figure 2).

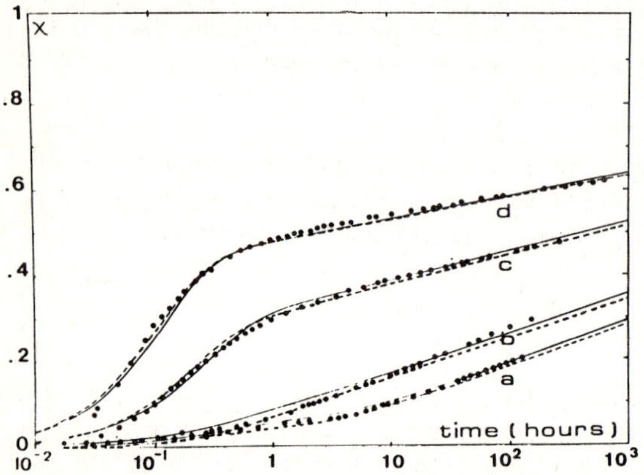

Figure 2

The helix amount χ versus time (logarithmic scale) in hours. (—) fit assuming independent mechanisms ; (---) fit assuming related mechanisms ; a) T = 28°C ; b) T = 26.5°C ; c) T = 20°C ; d) T = 10°C ; concentration 4.6% (for details see ref. [5])

The first step can be interpreted in terms of the phenomenological theory of crystallization, due to Avrami [4] and suggests that it represents the nucleation and unidirectional growth of fibers, according to GODARD et al. [3] . There are not at present direct measurements (X Ray, etc.) which would confirm this interpretation, nor any hypothesis concerning the microscopic interpretation of the second mechanism.

To this point, we can conclude that these systems *do not* reach an *equilibrium* state within a reasonable period of time, and that their structure, at a given temperature and a given time, depends on all the thermal history.

During helix formation, the solution loses its fluidity and becomes progressively a gel, with solid-like properties. The description of the rheological properties of these systems, in connexion with their structural parameters (number and size of the junctions, functionality, ...) is one of our goals. Here we report in Figure 3 the increase of the dynamic shear modulus $|G|$ as a function of time, for several quenching temperatures, which differ by only 1.5°C!

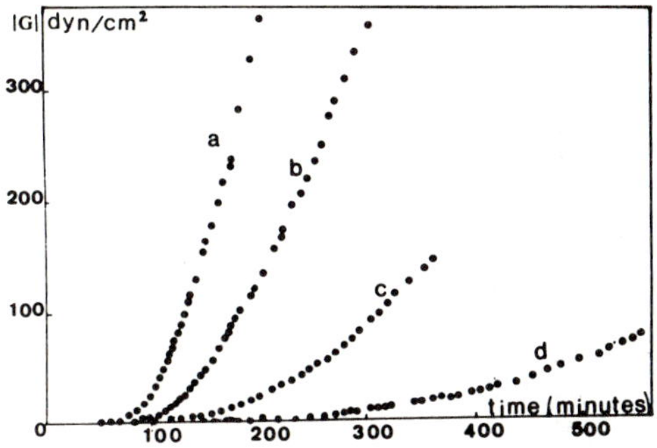

Figure 3

The shear modulus versus time during quenching experiments at a) T = 26.5°C ; b) T = 27°C ; c) T = 27.4°C ; d) T = 28°C ; concentration : 4.75%, frequency 0.5 Hz.

These experiments show that the mechanical properties are extremely sensitive to temperature variation in this range (26°C < T < 28°C). The striking point is that the helix amount χ for all these curves is of the same order of magnitude and rather low ($\chi \sim 10\%$ at t = 150 min. for curve a) and at t = 500 min. for curve d) on Figure 3). The sol-gel transition can be described by deriving the imaginary (G") and real (G') parts of the complex-modulus. Our preliminary results, on the kinetics of gelation [5] show a close analogy with the chemical gelation. A wide range of frequencies is needed (10^{-3} Hz < f < 10^2 Hz) in order to draw reliable conclusions from the rheological measurements. Special care must be taken to avoid breaking of the junctions under shear.

To conclude, the gelatin gels, as other physical gels, show various properties related to the network structure. The parameters of the network are sensitive to temperature variations. Whether these properties can be interpreted or not in the framework of the theoretical ideas concerning random systems is still an open question which is worth examining.

References

1. G.B. Ramachandran : Treatise on Collagen (Academic Press, N.Y., 1967) vol. 1.
2. W.F. Harrington and N.V. Rao, Biochem., 9, 3714 (1975).
3. P. Godard et al. J. Chem. Sci. Polym. Phys. Edn., 16, 1817 (1978).
4. M. Djabourov and P. Papon, Polymer, 24, 537 (1983).
5. M. Djabourov, J. Maquet, H. Theveneau, J. Leblond and Papon, Brit. Polym. Jour. to be published.

Clay Minerals: A Molecular Approach to Their (Fractal) Microstructure

H. Van Damme, P. Levitz, J.J. Fripiat, J.F. Alcover, L. Gatineau, and F. Bergaya
C.N.R.S., Centre de Recherche sur les Solides à Organisation Cristalline Imparfaite, F-45045 Orléans Cédex, France

1 Definition, Crystal Structure, Classification and Morphology

For most users, clays are just a class of mineral materials characterized by various properties such as a very small particle size, a strong adsorption capacity, a soft touch, or a plastic behavior when wetted. For geologists and soil scientists, clays are commonly defined as the fine fraction of rocks and soils, with an upper limit for the particle size at 2 μm. It turns out that this purely granulometric definition corresponds rather closely to a particular class of hydrous silicates with layer structures, which belong to the larger group of phyllosilicates.

The elementary polyhedra of layer silicates are tetrahedra and octahedra of oxygen ions. The coordinating cations, T, in the center of the tetrahedra are essentially Si, Al or Fe^{3+}. Si being the most frequent. The coordinating cations in the octahedra are usually Al, Mg, Fe^{3+} and Fe^{2+}. As described by G. BROWN [1], "the essential features of phyllosilicate structures are continuous two-dimensional tetrahedral T_2O_5 sheets, the basal oxygens of which form a hexagonal (in ideal configuration) mesh pattern". The tetrahedra are linked by sharing three corners while "the apical oxygen at the fourth corner, usually directed normal to the sheet, forms part of an immediately adjacent octahedral sheet in which octahedra are linked by sharing edges. The junction plane between tetrahedral and octahedral sheets consists of the shared apical oxygens of the tetrahedra and unshared OH groups...".

The unit formed by linking one octahedral sheet to one tetrahedral sheet is named a 1:1 layer (Fig. 1a), whereas the unit formed by sandwiching one octahedral sheet is named a 2:1 layer (Fig. 1b). The exposed surfaces of 1:1 layers consist of oxygen atoms on the tetrahedral layer side, and OH groups on the octahedral layer side. Only oxygen atoms are exposed on the basal surfaces of 2:1 layers.

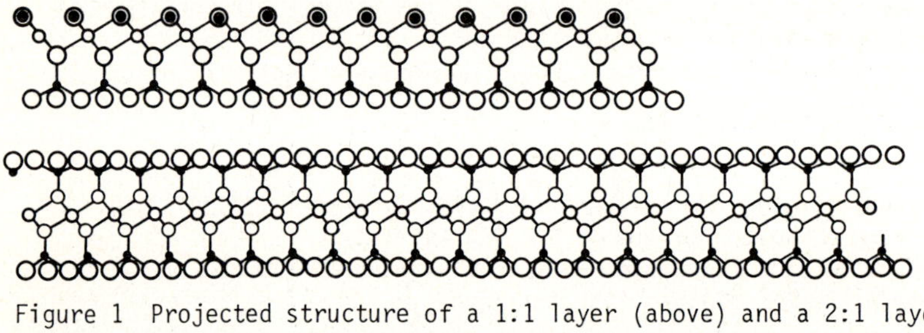

Figure 1 Projected structure of a 1:1 layer (above) and a 2:1 layer (below).

1:1 and 2:1 layers are not always neutral. Very often, as a result of isomorphic substitutions in their lattice, the layers bear a net negative charge which is balanced by interlayer material. This interlayer material can be individual cations, hydrated cations, or an additional octahedral hydroxide layer.

The layer type, the charge on the layer and the nature of the interlayer material provide the basis for the classification of clay minerals. We will not consider here all the varieties of clay minerals common in soils and sediments. An exhaustive compilation can be found in ref. [2]. A brief summary of the major groups is shown in Table 1. We will focus here only on three groups of clays, namely kaolinites, smectites and sepiolites. Each of them display very distinct and typical structural, morphological and physical properties.

Table 1. Classification of some clay minerals

Structure type	Group	Charge	Behavior in Water
1:1 layer	kaolin	∿ 0	non expanding
2:1 layer	talc	∿ 0	non expanding
	smectite	0.2 - 0.6	expanding
	vermiculite	0.6 - 0.9	expanding
2:1 inverted ribbons	sepiolite	∿ 0.1	-

Kaolinite is a typical 1:1 clay, with zero layer charge. There is no interlayer material (in the natural clays), and the elementary layers stack on each other, thanks mainly to the hydrogen bonding between adjacent layers. The resulting cristallites usually contain some degree of stacking faults [3]. Typical shapes and sizes are 0.5 μm hexagonal platelets, a few hundreds of Å thick. Water does not enter the interlayer space. This, together with the rigidity of the platelets, is of primary importance for understanding the properties of kaolins in water.

Smectites have properties almost opposite to those of kaolins. The layer charge is large, typically 0.5 negative charge per formula unit layer. Most important, the hydration energy of the interlayer cations allows water to enter the interlayer space (in clays with an even larger lattice charge the electrostatic attractive interaction becomes too strong to be overcome by the hydration forces). Smectites are therefore the prototypes of expanding clays. Typical members of the family are montmorillonites and hectorites. Morphologically, smectites appear by electron microscopy as *large* (up to two microns), *thin* (10 Å), and *flexible sheets*.

Sepiolite is a magnesian mineral with a 2:1 layer structure, like smectites, but the lateral extension of the layers is restricted to six tetrahedra, so that the layers become parallelipipedic ribbbons. Each ribbon is linked to four parallel ribbons, leading to an alternate ribbon-channel structure. The channels contain cations and water. Sepiolites have a fibrous morphology. Depending on the crystallinity, they may appear as rigid needles or more *flexible fibers*. Typical fibers have a diameter of 0.1 μm and are 2-3 μm long.

2 Clay-Water Systems: Classical Models and Electron Microscope Pictures

Among the best known properties of (some) clays are probably their ability to swell and to shrink upon addition or removal of water, and their ability to form gels or colloidal dispersions. Although they are not fully understood, swelling and gel formation are clearly related to the morphology of the elementary clay particles and to their electrical surface properties. Smectites (thin, large, flexible and charged layers) are able to form gels in water at extremely low concentrations (clay/water less than 0.01). Sepiolites (long, flexible, and weakly-charged fibers) are also able to form gels, but require larger concentrations. Kaolinites (rigid, very weakly charged platelets) do not form gels.

Because of their very particular properties, and because of their importance in nature, smectite-water systems have been extensively investigated in very broad range of clay/water (c/w) ratio, from the dry solid to the dilute suspension. Several classical concepts or models have been developed to describe these systems:

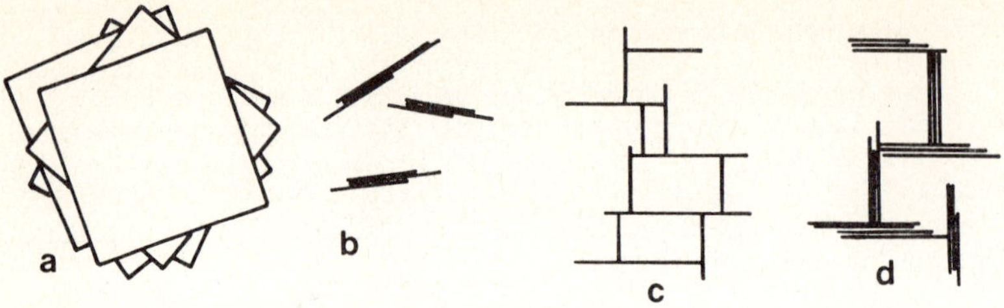

Figure 2 Schematic drawings of the turbostratic stack (a), tactoid (b), card-house (c) and book-house (d) models of smectites.

- In "dry" solids (c/w > 5), the most widely used model is undoubtedly an ensemble of independent elementary particles. Each particle is thought to result from the stacking of many layers, much like in kaolinites, but with some orientational (turbostratic) and translational disorder around the normal to the stack (Fig. 2a). This model emerged from the extensive X-ray diffraction studies of J. MERING and coworkers [4].
- In dilute suspensions, the most popular models are surprisingly similar to those of dry systems. The central concept here is the concept of tactoid [5], which is nothing else than a turbostratic micelle of a few parallel layers, in which the adjacent layers are separated by a small numbers of water layers (Fig. 2b). The average number of clay layers in the tactoid is dependent upon the nature of the interlayer cations and the history of the sample. In some exceptional cases, the tactoids might be limited to single layers.
- The classical model in gels has long been the card-house and book-house structure of H. van OLPHEN [6] (Fig. 2c and d). In these models, the gel properties are thought to be the result of the 3D arrangement of elementary layers (card house) or stacks (book-house), thanks to face-to-edge associations.

Recent electron microscopic and small angle X-ray diffraction studies have shown that this picture is most probably not correct [7,8]. Face-to-face associations seem to dominate the microstructure of smectite-water systems, in a broad range of conditions. The picture which has to be considered is a network of lenticular pores, the walls of which are connected to each other by flexible face-to-face associations of layers (Fig. 3a).

It is easy to understand that such a microstructure can hardly be achieved with the rigid platelets of kaolinites. As expected, electron microscopy shows that the elementary cristallites in kaolinite pastes touch each other only at very limited contact points [9] (Fig. 3b).

A still different arrangement is expected for the sepiolite fibers or laths. Thanks to the (limited) flexibility of the particles, one can easily imagine a highly entangled network, which is indeed observed by microscopy (Fig. 3c).

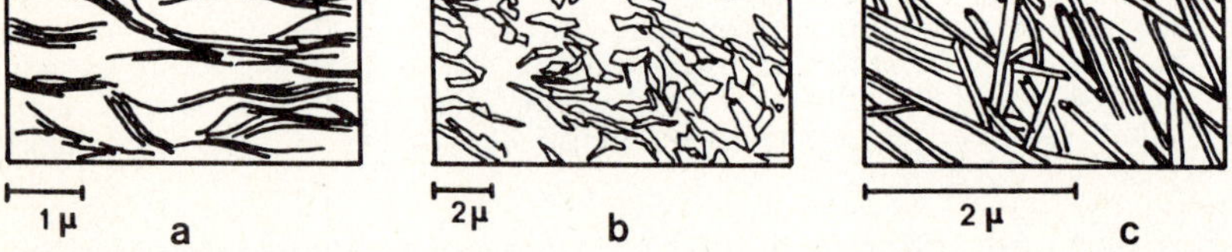

Figure 3 Schematic drawing of the particle arrangement in smectite-water (a), kaolinite-water (b) and sepiolite-water (c) systems, as revealed by electron microscopy.

3 Connectivity of the clay microstructure

One way to estimate the degree of connectivity of clay-water systems is to probe it with a molecule which is sensitive to the vicinity of the solid particles. This can be achieved by fluorescence decay spectroscopy (FDS) [10]. Indeed, consider a fluorescent probe molecule which is strongly adsorbed by the clay and which is quenched by some ions of the clay structure. The fluorescence properties of such a probe (quantum yield and excited state lifetime) will clearly be different if the molecule is merely adsorbed on *one* particle or if it is "sandwiched" in the contact zone between *two* adjacent particles. Hence, fluorescence should provide a way for estimating the fraction of total particle area involved in interparticular contact, or at least, for estimating the way this area changes upon varying physico-chemical factors such as c/w, temperature, or salt concentration.

This experiment was performed with a coordination compound of ruthenium(II) as fluorescent probe. This complex is strongly quenched by the Fe^{3+} ions, which are always present in variable amount in the clay lattice. The time law of the decay was found to be typical of a probe with very restricted (cluster limited) mobility. In particular, the long time limit of the decay was found to be totally independent of trap concentration. In such a situation, the observed decay is the ensemble average of the local decays over a distribution of trap concentrations. In the limit of low trap concentration, the decay law is:

$$P(t) = \exp(-k_o t)\ \exp\left\{-\bar{n}\ [1 - \exp(-k_q t)]\right\} \qquad (1)$$

where $k_o = 1/\tau_o$ is the unimolecular rate constant for decay in a trap-free environment, $\bar{n}$ is the average number of traps per cluster and k_q is the trapping probability per unit time in a cluster containing one trap.

$\bar{n}$ was measured for montmorillonite (the most common smectite), sepiolite and kaolinite in dilute suspension (c/w = 0.01) and in "dry" solid deposits, from c/w = 10 to c/w = 50). The domain of concentrated suspensions, gels and pastes has not yet been investigated. Nevertheless, the behavior of $\bar{n}$ (Fig. 4) closely reflects the electron microscopy pictures. In kaolinite, the contiguity is independent of the hydration level. This is what is expected for a disordered arrangement of rigid platelets, in which intimate contact over large area is hindered even by the smallest orientational disorder. In sepiolite, $\bar{n}$ increases somewhat upon drying the sample, as expected for an entangled network of fibers with some flexibility. In montmorillonite, the important increase of $\bar{n}$ upon drying is clearly consistent with the occurrence of large face-to-face associations, firmly connecting the adjacent particles.

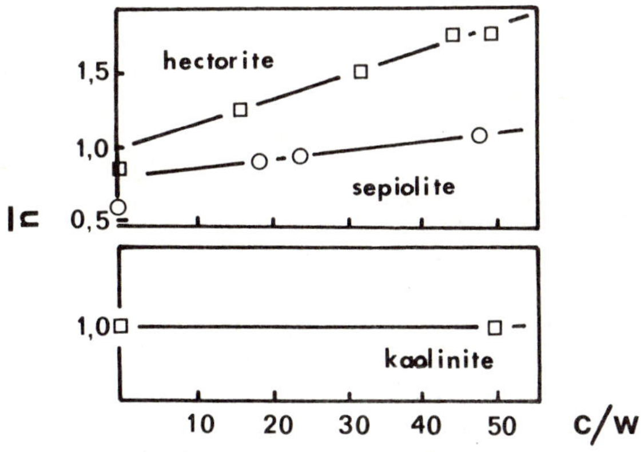

Figure 4 Average number of lattice Fe^{3+} traps accessible to an adsorbed fluorescent probe in various clay/water systems, as a function of c/w ratio.

As far as the microstructure of clay materials is concerned, the most interesting result is perhaps the continuity of the structural evolution upon wetting or drying, at least in sepiolite and in smectites. This can hardly be understood in terms of independent stacks of layers or in terms of independent bundles of fibers. An entangled system, in which drying would lead to a continuously increasing interdigitation, seems to be more consistent with the results. We suggest that self-similar networks might provide a suitable model for such an arrangement, and we will give additional arguments in what follows.

4 The Tactoids: Micellar Stacks or Fractal Aggregates ?

In the classical models of smectites (independent stacks of large and thin layers), the physical meaning of the concept of internal surface is clear: it corresponds to the interlayer space. The meaning of the external surface is obvious: it is just the sum of the upper and lower basal planes, plus the lateral surfaces of the stack (Fig. 5a). If the thickness of the stack is much smaller than their lateral extension, the external surface reduces to the upper and lower basal planes. The external surface, S_e, the internal surface, S_i, and the average number of layers per stack, N, are related by the simple relationship

$$S_t \,/\, S_e = N \qquad (2)$$

where $S_t = S_e + S_i$

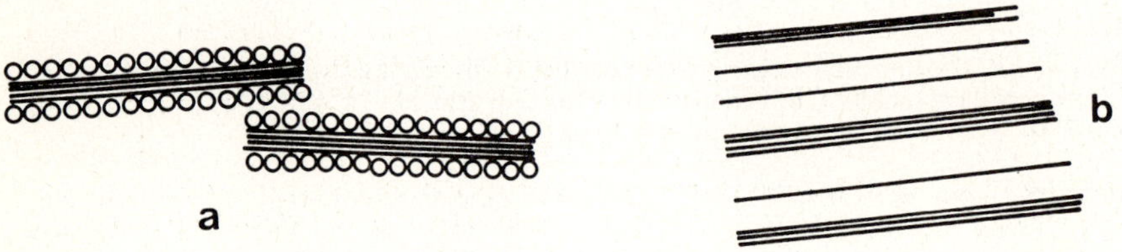

Figure 5 (a) Schematic drawing of the meaning of the basal external surface of a tactoid; (b) model for the arrangement of layers in a Na-montmorillonite suspension, derived from SAXS [8].

The total (basal) surface can easily be calculated, a priori, from the specific weight of the smectite ($\sim 2.65\ g/cm^3$). It is of the order of 800 m^2/g for a Na-montmorillonite. The external surface accessible of water in a smectite-water system is much more difficult to measure. Any adsorption process is indeed expected to perturb to some extent the microstructure of the system. However, NMR provides a means for measuring S_e by using the best molecular probe one could think of: the water molecule itself. FRIPIAT et al. [11] have shown that the spin-lattice relaxation time, T_1, of the first layers of water molecules ($\sim$ 3 layers, or about 10 Å) on the external surface of the tactoid (phase "b") is drastically different from the T_1 of the bulk water in the system (phase "a"). Since the two populations of molecules are in rapid exchange, the observed relaxation time is simply

$$T_1^{-1} = x_a \, T_{1a}^{-1} + x_b \, T_{1b}^{-1} \qquad (3)$$

where x_a and x_b are the mole fraction of phase a and phase b, respectively (the interlayer water is not in rapid exchange with the external water. It gives rise to a separate T_1 which can easily be separated from the T_1 of the external water). x_a is very close to 1 in dilute suspensions and gels, and x_b can thus be measured from the slope of the T_1 vs (c/w) relationship (T_{1b} is separately measured in clay powders). Taking into account the thickness of phase b, the external surface area of the tactoids is then easily calculated.

The results of these experiments, performed on hectorite/water systems ($10^{-3} < c/w < 0.18$) show that the average external surface area, S_e, is of the order of 80 m^2/g, in the whole concentration range which was explored. This was interpreted in terms of the classical tactoid model of Fig.5a, and was taken as evidence that a high degree of clay layers aggregation occurs, even in highly diluted suspensions, and that the average size of the tactoid ($\simeq$ 10 layers/tactoid) does not change significantly upon increasing the clay concentration by two orders of magnitude ; In other words, the classical interpretation of these results suggests a micellization process, in which any increase in clay concentration leads merely to an increasing number of independent tactoids, all identical to each other.

Serious (apparent) discrepancies appear between these NMR results and the small angle X-ray scattering results of PONS et al. for Na-montmorillonite [8]. Indeed, SAXS leads to the following conclusions : (i) in the diluted sol state, most of the layers are individual layers, with separation distances larger than 100 Å ; (ii) as the clay concentration increases up to $c/w \simeq 0.3$, the system gelified and "soft" stacks of layers with large internal separation distances, around 35 Å, start forming. At $c/w = 0.25$, the average number of layers per stack is around 8, and at $c/w = 0.40$, it is around 20 ; (iii) in the vicinity of $c/w = 0.40$, another characteristic transition occurs. The system goes into a hydrated solid state, in which "hard" stacks with internal separation distances around 31 Å are separated by "soft" stacks in a disordered interstratified arrangement. Drying further the system thickens the hard stacks with layers from the soft stacks.

In fact, it seems that these discrepancies can be partially resolved by considering fractal aggregation models, at least in some dimension range; In order to see to what extend such models might account for the main trends of the experimental results obtained with smectites, we have performed [12] a two-dimensional diffusion-limited aggregation (DLA) [13] of anisotropic particles with aspect ratios up to 19, all parallel to each and to the seed. DLA is probably not the best aggregation model in this case. Clustering of clusters would be more realistic, but DLA should show the main features. The morphological, structural and surface properties of the aggregates can be summarized as follows :

(i) the aspect ratio of an aggregate is, within the range of the fluctuations, the same as the aspect ratio of the particles it is made of .

(ii) the specific external perimeter of the aggregates (S_e/N) reaches very rapidly a constant value, for a given aspect ratio. For instance, for an aspect ratio of 9, $S_e/N = 0.4$ (normalizing the perimeter ofone particle to 1). Hence, in the classical tactoid model, this would correspond to an ensemble of independent stacks of 2.5 layers, on the average, although the real aggregate may contain more than 10^3 particles !

(iii) the interference function, $I(0,k)$ in the direction perpendicular to the long axis of the aggregate, i.e. in the $(0,k)$ direction of reciprocal space, gives, as expected, strong $(0,k)$ reflexions corresponding to adjacent particles with period d, but also weak ($I/I_0 \simeq 0.03$) modulations corresponding to much larger periods (for instance, for an aggegate of 1000 particles with aspect ratio = 9, the maximum of the modulation appears at $s = 0.04/d[0;1], Å^{-1}$). A cross-sectional analysis of the mass distribution within the aggregate suggests that these large periods might be related to the distribution of "fingers" of the aggregate (Fig.6);

(iv) finally, the morphology of an enlarged picture, showing the particle entanglement in the aggregate, is surprisingly similar to the transmission electron micrographs obtained with real smectites [7], although the aggregation procedure did not allow for particle flexibility.

When taken together, these features clearly suggest that fractal models might improve our insight of clay microstructures (at least for smectites and sepiolites). The model proposed here should merely be considered as an exploratory attempt. The details of the aggregation process, the dimension range of self-similarity, and the dependence of the associated fractal dimension upon physico -chemical factors, are all important points which are to be investigated before a critical evaluation could be performed.

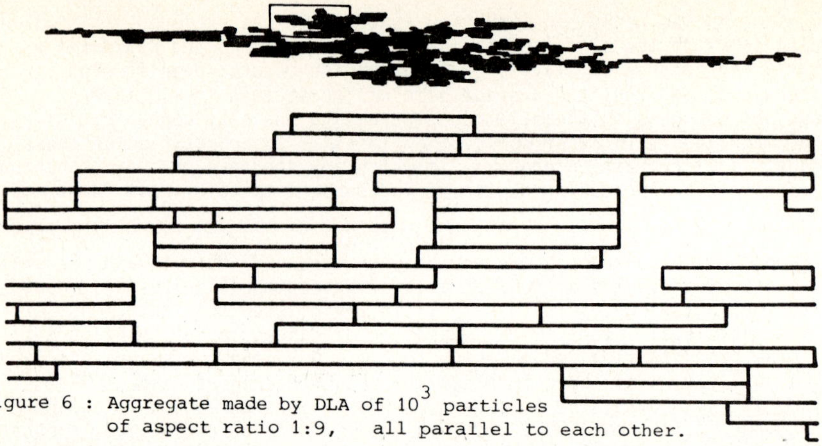

Figure 6 : Aggregate made by DLA of 10^3 particles of aspect ratio 1:9, all parallel to each other.

1. G. Brown : Clay Minerals : Their Structure, Behavior and Use : London, Royal Society, pp 1-20 (1984).
2. S.W. Bailey : Crystal Structures of Clay Minerals and their X-ray Identification, London, Mineralogical Society, Chap. 1 (1980).
3. C. Tchoubar : Clay Minerals : Their Structure, Behavior and Use : London, Royal Society, pp.99-50 (1984).
4. J. Mering : Acta Cristallogr., 2, 371 (1949).
5. A. Banin and N. Lahav : Isr. J. Chem. 6, 235 (1968).
6. H. van Olphen : J. Colloid Interf. Sci. 19, 313 (1964).
7. D. Tessier and G. Pedro : Developments in Sedimentology, 35, 165 (1982).
8. C.H. Pons, F. Rousseaux and Tchoubar : Clay Minerals, 16, 23 (1981).
9. D. Tessier, G. Pedro and L. Camara : C.R. Acad. Sci. Paris, 290D, 1169 (1980).
10. A. Habti, D. Keravis, P. Levitz and H. van Damme : J. Chem. Soc. Faraday Trans. 2, 80, 67 (1984).
11. J. Fripiat, J. Cases, M. François and M. Letellier : J. Coll. Interf. Sci., 89, 378 (1982).
12. J.F. Alcover, L. Gatineau, P. Levitz and H. van Damme : in preparation.
13. T.A. Witten and L.M. Sander, Phys. Rev. Lett., 47, 1400 (1981).

Structure of Random Materials

D.W. Schaefer, J.E. Martin, A.J. Hurd, and K.D. Keefer
Sandia Natinal Laboratories, Albuquerque, NM 87185, USA

Light scattering and small angle x-ray scattering results are reported for a variety of random materials. Random processes such as polymerization and aggregation account for the structure of these materials. Materials studied include linear and branched polymers, colloidal aggregates (prepared in solution, in flames and at an air-water interface), and composites. Although the concept of fractal geometry is essential to interpretation of the scattering curves, not all the materials show fractal character.

In spite of the enormous interest in the theoretical physics community in random growth processes, there have been few experimental studies of the products of random growth, namely random materials. In this paper we report small-angle x-ray and light scattering studies of a variety of random materials including porous silica aerogels, carbon black composites, linear and branched polymers, and colloidal aggregates prepared in solution, in flames, and at an air-water interface.

Scattering principles

To interpret scattering curves it is convenient to distinguish between structures for which the density is or is not dependent on the mass. For systems in the former category, such as polymers in solution, the scattered intensity, I, is often power-law in the magnitude of the scattering vector, K.

$$I(K) \sim K^{-D} \tag{1}$$

Here $K = 4\pi\ \lambda^{-1}\ (\sin\theta/2)$, where λ is the incident wavelength and θ is the scattering angle. D is the fractal dimension which relates the mass, M, to radius, R.

$$M \sim R^{D} \tag{2}$$

For random objects, D may be nonintegral and thus D is known as the fractional or fractal dimension [1]. Equation (1) is valid only for the "Porod" region of the scattering curve, which is defined such that K^{-1} is smaller than the correlation range ξ of the system [2].

Objects described by eqs. (1) and (2) are called "mass" fractals. Examples include polymers [3] diffusion-limited aggregates [4], and percolation clusters [5]. Table 1 gives the Porod exponents for various model objects.

In contrast to mass fractals, the interior of "surface" fractals is uniform, so scattering occurs only from the surface. Surface fractals are part of a more general class of objects for which the density is independent of the mass. Solid spheres belong to this class. For this case, Bale and Schmidt [6] show,

$$I(K) \sim K^{D_s-6} \tag{3}$$

where D_s is the "surface" fractal dimension, which relates surface area, S, to particle radius

$$S \sim R^{D_s} \tag{4}$$

Fractal surfaces are rough on all length scales up to some surface correlation range, ξ_s, beyond which they are likely to be smooth or otherwise non-fractal ; a cross-over to smooth texture may also occur on small scales. Hence, eqs. (1) and (3) are applicable only in the Porod regime, where scattering probes length scales small compared to ξ or ξ_s but large compared to atomic dimensions.

Table 1 distinguishes the two classes of fractal objects according to the observed Porod slope. Mass fractals have slopes between -1 and -3, and surface fractals have slopes between -3 and -4. Cognizance of these categories is essential to interpretation of the scattering from random materials.

Table 1

Porod slopes for various structures

	Slope	
Linear ideal polymer[22] (random walk)	-2	
Linear swollen polymer[22] (self-avoiding walk)	-5/3	
Randomly branched ideal polymer[22]	-16/7	Mass Fractals
Swollen branched polymer[22]	-2	Slope = -D
Diffusion limited aggregate[4]	-2.5	
Multiparticle diffusion limited aggregate[17]	-1.8	
Percolation[5] (single cluster)	-2.5	
Fractally rough surface[6]	-3 to -4	Surface Fractals
Smooth surface[6]	-4	Slope = D_s-6

Polymers

Figure 1 shows the measured scattering curves for a variety of polymers. The polystyrenes(PS) are linear polymers dissolved in cyclopentane, which is a theta solvent for PS at 20°C, As expected from Table 1, the slope in the Porod region is -2 at 20°C, consistent with the well-known result that theta polymers display random-walk statistics. At higher temperatures, the chain swells and scattering slopes approach the value for the self-avoiding walk. It is interesting to note that at large K and intermediate temperatures, intermediate slopes are observed [3] with no indication of a cross-over to -2 at a large K. In contrast with the so-called "blob" model [7] of chain statistics, these chains do not display ideal statistics on small length scales.

Several silicate polymers are also shown in Fig.1. These polymers are all prepared by condensation polymerization of silicon alkoxides. Depending on polymerization conditions, a variety of structures is possible [8]. The commercial colloidal silicate, LUDOX™, is prepared by polymerization of $S_1(OH)_4$. The observed Porod slope of -4 confirms that the colloid surface is smooth and non-fractal [2]. The smooth surface is the result of the tetrafunctional nature of the silisic acid monomer.

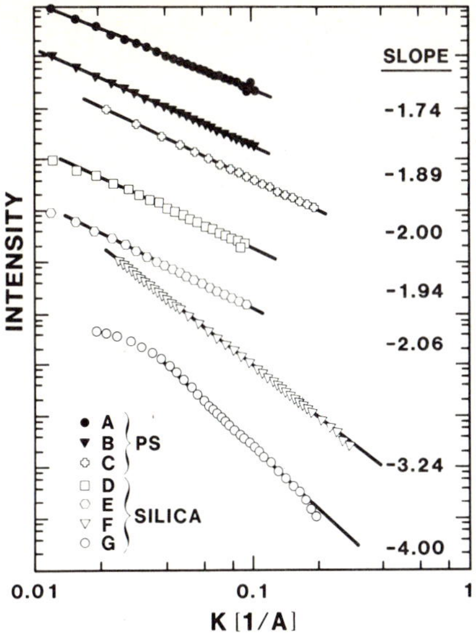

Figure 1. Scattering curves for various polymers. Curves A-C are for polystyrene in cyclopentane at and above the theta temperature [A=65°C, B=30°C and C=theta=20°C]. Curves D-G are silicates polymerized under different conditions. Curve G represents the most aggressive hydrolysis conditions which leads to smooth colloidal particles. Curve F is a fractally rough colloid prepared at near stoichiometric H_2O/Si. Two-step polymerization leads to branched polymers in D and E. D is acid-catalyzed and E is base-catalyzed.

Under mild hydrolysis conditions (near stoichiometric H_2O/Si), where the polymerizing monomers are not all tetrafunctional, a new and unexpected structure is formed [9]. The data for this case show a Porod slope of -3.3, which, from Table 1, indicates a fractal surface. We call these structures rough colloids. To our knowledge such structures have not been seen previously.

Since the fractal surfaces are realized under mild hydrolysis conditions, we believe they arise because of random polymerization of partially hydrolyzed monomers. We simulate such a process [10] by Eden-like [5] growth on a lattice. In this model, growth starts from a seed. As new sites are occupied (chosen at random from available perimeter sites), a specified number (one to four, corresponding to tetrafunctional silicon) of sites are blocked or "poisoned". Depending on the details of the poisoning, structures with fractal hulls (exterior surfaces) may be generated. Figure 2 shows a cluster which was generated from an equal mixture of 2-, 3- and 4-functional monomers. Although this cluster has mass as well as hull-fractal character, we believe that subsequent uniform depoisoning (ripening) will lead to uniform cores with fractally rough surfaces remaining.

If the silicates are prepared by a two-step polymerization, true branched polymeric structures result [11, 12]. In the first step, a weakly branched prepolymer is formed under mild hydrolysis conditions and sub-stoichiometric H_2O/Si. Upon subsequent linking of these prepolymers, a non-colloidal polymeric structure is produced. Consistent with Table 1 — and the observed slopes near -2 in Fig.1 — we believe these structures are best modeled as lattice animals. Dilution studies [11] of the branched silicates show that base catalysis leads to more compact, more highly branched polymers. Nevertheless, slopes near -2 are found regardless of the degree of branching. This result was predicted by Family for asymptotically large polymers [13].

Colloidal aggregates

Starting with the polymeric and colloidal structures in Fig.1, it is possible to create a variety of new structures by aggregation, gelation and drying. First consider the colloidal aggregates. If a stabilized colloidal silica solution is destabilized by reducing the pH and increasing the ionic strength, highly ramified colloidal aggregates are formed. The lower curve of Fig.3 is the scattered intensity from a silica aggregate [14]. Two regimes are found, with Porod slopes of -4 and -2. The slope of -4 indicates smooth surfaces for dimensions smaller than the size of the unaggregated particles, and the slope of -2 indicates a polymer-like fractal structure on larger length scales. Light scattering studies [14] reveal a slope of -2.1 ± .05 down to K = .0001 $Å^{-1}$. We conclude that these aggregates are branched, chain-like structures between 100 and 10000 Å.

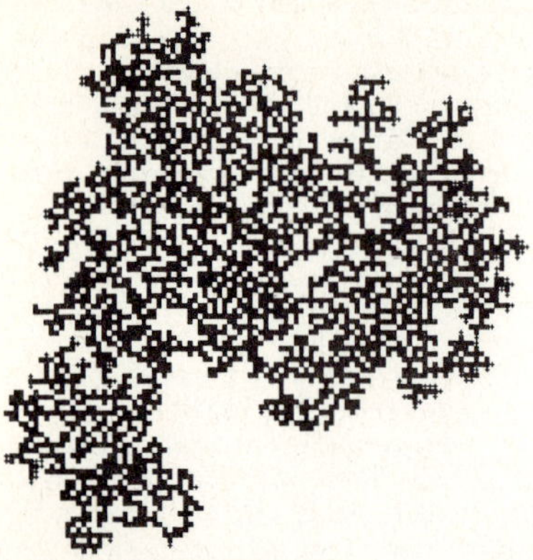

Figure 2. A cluster with a fractally rough hull. The cluster was generated by random addition of 2-, 3- and 4-functional monomers to a seed on a square lattice.

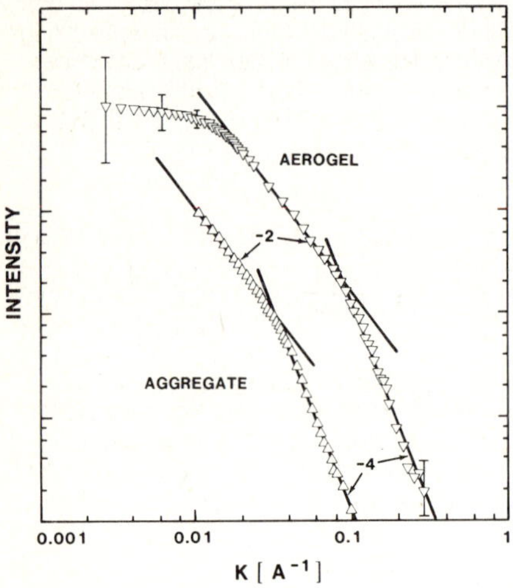

Figure 3. Scattering curves for a colloidal silica aggregate (lower) and a porous silica aerogel (upper).

Two-dimensional aggregation was also studied by trapping 3000 Å silica particles at an air-water interface and destabilizing the system by increasing the ionic strength of the aqueous substrate [15]. The resulting aggregates were then studied by optical microscopy. A photomicrograph is shown in Fig. 4. The fractal dimension of these aggregates was obtained by analysis of the photomicrographs, using eq. (2), with the result D = 1.20 ± 0.15. This value is smaller than any known model for two-dimensional growth. We attribute the low D to biased growth caused by anisotropic repulsive forces at the tip of a growing "dendrite".

Colloidal particles and aggregates can also be prepared in flames. Fumed silica, for example, is prepared by burning $SiCl_4$. Depending on flame conditions, highly ramified colloidal clusters result. We have studied the structure of a fumed silica (Cabosil M5) by a combination of x-ray and light scattering, and the results [16] are combined in Fig.5. The data show two power-law regimes with slopes of -1.8 and -4.0. Like their solution counterparts, these clusters are smooth on scales below 100 A and are mass fractals on scales above 1000 A. The most reasonable model explaining the fractal nature of these clusters is multiparticle diffusion-limited aggregation [17]. This result differs from an earlier study [18] of this system in which the intermediate slope in the cross-over regime near K = 0.01 $Å^{-1}$ was taken as evidence for a mass fractal with D = 2.5. The data in Fig.5 show that fumed silica is non-fractal on scales below 100 Å. This result is consistent with the recent studied of Avnir et al. [19] who explain scale-dependent adsorption in terms of fractally rough surfaces. They find, however, that D_s = 2 for fumed silica.

Carbon blacks are a second class of gas-phase aggregates that display ramified structures in the electron microscope [20]. Carbon black is prepared by pyrolysis of hydrocarbons in an oxygen-deficient flame. Two blacks were studied : Spheron 6 and Vulcan 3 from Cabot corporation. The data are shown in Fig.6. The more extensive data on Spheron 6 shows non-fractal, smooth surfaces below 100 Å. This result is consistent with adsorption studies [19]. For $0.002 < K < 0.01$, the window between the x-ray and light scattering data precludes definitive statements. It appears, nevertheless, that these structures are non-fractal or at best fractal over a limited regime with D = 2.1.

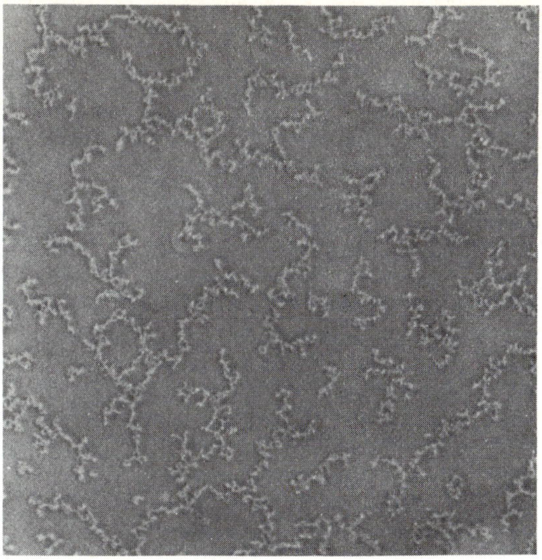

Figure 4. A two-dimensional silica aggregate prepared at an air-water interface.

Carbon black composites

Because of the relatively high conductivity of graphitic carbon, porous carbon black electrodes are often used in electrochemical applications. We studied two composites made with teflon and polyethylene binders. These materials are manufactured by Honeywell and Raychem, respectively. The data are shown in Fig.7. Although the curves are nearly identical for the two materials, they are distinctly different from the Spheron 6 in Fig.6. Both materials have slopes between -3 and -4, and thus appear to be surface fractals below 50 Å. Since carbon black itself is non-fractal, the origin of the fractal surface is uncertain : one possibility is variability among carbon blacks depending on flame conditions. Unfortunately we do not have the actual blacks used in the composites. Another possibility is that the porosity of these materials is not determined by the surface structure of the filler. Since power-law pore size distributions [6] can also give slopes between -3 and -4, it is quite possible that the observed curves reflect a fractal distribution of pore sizes. Such a distribution is equivalent to a fractally rough surface. Electrochemical studies [21] indicate D_S = 2.7 for the teflon composite.

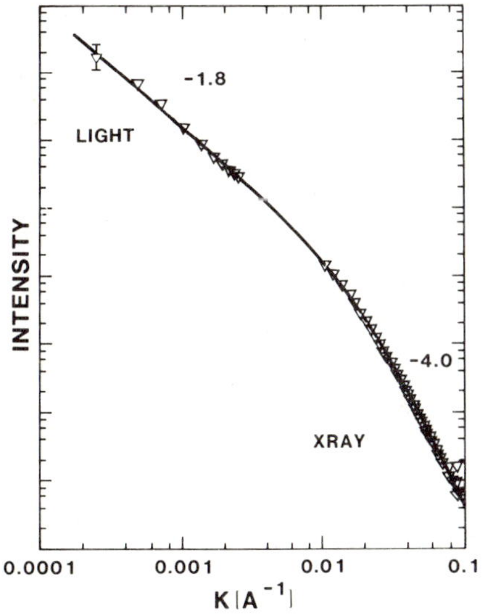

Figure 5. Scattering curve for fumed silica (Cabosil M5).

Porous silica gels

We studied two porous silicas made by drying alcoholic silicates. The two materials have radically different structure, which we attribute to differences in polymerization conditions in the solution precursor.

A silica aerogel with a density $\rho = 0.088 g/cm^3$ was obtained from Airglass AB, Sjobo, Sweden. This material is prepared by base-catalyzed hydrolysis and condensation of silicon tetramethoxide in alcohol. The condensation ultimately leads to a gel which is critical-point dried to yield a nearly transparent, fragile solid. The scattering data for this material are shown in the top curve of Fig.3. The flat region of the curve for $K < 0.01$ Å^{-1} shows that the material is uniform for lengths larger than 100 Å. The rest of the curve looks very much like the lower curve for the colloidal

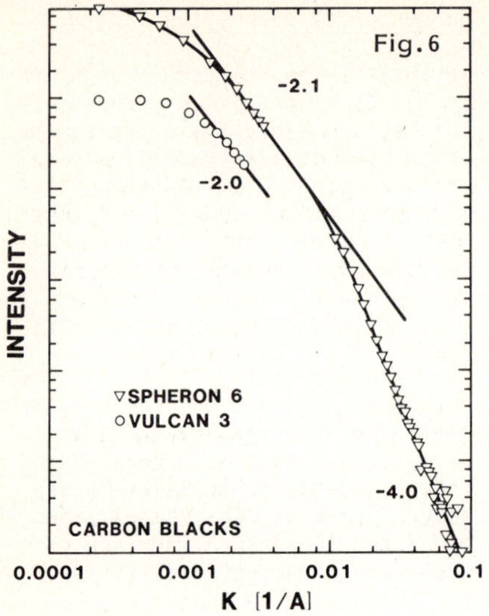

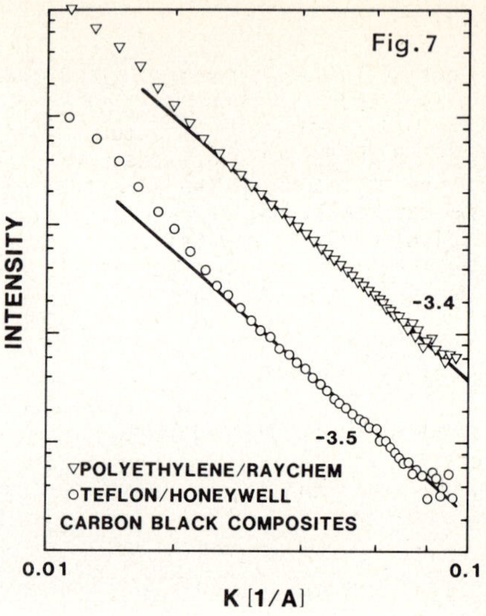

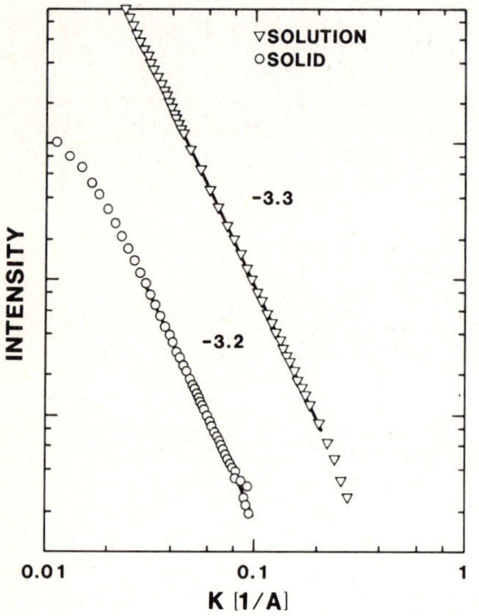

Figure 6. Scattering curves for carbon blacks.

Figure 7. Scattering curves for carbon black composites.

Figure 8. Scattering curves for silica polymerized in solution (▽) with $H_2O/Si = 2$ and for the porous solid made by air drying this solution. Both curves indicate surface fractals.

aggregate. Both show slopes of -2 and -4. Because of the similarity of the two curves, we conclude that the structure of the material is identical to the aggregates. The solid is composed of branched chain-like struts of colloidal "monomers". This unique structure demonstrates that the common notion of porosity as interconnected bubbles is completely wrong for this material.

Another porous silica was made by air drying the rough colloidal particles studied in Fig.1. The scattering data [9], along with those for the solution precursor from Fig.1, are plotted in Fig.8. In the intermediate regime, the power-law exponent for the two curves is nearly identical, corresponding to a fractal surface with $D_s = 2.7$. This solid then is composed of fractally rough colloidal particles in a randomly packed array.

Conclusion

The scattering data presented here indicate that a wide variety of different structures is found in random materials, and that an equally wide variety of growth processes determines these structures. Although the emphasis here is on polymerization and aggregation processes, phase separation and ripening are other important processes that control the structure of materials. In view of the prevalence of random materials, natural and synthetic, it seems that an entire materials science discipline could be devoted to the study of the control of material properties through chemical and physical growth processes. We hope that our work convinces the reader of the value of such studies.

Acknowledgments

We thank George Wignall for the polyethylene composite. Most of the x-ray data were taken at the National Center for Small Angle Scattering Research at the Oak Ridge National Laboratory. We thank Steve Spooner and J.S. Lin for help in collecting these data. This work performed at Sandia National Laboratories supported by U.S. Dept. of Energy under Contract N° DE-AC04-76DP00789.

References

1. B.B. Mandelbrot, Fractals, Form and Chance, Freeman, San Francisco, 1977.
2. D.W. Schaefer and K.D. Keefer, Mat. Res. Soc. Symp. Proc., 32, 1 (1984).
3. D.W. Schaefer and J.G. Curro, Ferroelectrics, 30, 49 (1980).
4. T.A. Witten and L.M. Sander, Phys. Rev. Lett., 47, 1400 (1981).
5. H.E. Stanley in Structural Elements in Particle Physics and Statistical Mechanics, K. Fredenhagen, and J. Honerkamp, Eds., Plenum, New York, 1982.
6. H.D. Bale and P.W. Schmidt, Phys. Rev. Lett., 53, 596 (1984).
7. B. Farnoux, F. Boué, J.P. Cotton, M. Daoud, G. Jannink, M. Nierlich and P.G. de Gennes, J. Phys. (Paris) 39, 77 (1978).
8. K.D. Keefer, Mat. Res. Soc. Symp. Proc., 32, 15 (1984).
9. K.D. Keefer and D.W. Schaefer, to be published.
10. K.D. Keefer, J.E. Martin and D.W. Schaefer, to be published.
11. D.W. Schaefer and K.D. Keefer, Phys. Rev. Lett., 53, 1383 (1984).
12. C.J. Brinker, K.D. Keefer, D.W. Schaefer, R.A. Assink, B.D. Kay, and C.S. Ashley J. Non Cryst. Solids, 63, 45 (1984).
13. F. Family, J. Phys. A, 13, L325 (1980).
14. D.W. Schaefer, K.D. Keefer and J.E. Martin, J. Phys. (Paris) Colloq. (1985).
15. A.J. Hurd and D.W. Schaefer, Phys. Rev. Lett., 54, 1043 (1985).
16. J.E. Martin, D.W. Schaefer and A.J. Hurd, to be published.
17. P. Meakin, Phys. Rev. Lett., 51, 119 (1983); M. Kolb, R. Botet and J. Jullien Phys. Rev. Lett., 51, 1123 (1983).
18. S.K. Sinha, T. Freltoft and J. Kjems, in Kinetics of Aggregation and Gelation, F. Family and D.P. Landau, Eds., Elsevier-North Holland, Amsterdam, 1984.
19. D. Avnir, D. Farin and P. Pfeifer, J. Chem. Phys. 79, 3566 (1983).
20. A.I. Medalia and F.A. Heckman, Carbon, 7, 567 (1969).
21. A. Le Méhauté and G. Crepy, Solid State Ionics, 9 and 10, 17 (1983).
22. M. Daoud and F. Family, J. Phys. (Paris), 42, 1359 (1981).

Part II **Microemulsions**

Ionic Microemulsions

P. Pincus, S. Safran, and S. Alexander
Exxon Research and Engineering Company, Annandale, NJ 08801, USA
D. Hone
Department of Physics, University of California, Santa Barbara, CA 93106, USA

We review some of the theoretical models for microemulsion behavior, based on the Schulman conjecture for surfactant saturated interfaces. Emphasis is placed on the Safran-Turkevich application of Helfrich curvature energy concepts to globular shapes, phase transitions, and interfacial tensions. Extension of this model to include the Coulomb forces associated with ionic surfactants suggests that the electrostatic effects renormalize the curvature energies to favor water external phases.

I. Introduction

Microemulsions are equilibrium phases of two immiscible fluids, e.g. water and oil, emulsified by surfactants. Over the last few years, a great deal of experimental and theoretical effort has been applied to uncover the structures and properties of these exotic phases. Technological impetus to study such complex fluids arises, e.g. from their applications to enhanced oil recovery. Their intellectual scientific interest derives from several intriguing problems and possibilities, e.g. what determines whether the phases will be low viscosity microemulsions or ordered elastic structures similar to the lyotropic phases constructed with surfactants and a single solvent? Often the microemulsions may be described in terms of swollen micelles which encapsulate one solvent while the other solvent forms an external phase. As a function of a control parameter, such as ionic strength, the micelles may invert. What is the nature of the solution in the inversion region? Is there a bicontinuous phase formed of two randomly interleaved channels? These and other questions have been discussed theoretically over recent years.

An idealized picture of the essential features of the microemulsion phases has been emerging. In particular, if the surfactants are assumed to be insoluble in either fluid, SCHULMAN [1] has suggested that they accumulate and form interfaces which have the property of ultra-low interfacial tension. It is the organization of these oil-water boundaries which seem to be at the heart of the problem. DE GENNES and TAUPIN [2] have elegantly discussed the interfaces in terms of HELFRICH'S [3] ideas on curvature energies. They consider both globular micelles and fluctuating lamellar phases. If one considers an instantaneous snapshot of a microemulsion, two possible situations emerge from the de Gennes-Taupin viewpoint: (i) GLOBULAR phase composed of one of the fluid components encapsulated by the surfactant saturated interfaces embedded in a continuum made up of the other solvent and (ii) BICONTINUOUS phase which may be imagined to be derived from strongly fluctuating lamella. Of course, when dynamic considerations are included, it is conceivable that this convenient classification loses meaning. In this school, the present experimental situation will be reviewed by Anne-Marie Cazabat and Christianne Taupin. In particular, Cazabat will focus on the globular structures and the interactions between them as determined by transport measurements. Taupin's presentation concentrates on the bicontinuous picture and the various experimental facts related to its characteristics.

In a series of papers, TALMON and PRAGER [4-6] emphasized the role of mixing entropy in order to describe the bicontinuous phases. This work was simplified and extended by JOUFFROY, LEVINSON and DE GENNES [7]. Further elaboration and detailed considerations were carried out by WIDOM [8] who added the further feature of a microscopic cutoff, presumably associated with surfactant dimensions.

On the other hand, SAFRAN and TURKEVICH [9] have concentrated their attention on interfacial curvature energy aspects. In these terms, they are able to derive phase diagrams that appear to correspond to some experimentally observed cases and to extend their model to predict globular shape transitions [11]. In the next section, we shall briefly review this approach and recall some of its consequences.

The purpose of the present work is to further develop the curvature energy approach with the inclusion of Coulombic effects associated with ionic surfactants. In Section III, basing our analysis on the Boltzmann-Poisson equation, we suggest that the principal contribution of the ionizable groups is to bias the system toward water external phases. This ultimately results in the renormalization of the parameters which characterize the curvature effects, as suggested by DE GENNES and TAUPIN [2].

Let us briefly recall the SCHULMAN [1] argument. Assume that the system contains n_s surfactant molecules which are insoluble in either solvent. Consequently, they reside at the interface (total area A) which we take to be of arbitrary shape. Then neglecting curvature effects and interactions between different pieces of interface, the interfacial contribution to the free energy may be expressed as

$$f = \gamma_0 A + n_s G(\Sigma) \qquad \text{I-1}$$

where γ_0 is the bare oil-water interfacial tension and $G(\Sigma)$ is the surfactant free energy which depends only on the area per molecule $\Sigma = A/n_s$. Minimizing f with respect to Σ at fixed n_s, we obtain the condition

$$\gamma_0 + dG/d\Sigma = 0 \qquad \text{I-2}$$

which is equivalent to the vanishing of the actual surface tension, $\gamma = \partial f/\partial A$. Therefore, the system will find an optimal area per surfactant Σ^* which is implicitly given by the solution to I.2. The total interfacial area is then set by the number of surfactants to be $A^* = n_s\Sigma^*$.

II. Curvature Effects

The SAFRAN-TURKEVICH [9] approach is then to start with the Schulman condition and augment the interfacial free energy with the Helfrich curvature contribution which may generally be written as

$$f_b = (1/2)K\int(R_1^{-1} + R_2^{-1} - 2\rho_0^{-1})^2 dS + (1/2)K'\int(R_1^{-1} - R_2^{-1})dS \qquad \text{(II.1)}$$

where $R_{1,2}$ are the two local radii of curvature, ρ_0 is a natural radius of curvature which has its origin in the intrinsic asymmetry of the surfactant [12], K is the interfacial rigidity and K' is the saddle splay constant. The spontaneous radius of curvature ρ_0 may be either positive or negative. We define $\rho_0 > 0$ when the interface bends toward the water. This contribution must be augmented with entropic effects arising from interfacial fluctuations and constraints associated with: (i) incompressibility of the solvents and (ii) insolubility of the surfactants in the either solvent. These may be expressed in terms of x, the volume fraction of water in the system and v_s, the volume fraction of surfactant.

If we assume a concentration c of monodisperse spherical globules of radius ρ, the constraints become

$$x = (4/3)\pi\rho^3 c \ ; \ v_s = 4\pi\rho^2 \delta c \qquad \text{II.2}$$

where δ is a characteristic surfactant length. Then the bending free energy density becomes

$$F_b = 2k(1 - \rho_0/\rho)^2 \qquad \text{II.3}$$

where $k = Kv_s(\rho_0{}^2\delta)^{-1}$ is a rigidity density which is constant at fixed surfactant concentration. As the concentration of water is increased at fixed v_s, ρ grows linearly until $\rho = \rho_0$. At this point the chemical potential of the water, $\partial F_b/\partial x$, is zero and the added water prefers to phase separate into a nearly pure excess phase. This two-phase coexistence has been called emulsification failure. This discussion neglects both the entropy of mixing of the swollen micelles and their mutual interactions. SAFRAN and TURKEVICH [9] have shown that if such terms are included and if the intermicellar forces are attractive, as might be expected for non-ionic surfactants with a water internal phase [10], another phase equilibrium may be provoked with a coexistence of dilute and dense microemulsions. This is completely analogous to the standard phase separation of a binary fluid mixture with a miscibility gap. Both emulsification failure and the "liquid-gas" phase separation may simultaneously occur yielding a possible picture for the well-known middle phase microemulsions [9].

The application of microemulsions to tertiary oil recovery depends crucially on the ultralow interfacial tensions observed between the middle and both upper and lower phases. It is relatively easy to rationalize this behavior in terms of the S-T model. For the interface between pure water and the micellar "gas", the flat meniscus will have an effective tension determined by the curvature energies and should be approximately [2,13] given by

$$\gamma = K/2\rho_0{}^2 \; . \qquad \text{II.4}$$

For ergs $K = 10^{-13}$ and Angstroms $\rho_0 = 100$ Angstroms, we find γ of order 1/10 dynes/cm, about 10^2-10^3 times smaller than the bare surface tension.

For the "liquid-gas" boundary, typical interfacial energies are given by $\gamma = T_c/\xi^2$ where T_c is the critical temperature in energy units and ξ is the correlation length. Far from the critical point, ξ tends to a molecular dimension which in this case is the swollen micellen radius ρ_0. Thus, again we find low surface tensions which may become ultralow in the vicinity of the critical point where ξ diverges. This is in accord with observations by KIM, BOCK and HUANG [14].

The expressions for the constrained bending energy (II.1, II.2) have also been used [11,12] to discuss other possible globular shapes. For flat lamellae, $F_b = 2k$ and at small water concentrations ($\rho \leq (1/2)\rho_0$), a "bicontinuous" lamellar phase is more stable than swollen micelles. Thermal fluctuations will more or less wrinkle the layers yielding, on large length scales, a fairly random interpenetrating system of oil and water channels. Such a structure will be electrically conducting via the ions dissolved in the aqueous regions. Thus, with increasing water fraction, we may obtain the anomalous situation of a transition from a conducting to a non-conducting micellar phase.

For sufficiently small saddle splay, K', a cylindrical phase may intercalate between the lamellar and spherical phases. Thermal fluctuations will cause the cylinders to obtain a random walk conformation very similar to worm-like polymers. Indeed there is growing evidence for the existence of worm-like micelles and their polymeric properties [15,16].

III. Coulomb Effects

In practice one is often dealing with ionic surfactants. Under high ionic strength conditions, where the Debye screening length is short compared to other distance scales, the effects of electrostatic interactions may be lumped with the other curvature parameters already discussed. However, at low salt concentrations, we may expect long-range Coulomb forces to play an important role in the interactions between oil internal globules embedded in an aqueous matrix. Here we would like to speculate that the electrostatics may also vigorously compete with the curvature effects in determining globular shapes.

For oil-water interfaces decorated with ionic surfactants, we are dealing with charged surfaces having the neutralizing counterions occupying the adjacent aqueous phase. The counterion distribution, in mean field theory, is determined from the Boltzmann-Poisson (BP) equation,

$$-\varepsilon\nabla^2\phi = 4\pi n$$
$$n = n_o \exp-(e\phi/T), \qquad \text{III.1}$$

where n is the counterion density, ε is the dielectric constant of water (= 80), and n_o is fixed by the area per surfactant. Linearization of these equations leads to Debye-Huckel theory of electrolytes. However, in the vicinity of a highly charged surface, the non-linear effects become dominant and control the charge distribution on all scales. For a planar interface, the BP equations were exactly solved by GOUY [17] and CHAPMAN [18] to yield

$$n(x) = [2\pi\ell(x + \lambda)^2]^{-1} \qquad \text{III.2}$$

where ℓ is the Bjerrum length ($\ell = (e^2/\varepsilon T$ = 6 Angstroms) and $\lambda = \Sigma/2\pi\ell$ is an electrostatic surface length. If the water concentration is not too small, the electrostatic free energy density derived from III.1-2 in our microemulsion notation is

$$F_e = -(Tv_s/\delta\Sigma)\ln(\Sigma\lambda/v_c) \qquad \text{III.3}$$

where v_c is the volume occupied by one counterion. This is essentially the entropy of mixing of n_s counterions confined to a volume $n_s\lambda\Sigma$. Comparing F_e to the bending energy (II.3), we note that their ratio is approximately $(T/K)(\rho_o^2/\Sigma)$ which is typically of order 10-100. Thus, the electrostatic effects are <u>not</u> negligible and renormalize Σ^*.

For curved interfaces, there is a strong asymmetry between water internal and external cases. This arises from the different qualitative nature of the electric field distribution in the vicinity of charged surfaces. For example, consider a charged spherical shell of radius ρ embedded in a medium of dielectric constant ε. Outside the shell Gauss' Law gives a field $E = (Ze/\varepsilon r^2)$ while in the interior E = 0. Thus, for our case with neutralizing counterions, we expect weak concentration gradients for water internal phases but stronger variations for water external situations. In particular, for spherical W/O globules, the B-P equations may be linearized, and the resulting Debye-Huckel equations are easily solved to give an approximate free energy density (appropriate to the highly charged microemulsion interfaces)

$$F_s = F_e + (3/8\pi)k(T/K)\ (\rho_o^2/\ell\rho). \qquad \text{III.4}$$

Note that the correction to the planar interface result, F_e, is positive, reflecting the increased charge localization and comparable to the $4k\ \rho_o/\rho$ contribution to the bending energy (II.3). A positive steric ρ_o may be effectively driven negative by the Coulomb interactions.

For water external phases the situation is somewhat more complicated. The strong electric fields near the interface cause condensation of a fraction of the counterions into the Stern layer for both cylindrical [19] and spherical [20] globules. For both geometries, a distant test charge experiences an E field associated with an effective surface charge density $\Sigma^* = m\rho\ell$ where m is a shape-dependent constant of order 5. Thus, for a spherical swollen micelle with typical microemulsion parameters, only about 10% of the ionic groups contribute to the effective charge. This is the charge renormalization effect discussed in reference 20. However, it is precisely these counterions which may escape from the sphere, i.e. desorb, that contribute to lowering the free energy of the water external phase. This effect amounts to reducing the free energy relative to the planar reference state to

$$F_s' - F_e = N^* cT \, \ell n(N^* c\Sigma\ell) \qquad \text{III.5}$$

where N^* is the number of effective counterions per micelle, i.e. $N^* = 4\pi\rho^2/\Sigma^*$. In the notation of III.4, this becomes

$$F_s' = F_e - gk(T/K)(\rho_0^2/\ell\rho) \qquad \text{III.6}$$

where the constant g is of order 10 and derives from logarithmic terms in III.5 and Σ^*. Thus, the tendency for ionic surfactants to be water external persists, being driven by the increased counterion entropy of mixing.

We have seen that electrostatic effects arising from ionic surfactants may profoundly alter the interfacial shapes of microemulsion phases. A more detailed treatment including the cylindrical phases and effects of pH and ionic strength will be given in a forthcoming publication. With pH, the charge per surfactant may be controlled to engender phase-inversion transitions. The major effect of salts is to screen the electrostatic interaction between charged globules in the water external phases.

In summary, we have shown that electrostatic forces are comparable to steric curvature effects and are therefore important ingredients in determining the interfacial structure of microemulsions. In the future, we plan to develop the ideas presented here with the expectation of making contact with experiment in order to definitely fix the scale of the Coulombic interactions in the context of microemulsions.

References

1. J.H. Schulman and J.B. Montagne: Ann. N. Y. Acad. Sci. 92, 366 (1961)
2. P.G. de Gennes and C.Taupin: J. Phys., Chem. 86, 2294 (1982)
3. W. Helfrich: Z. Naturforsch, 28, 6693 (1973)
4. Y. Talmon and S. Prager: Nature 267, 333 (1977)
5. Y. Talmon and S. Prager: J. Chem. Phys. 69, 2984 (1978)
6. Y. Talmon and S. Prager: J. Chem. Phys. 76, 1535 (1982)
7. J. Jouffroy, P. Levinson and P.G. de Gennes: J. Phys. (Paris) 43, 1241 (1982)
8. B. Widom: J. Chem. Phys. 81, 1030 (1984)
9. S.A. Safran and L.A. Turkevich: Phys. Rev. Letters 50, 1930 (1983)
10. B. Lemaine, P. Bothorel and D. Roux: J. Chem. Phys. 80, 1023 (1984); J.S. Huang, S.A. Safran, M.W. Kim, G.S. Grest, M. Kotlarchyk and N. Quirke: Phys. Rev. Letters 53, 592 (1984)
11. S.A. Safran, L.A. Turkevich and P. Pincus: J. Physique Lettres 45, L69 (1984)
12. L.A. Turkevich, S.A. Safran and P. Pincus: to be published in "Surfactants in Solution," Bordeaux (1984)

13. M.L. Robbins: "Micellization, Solubilization and Microemulsions," Vol. 2, K. Mittal, Ed., Plenum, New York (1977)

14. M.W. Kim, J. Bock and J.S. Huang: in "Waves on Fluid Interfaces," p. 151 (Academic Press, 1983); M.W. Kim, J.S. Huang and J. Bock, SPEJ, p. 203, April 1984

15. J. Appell, G. Porte, Y. Poggi: J. Physique Lettres 44, 689 (1983)

16. S.J. Candau, E. Hirsch and R. Zana: J. Physique, to be published

17. G.J. Gouy: J. Phys. 9, 457 (1910); Ann. Phys. 7, 129 (1917)

18. D.L. Chapman: Phil. Mag. 25, 475 (1913)

19. G.S. Manning: J. Chem. Phys. 5, 924, 934, 3249 (1969)

20. S. Alexander, P.M. Chaikin, P. Grant, G.J. Morales, P. Pincus and D. Hone: J. Chem. Phys. 80, 5776 (1984)

The Structure of Microemulsions

A.M. Cazabat, D. Chatenay, P. Guering, D. Langevin, J. Meunier, and W. Urbach
Groupe "Microémulsions", Laboratoire de L'Ecole Normale Supérieure, 24, rue Lhomond, F-75231 Paris Cedex 05, France

Microemulsions [1] are transparent, isotropic, low viscosity dispersions of oil and water, stabilized by surfactant molecules.

The transparency of these media indicates that the length scale of the dispersion is small compared to optical wavelengths (∿ 100 Å).

The surfactant molecules are located at the interfaces between oil and water, (polar heads on the water side and aliphatic chains on the oil side), therefore considerably decreasing the interfacial energy. The entropy of mixing is then strong enough to induce spontaneous emulsification [2] .

Several models have been proposed for describing these phases : for example, at low water content, water droplets sheathed by a film of surfactant molecules are dispersed in the continuous oil phase : this droplet model was first proposed by Schulman [1] and explains most of the experimental results obtained on dilute systems (low oil or water content). But it is notsuitable if oil and water amounts are comparable : other models, assuming symmetrical roles for oil and water, are then needed [3] [4].

In the following, the main features of the various models are recalled and compared to experiment. However, it must be kept in mind that these models are static ones. The low viscosity of microemulsions suggests that the structures are continuously changing. It allows to explain most of the apparent discrepancies between models and experiments.

The droplet model [1]
The dispersed phase is supposed to be composed of identical spherical droplets, the size of them being independent of their volume fraction ϕ

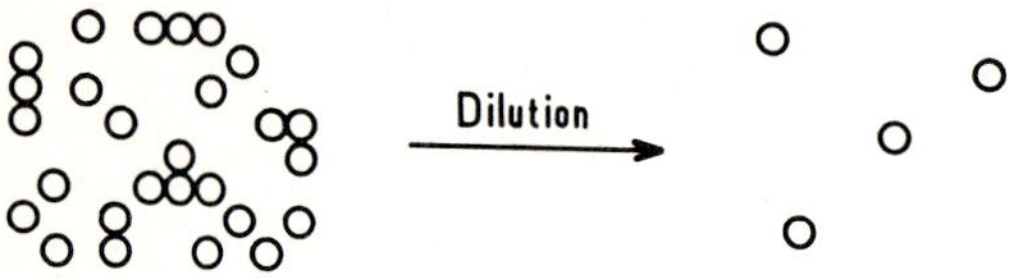

Figure 1

The system can be diluted by adding increasing amounts of continuous phase which allows to use scattering techniques in order to find the size of droplets and the interactions between them.

At low ϕ ($\phi < 0.05$) this model agrees with X rays [5] and neutron scattering [6] results. In some cases, it can be used up to very large ϕ ($\phi \sim 0.20$ [7]). Its validity can be checked using variable contrast methods [8].

Light scattering has been widely used for systematic studies of microemulsions [9] [10] [11] . The interaction between droplets is well described by the sum of a hard sphere term [9] (accounting for steric exclusion) and a perturbation (screened Coulombic repulsion in water continuous systems, or Van der Waals attraction). In oil continuous systems, a quantitative agreement with experiment is obtained if some overlapping of droplets interfaces during collisions is allowed.

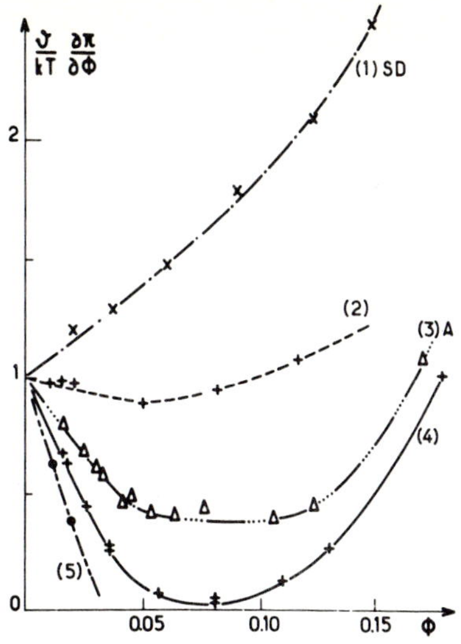

Figure 2 :
Water in oil droplets
$\mathcal{V}$ droplet volume
Π osmotic pressure
k Boltzmann constant
T absolute temperature

$\frac{\mathcal{V}}{kT}\frac{\partial\Pi}{\partial\phi}$ reduced osmotic compressibility

Typical results for the reduced osmotic compressibility can be found on the figure 2

The slope of the curve (1) at $\phi=0$ is close to +8, indicating hard sphere-like behavior (SD). The slopes of the other curves are lower, therefore some attractive interaction is present. The balance is weakly (2), moderately (3) or strongly (4) (5) attractive. In the latter cases, a critical behavior ((4) $\phi \sim 0.08$) or even a demixion ((5) $\phi \sim 0.03$) are observed.

The strength of the attractive potential is governed by the volume allowed for interfaces overlapping [11], therefore by geometrical and dynamical properties of the interfaces.

Random structure models [4] [5]
The droplet model is not suitable if water and oil are in comparable amount and play symmetrical roles. Scriven [4] and Friberg [5] suggested models with random filling of the space with oil and water domains [12]. A simpler formulation using a lattice model has been given by de Gennes [13].

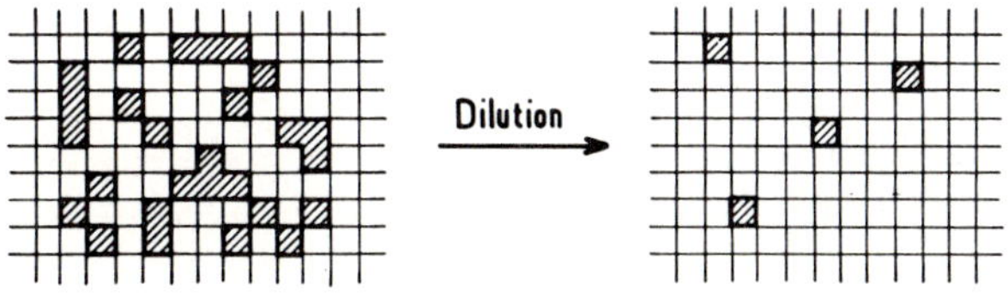

Figure 3

The structure of the dispersed phase is strongly polydisperse and changes during dilution. This model has interesting connectivity properties : at low water content ($\phi_W < 0.15$) the water phase is discontinuous (the same for oil phase at low oil content). But in a large concentration range both water and oil phases are continuous [12]. The system is said to be bicontinuous and exhibits very peculiar transport properties.

Comparison between the two models
The basic difference between the two models can be seen when two droplets (or two identical "cubes") come into contact.

Nothing happens in the first model. In the second model there is merging and the interface disappears. In other words, the probability of interface opening during a collision between droplets is zero in the first case, one in the second.

One may expect real systems to be closer to the first case if the interactions between droplets are repulsive (ex : hard spheres) and to the second case if they are strongly attractive. (Some droplets overlapping actually occurs in the latter case).

Therefore the microemulsion structure depends on the interfacial properties of droplets (through the opening probability) and on their volume fraction ϕ (collision frequency).

Hence, structural informations can be obtained by studying collision processes (through ultrasonic absorption [14] or fluorescence decay [15] techniques) or transport properties of the system (like viscosity [16], electric conductivity [17], diffusion coefficients).

Hard spheres and moderately attractive systems

Let's discuss some properties of systems (1) (hard spheres, referred to as SD) and (3) (attractive droplets, referred to as A), previously characterized using light scattering.

The self-diffusion coefficient D_s of the various components is deduced from tracer [18] or NMR [19] experiments. The results can be found on the figure 4 [19] These systems are water in toluene microemulsions, stabilized by SDS (sodium dodecyl sulfate). A short chain alcohol (pentanol or butanol) is needed as cosurfactant to get stability, and is partitioned between interface and continuous phase.

- the toluene molecules move freely in the continuous phase;
- the mobility of SDS molecules is low, indicating that they are not free, but included in a larger particle. Extrapolating D_s to $\phi = 0$, we get a value in perfect agreement with the diffusion coefficient D_0 of the droplets, separately measured using dynamic light scattering.

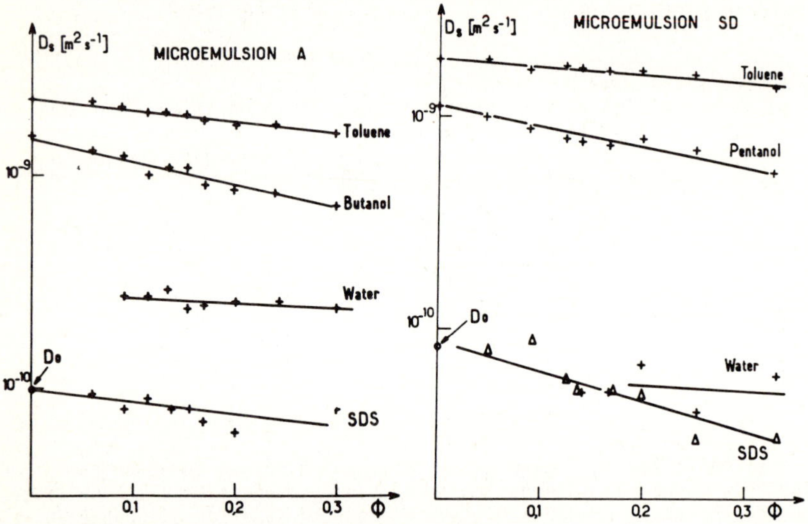

Figure 4

Self diffusion coefficients in W/O microemulsions

- In the first system (SD) water is bound to droplets up to $\phi \sim 0.25$. On the contrary, in the system (A), water moves faster than the droplet (but slower than free water). In this case, droplets can exchange water molecules, which indicates interface opening during collisions.

Ions can also been exchanged. As a result, an electric percolation process [17] is observed when a continuous path of connected water cores is formed. In the non-connected systems, the electric conductivity K stays low : typical results on systems A and SD can be found on the figure 5, together with collective diffusion coefficients D, deduced from dynamic light scattering measurements. Therefore, we can imagine a picture for the structure of hard sphere-like or moderately attractive systems : at low ϕ, the droplets are isolated. With increasing ϕ, the number of collisions increases, and the probability of transient merging increases with the strength of the attractive potential. However, the memory of the basic droplet structure is retained [20].

Strongly attractive systems. Phase separation [21]

In strongly attractive systems, critical behavior and phase separation are observed. A suitable system for experimental study is a mixture containing equal amounts of oil and water. The overall composition is fixed, the varying parameter being the water salinity. At low salinity, a 2-phases equilibrium is observed between a water-continuous microemulsion and oil (I). With increasing salinity, the screening of repulsive interactions increases, the interaction balance becomes attractive, leading to phase separation (II); a 3-phases system with microemulsion in equilibrium with both oil and water is obtained.

The situation II can also be reached by starting at very high salinity : III : oil continuous microemulsion in equilibrium with water, and decreasing salinity. This makes the interactions between water droplets more attractive, leading again to phase separation (III).

The self diffusion coefficients of the components can be found in the fig.6 [19] The droplet model, with enhanced exchanges when phase boundaries are approached, accounts well for the 2-phases situations I, III.

In the 3-phases domain, the memory of droplet structure remains probably near the boundaries [22] but is lost in the middle part where water and oil move freely and play symmetrical roles.

By contrast, the SDS mobility stays low. That means that oil and water domains are separated by well-defined interfaces. No molecular mixture of the components is observed.

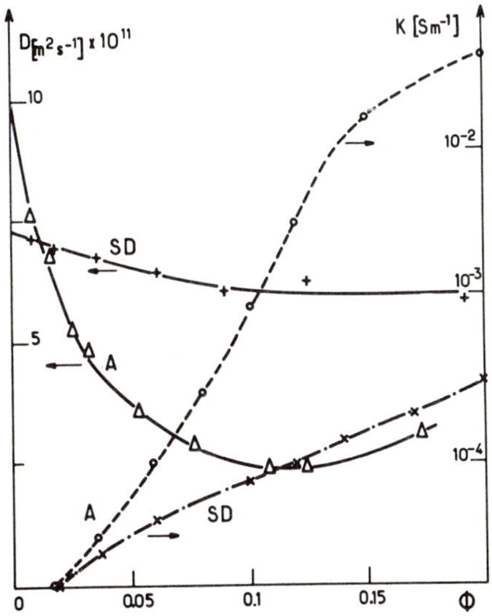

Figure 5 : Self-diffusion coefficient DS and electric conductivity K for A and SD microemulsions

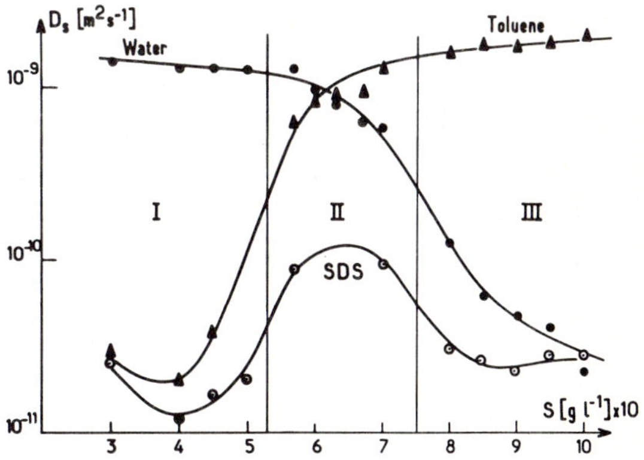

Figure 6
Self-diffusion coefficients versus water salinity .

Electric conductivity measurements in this range show a linear dependence on the water content of the microemulsion [16] [20] , in good agreement with the assumption of a bicontinuous structure.

Some concluding remarks

This paper proposes a coherent picture of microemulsions structure, with a progressive change from droplet to random model when the dispersed phase concentration and the attractive interactions increase. Although quite frequent, this scheme is not always followed : without cosurfactant, lamellar phases often occur as ϕ increases due to large interface rigidity [23] . On the contrary, with large amount of cosurfactant, the characteristic times for structural changes become so short that the system is almost a molecular mixture.

1. J.H. Schulman and J.A. Friend: J. Coll. Int. Sci. 4, 497 (1949)
2. E. Ruckenstein and J.C. Chi: J. Chem. Soc. Far. Trans. II, 71, 1690 (1975)
3. S. Friberg, I. Lapcynska, G. Gillberg: J. Coll. Int. Sci. 56, 19 (1976)
4. L.E. Scriven: Nature (London), 263, 123 (1976)
5. J.H. Schulman, W. Stoeckenius and L. Prince: J. Phys. Chem. 63, 1677 (1959)
6. M. Dvolaitzky, M. Guyot, M. Lagües, J.P. Le Pesant, R. Ober, C. Taupin: J. Chem. Phys. 69, 3279 (1978)
7. M. Kotlarchyk, S.H. Chen, J.S. Huang, M.W. Kim: Phys. Rev. Lett. 53, 941 (1984)
8. M. Dvolaitzky, M. Lagües, J.P. Le Pesant, R. Ober, C. Sauterey, C. Taupin: J. Phys. Chem. 84, 1532 (1980)
9. A.A. Calje, W.G.M. Agterof and A. Vrij in "Micellization, Solubilization and Microemulsions", vol. II, Plenum, N.Y. 1977
10. A.M. Cazabat, D. Langevin: J. Chem. Phys. 74, 3148 (1981)
11. B. Lemaire, P. Bothorel, D. Roux, J. Phys. Chem. 87, 1023 (1983)
12. Y. Talmon and S. Prager, J. Chem. Phys. 69, 2984 (1978)
13. J. Jouffroy, P. Levinson, and P.G. de Gennes, J. Phys. (Paris), 43, 1241 (1982)
14. R. Zana, J. Lang, O. Sorba, A.M. Cazabat, D. Langevin, J. Phys. Lett. 43, 289 (1982)
15. P. Lianos, R. Zana, J. Lang, A.M. Cazabat, to be published
16. - K.E. Bennet, J.C. Hatfield, H.T. Davis, C.W. Macosko and L.E. Scriven: in "Microemulsions" (I.D. Robb Ed., N.Y. 1982)
    - A.M. Cazabat, D. Langevin, O. Sorba, J. Phys. Lett. 43, 505 (1982)
17. - M. Lagües, R. Ober, C. Taupin, J. Phys. Lett. 39, 487 (1978)
    - A.M. Cazabat, D. Chatenay, D. Langevin, A. Pouchelon, J. Phys. Lett. 41, 441 (1980)
18. B. Lindman, N. Kamenka, T.M. Kathopoulis, B. Brun, P.G. Nilsson, J. Phys. Chem. 84, 2485 (1980)
19. - D. Chatenay, P. Guering, W. Urbach, A.M. Cazabat, D. Langevin, J. Meunier, L. Léger, B. Lindman, to be published
    - P. Guering, B. Lindman, to be published
20. A.M. Cazabat, D. Chatenay, D. Langevin, J. Meunier, Far. Disc. Chem. Soc. 76 291 (1983)
21. P.A. Winsor, Trans. Far. Soc. 44, 376 (1948)
22. - M. Drifford, J. Tabony, A. De Geyer, Chem. Phys. Lett. 96, 119 (1983)
    - L. Auvray, J.P. Cotton, R. Ober, C. Taupin, J. de Phys. 45, 913 (1984)
23. C. Taupin, this volume.

Random Bicontinuous Microemulsion Predictions of Talmon-Prager and de Gennes Models. Structure Experiments and Determination of the Flexibility of the Interface

C. Taupin, L. Auvray, and J.M. di Meglio
Collège de France, Laboratoire de Physique de la Matière Condensée[1],
11, Place Marcelin-Berthelot, F-75231 Paris Cedex 05, France and
Greco "Microémulsions" du C.N.R.S.

1. Introduction

For several years, theoretical papers [1,2] have presented a new aspect of microemulsions based on the main idea that the usual energy parameters which are predominant in determining the structure in these systems could be unusually small, leading to a very subtle competition. These parameters are :

- interfacial tension ;
- curvature energy (including spontaneous curvature and interfacial rigidity as introduced by HELFRICH [3]) ;
- entropy of dispersion of oil and water (recall that the scale of dispersion is 100 Å) ;
- interactions (Van der Waals or structural forces).

A particular emphasis was given to the highly flexible films [2] where one can define a persistence length ξ_K :

$$\xi_K = a \exp\left[\frac{2\pi K}{kT}\right] \qquad (1)$$

where a is a molecular length, and K the rigidity coefficient [3] . At scales smaller than ξ_K the lamella is stiff but it becomes wrinkled at scales larger than ξ_K.

In this case of highly flexible films and very small energy parameter, it has been suggested that random bicontinuous structures could exist.

The main predictions of the "Talmon-Prager" and "de Gennes" models concern :

- the existence of an interfacial film ;
- the relationship of the length scale ξ of the system (which is ξ_K in ref. [2]) with the volumic fractions ϕ_W, ϕ_0 and the interfacial area $n_s\Sigma$ (n_s number of surfactant molecules per unit volume and Σ area per molecule) ;

$$\xi = \frac{6\phi_W\phi_0}{n_s\Sigma} \qquad (2)$$

---

[1] Unité Associée au C.N.R.S. (U.A. n° 792).

This relation is characteristic of bicontinuous model, by contrast to the well known

$$R_W = \frac{3\phi_W}{n_S \Sigma} \tag{3}$$

of water droplets of radius R_W.

The introduction of ξ_K was shown to be very useful in explaining the transition from the usual highly organized lamellar phases towards the disordered microemulsion phases [2].

Two sets of experiments are presented. The first concerns the measurement of the rigidity coefficient K and the second the experimental evidence for random bicontinuous structures.

2. Flexibility of the Interface

The birefringent microemulsions which are lamellar phases in the vicinity of microemulsions in the phase diagram, and which contain approximately half the amount of cosurfactant, exhibit a very labile behavior and are good systems to determine the rigidity coefficient of the interfacial film.

The first observation [4] was the presence of residual undulations even in flat glass cell oriented samples. This observation was made by means of the local spin label technique, which is a very sensitive detector of molecular orientations. The angular amplitude of these undulations $< \theta^2 >$ increases with the degree of swelling (associated with increasing reticular distances up to a few hundred Å).

Theoretical calculations [5] have shown that $< \theta^2 >$ is due to a competition of the tendency to curvature due to a high flexibility and the steric repulsion between undulated lamellae following :

$$< \theta^2 > = kT/\pi K \ \ \mathrm{Log}(\xi_u/a) \tag{4}$$

where a is a molecular length (taken around 10 Å) and ξ_u a length determined by $\xi_u = (K/U")^{1/4}$ with U" is the curvature of the interaction potential u between the lamellae. In this case of highly flexible interface, the predominant term in U" is due to the steric repulsion potential of HELFRICH [3]

$$U" = 5.04\ (kT)/(Kd^4) \tag{5}$$

with d the distance between lamellae. Thus one gets

$$< \theta^2 > = \frac{kT}{\pi K} \ \mathrm{Log}\ d + \mathrm{cste} \tag{6}$$

By determining $< \theta^2 >$ using a classical procedure of synthesis of the EPR spectra, the study of $< \theta^2 >$ either as a function of temperature or swelling leads to two experimental determinations of K [5].

$K_T = 6.4 \cdot 10^{-15}$ erg $\qquad$ $K_d = 4 \cdot 10^{-14}$ erg

Let us recall that the value for lecithin bilayers is $K \sim 2 \cdot 10^{-12}$ erg evidencing thus the high flexibility of microemulsion films.

In these experiments, the role of cosurfactant as decreasing K was demonstrated, and the observation of dynamical effects due to the lateral diffusion of the spin labelled molecule on curved surfaces lead to evaluations of ξ_K in very good agreement with the model.

Moreover, these systems exhibit hydrodynamical collective modes of lamellar phases which were predicted in a paper by F. BROCHARD [6] . In particular, a pure undulation mode appears at a characteristic time τ :

$$\tau^{-1} = \frac{K}{\eta d} q^2 \tag{7}$$

if the wave vector q is parallel to the lamellae. This mode was observed by quasi-elastic light scattering [7] and led to a value of K in the same range ($K_s \sim 2 \cdot 10^{-15}$ erg).

3. Random Bicontinuous Structures

Another particular case of microemulsion system is "the Winsor III Microemulsions" which appear when an oil, surfactant and brine system separates in three phases. The "middle phase" microemulsions are associated with extremely low interfacial tensions ($\sim 10^{-3}$ dynes/cm) and have many industrial applications in enhanced oil recovery and phase transfer. Their structural organization is not well understood. Several features are particularly interesting :

- they are associated with extremely low interfacial tensions ;
- they appear in a range of salinity of the brine where the spontaneous curvature of the amphiphilic film (which is strongly dependent on electrostatic repulsions) could be very low ;
- if they also correspond to highly flexible films they could be good candidates for testing the random bicontinuous models which were proposed recently [1,2].

A systematic study of their structure,which was performed by means of small angle X-ray scattering and neutron scattering using variable contrasts with deuterated molecules,led us to the proof of three facts :

- existence of a well-defined interfacial film even in "critical microemulsions" ;
- the characteristic size of the structure determined either by the radius of curvature of the film or the volume of the oil or water elements, obeys relation which is characteristic of random bicontinuous media ;
- in the case $\phi_u = \phi_0$ the mean curvature of the film is evidenced to be zero.

Moreover, the existence of volumic correlations is in favor of a finite elementary volume in the system, which correlates well with the existence of the persistence length ξ_K.

References

1. Y. Talmon and S. Prager: J. Chem. Phys. 69 , 2984 (1978) and 76 , 1534 (1982)
2. P.G. de Gennes and C. Taupin: J. Phys. Chem. 86 , 2294 (1982)
3. W. Helfrich: Z. Naturforsch. 28 C , 693 (1973) and 33 a , 305 (1978)
4. J.M. di Meglio, M. Dvolaitzky, R. Ober, C. Taupin: J. Physique (Lettres) 44 , L-229 (1983)
5. J.M. di Meglio, M. Dvolaitzky, C. Taupin: J. Phys. Chem. 89 , 871 (1985)
6. F. Brochard and P.G. de Gennes: PRAMANA 1 , 1 (1975)
7. J.M. di Meglio, M. Dvolaitzky, L. Léger, C. Taupin: Phys. Rev. Lett. (under press)
8. L. Auvray, J.P. Cotton, R. Ober and C. Taupin: J. Phys. (Orsay, France) 45 , 913 (1984)
9. L. Auvray, J.P. Cotton, R. Ober and C. Taupin: J. Phys. Chem. 88 , 4586 (1984)
10. L. Auvray: J. Physique (Lettres) 46 , L-163 (1985)

Part III **Packings**

The Relation Between the Structure of Packings of Particles and Their Properties

John Dodds and Muriel Leitzelement
LSGC-CNRS, ENSIC, 1, rue Grandville,
F-54042 Nancy, France

NOMENCLATURE

- A cross-sectional area
- d diameter
- f friction factor defined in (4.3)
- h Kozeny constant
- K permeability
- L length
- m hydraulic diameter
- P pressure
- r radius
- Re Reynolds number
- R resistance to flow
- S surface per unit bed volume
- S_0 surface per unit solid volume
- T surface tension
- v velocity
- x number fraction
- ε porosity
- θ contact angle
- μ viscosity
- τ tortuosity
- ρ density
- ϕ shape factor

SUBSCRIPTS

- t tube
- p particle
- c capillary

11 INTRODUCTION

A porous medium is one containing holes ; that is, composed of two complementary spaces, the solid phase and the void phase, forming a heterogeneous system in which either phase can be continuous or discontinuous, isotropic or have a preferential direction. The "solid" phase can be rigid, plastic or even fluid as in the case of foams.

This general definition opens a vast domain, which in chemical engineering includes columns packed with raschig rings, filter papers, catalyst pellets... A useful classification is by the size of the pores which can be either : *macroscopic* pores where the pore walls have only an insignificant effect on phenomena taking place ; *capillary* pores where the proximity of the walls is determinant in the phenomena occurring but does not bring the molecular structure of the two phases into account ; *molecular* pores, considered as force spaces in which the molecular structure of the two phases is dominant.

What follows is concerned with capillary systems with a characteristic dimension of the order of between a centimeter and a nanometer, in which the geometry of the structures controls the processes. This is mainly dealt with from the point of view of fluid flow phenomena and for unconsolidated packings of grains. This latter restriction is due to the fact that there is a direct relationship between the structure of the solid phase and the structure of the void phase only in these systems, and that in addition, such systems are very important in practice e.g. bulk powders, filter cakes, soils...

The principal feature of the structure of porous media is its complexity and the difficulty of making any description of the geometry involved. It is possible to imagine a full knowledge of a packing of spheres by having the 3 dimensional coordinates and diameter of every sphere in a packing. Such information can easily be obtained by computer-simulated packings. But this raw data is quite useless for any practical purpose of relating the structure of the packing to its properties.

It is necessary to extract the main features important for the problem under study, and use some simplified conceptual model of the structure to relate measurable quantities to measurable properties.

12 STRUCTURAL PARAMETERS

121 POROSITY (ε)

This is the most obvious and fundamental property of a porous medium and is defined as the ratio of the volume of the void space to the total volume of the porous medium

$$\varepsilon = \frac{V_{void}}{V_{void} + V_{solid}} \qquad (1)$$

This is not an unambiguous definition, as the total pore volume can be divided into connected or non-connected pores, micro or macro pores. In fluid flow phenomena, we are only concerned with the connected pore space open to flow. Porosity is independent of the absolute size of the pores and in an isotropic, homogeneous, random packing the free area of any representative cross-section is equal to the porosity. It is important that the four conditions given here be observed.

122 SPECIFIC SURFACE

The surface exposed in a porous medium is expressed in terms of either the surface per unit volume of solid, called the specific surface (S_0) or the surface per unit volume of the packing (S). As with porosity, these can include micropores or sealed-off pores, but unlike porosity the specific surface depends on absolute size. For example, the specific surface of an assembly of x spheres is

$$S_0 = \frac{x\pi d^2}{x\pi d^3/6} = \frac{6}{d} \qquad (2)$$

$$S = S_0(1-\varepsilon) \qquad (3)$$

123 PARTICLE SIZE, SHAPE and DISTRIBUTION

In a porous medium formed by packing particles, it is possible to make measurements on the solid phase, notably the size and shape of the grains. In general, practical problems involve irregular particles with a distribution in particle size which it is useful to represent by some average quantities. The book by ALLEN (1) gives an extensive discussion of these problems.

For example, a simple shape factor is the sphericity of a particle defined as

$$\psi_w = \frac{\text{surface area of a sphere having the same volume as the particle}}{\text{surface area of the particle}}$$

Other methods can involve characterising shape by Fourier analysis of particle outlines or determining fractal dimensions.

Expressing a particle size distribution, even of spheres, in terms of an average particle size is not unambiguous. For example if we have a set of 10 spheres of diameter d_i = 1, 2, 3 ... 10, we may define several substitute sets of uniform spheres. 10 spheres of diameter 5.5 will have the same number and total length of the original distribution, $\bar{d} = \Sigma x_i d_i / \Sigma x_i$. 10 spheres of diameter 6.21 will have the same total surface area as the original set, $\bar{d} = \Sigma x_i d_i^3 / \Sigma x_i$. A set of spheres with the same specific surface, that is total surface and total volume, is given by

$$\bar{d} = \Sigma x_i d_i^3 / \Sigma x_i d_i^2$$

In addition, the way particles contact one another is an important structural parameter. The various ways of defining such a coordination number will be discussed more fully in the next section.

124 PORE SIZE, SHAPE and DISTRIBUTION

The porosity and specific surface are overall properties of a porous medium. The pore size is a microscopic quantity, and the notion of pore size distribution implies a more detailed description of the pore texture. Nevertheless, it is not possible to define the size of a pore as it requires a discretisation of the whole pore space which cannot be done on a rigorous basis. It is easy to grasp intuitively what is meant by a pore, but in a porous medium the effective void volume is really a single pore of complex shape. In practice, the definition of pore size and distribution is made by the method of measurement, and will be discussed on the section on capillary pressure. Two simplifications for describing pore texture are tortuosity and connectivity.

If we consider an element of fluid flowing through a porous medium, it is easy to imagine that the length of the flow path followed by this element in passing through the porous medium will be longer than the apparent length from the inlet to the outlet. *The tortuosity* is the ratio of the real path length to the apparent path length

$$\tau = \frac{L_{effective}}{L}$$

This is not an absolute parameter, and the overall definition cannot be expressed as a sum of the microscopic effects as they cannot be measured. It is also easy to see that the tortuosity is a sub-factor of what might be called the texture of the porous medium. It should, however, be independent of pore size in the same way as porosity. If a fluid flowing through a porous medium is considered to be subjected to a series of deviations of angle θ to the apparent direction, then the tortuosity is

$$\tau = \frac{L}{L\langle\cos\theta\rangle_{av}}$$

and if θ is taken to be 45° this gives a value of $\sqrt{2}$ for the tortuosity.

The connectivity of pore space can be defined as the number of pores meeting at a point. In porous media formed by regular packing, it is possible to imagine a value for this. In random porous media, it becomes less obvious. In addition, as we cannot define a pore, the difficulties become unsurmountable. It is, however, an extremely important property and essential in differentiating real porous media from most models. Tortuosity can be considered as a sub-factor of connectivity.

13 MEASURABLE PROPERTIES

The aim of a structural description of a porous medium is to relate the quantities listed above to measurable properties. Here we restrict our attention to two features involving single-phase fluid flow. Other important properties such as thermal conductivity, mechanical strength and two-phase flow are also very dependent on structure but will not be dealt with.

131 PERMEABILITY (K)

The permeability of a porous medium is the conductivity to fluid flow. It is defined by Darcy's law

$$\nu = \frac{K}{\mu}\frac{\Delta P}{L}$$

It is a geometrical factor having units of (metres)^2 which depends on the structure of the porous medium. In anisotropic media the permeability is a directional quantity.

132 DISPERSION (D)

Fluid flowing in a porous medium is subjected to a mixing process by the configuration of the pore space. This can be described by a dispersion coefficient measured by a step change in a miscible fluid flowing in a porous medium and analysed as if it was a true diffusion process. This factor is made up of several different contributions, one of which is structure-dependent caused by mixing effects between streamlines.

2 PACKINGS of PARTICLES

Systems formed by packings of particles have been studied for at least 200 years and it is not possible to give a full review of the many different approaches and the results they have achieved. The vast majority of the work has been on the packing of equal spheres, very little on mixtures of sizes and even less on real systems with irregular particles. An excellent review is available by HAUGHEY and BEVERIDGE (2). Here 3 aspects will be covered. The properties of ordered assemblies of equal spheres, the structure of random packings of equal spheres and a presentation of an idealized model of the packing of spheres with different sizes.

21 ORDERED PACKINGS of EQUAL SPHERES

Ordered packings of equal spheres have been studied in detail for crystallography, and give some idea of the characteristics of real particle packings. As shown on Fig. 2.1, different structures can be obtained by superposition of layers in square contact or triangular contact. The Table 2.1 gives the characteristics of these different packings.

Table 2.1 : Ordered packings of equal spheres

PACKING TYPE	POROSITY	C	FREE AREA
Cubic (1)	0.476	6	1.0 -0.2
Cubic tetrahedral (4)	0.395	8	1.0- 0.0
Orthorhombic (2)	0.395	8	0.64-0.25
Tetragonal spenoidac (5)	0.302	10	0.58-0.09
Cubic close packing (3)	0.260	12	0.35-0.2
Tetrahedral close pack (6)	0.260	12	0.46-0.09

It can be seen that the porosity varies from 0.476 to 0.2595 and the coordination numbers vary from 6 to 12. This is comparable with values found in practice on irregular packings. Important differences from these ideal cases are that the coordination number is unambiguous, in that all spheres do actually touch and the fact that the mean free area of a cross-section is not equal to the porosity.

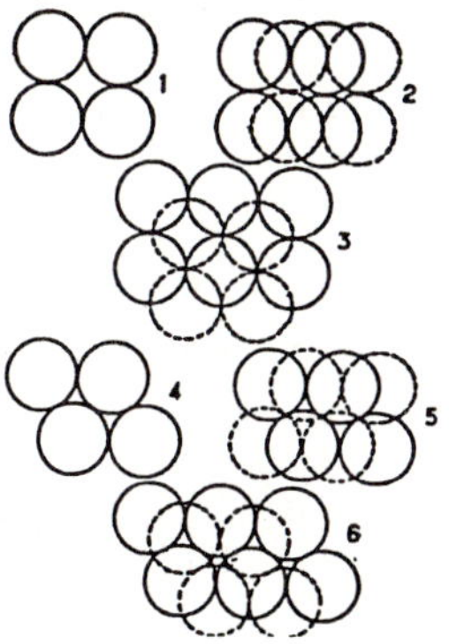

Fig. 2.1 : Ordered packings of equal spheres

This latter feature can be seen from the last columns of Table 2.1 where maximum and minimum values are given which depend on where the section plane is cut.

Concentrating on the packings n° 1 and 6 that is the loosest and the tightest, we may obtain some idea of the configuration of the voids. In *cubic packing*, the pore space may be taken to be composed of a central void in each unit cell which can just contain a sphere of diameter 0.732 D_p. This void connects with similar voids in corresponding unit cells by 6 outlets which will just allow a sphere of diameter 0.414 D_p to pass. In *close packing*, there are two types of cell, a rhomboid shape formed by eight spheres which can contain a sphere 0.29 D_p and a smaller tetrahedral void which can contain a sphere of 0.225 D_p. Thus the four corners of the tetragonal cells connect with the 8 corners of the rhomboidal cells, there being twice as many tetragonal cells as rhomboidal, and communication being always between cells of different kinds. The waists which allow communication are all of the same size, formed by 3 spheres and can just allow a sphere of size 0.155 D_p to pass. The total void volume is divided between rhomboidal and tetragonal cells in the ratio 6:13. Ordered assemblies therefore give at least some idea of the complexity of pore space in sphere packings, and especially the fact that it is essentially cellular in nature.

22 RANDOM PACKINGS of EQUAL SPHERES

The greatest difference between ordered assemblies of spheres and real packings is in the introduction of randomness. This is less easy to define, but the very reproducibility of random packings makes an experimental study possible. If equal spheres are poured into a container they give a *loose random* packing with a porosity of about 0.40. If this is then tapped or vibrated, we obtain a *close random* packing with a porosity which is now well established as 0.359. It is not possible to reduce the porosity further except by ordering. Another reproducible packing state is that of *minimum fluidisation* with a porosity of 0.46 which is obtained by flowing fluid upwards through a bed, which expands until there is a state of incipient fluidisation.

Close random packing is the most well defined of these and has been extensively studied as a model of the liquid state by BERNAL (3) and FINNEY (4). Real packings have been dismantled sphere by sphere, and computer simulations have both given the coordinates of all the spheres in a packing and have established the radial distribution functions and the coordination numbers of the spheres.

It is necessary to define the meaning of a contact. This can be a real contact where two sphere centres are a distance D apart or a close contact with a separation of 1.1D or just be near neighbours with centres at a distance of say $\sqrt{2}D$. Values given are

	Contacts		
	Real	Close	Neighbours
Close packing	6-7	8-9	13-14
Loose packing	5-6	7-8	13-14

More complete information is given by the radial distribution of sphere centres relative to given sphere. This shows that there is some order at close range in a random packing, but beyond about 3 sphere diameters there is a uniformity.

23 PACKING of DIFFERENT SIZE PARTICLES

This experimental approach to the study of packings is extremely onerous, and it is not possible to extend the method to a general investigation of multicomponent packings. To determine a porosity diagram of a ternary mixture of sizes would require the preparation of at least 50 different mixtures. To then dismantle each of these to determine the contacts between particles would be unthinkable. To then repeat this procedure for another different set of three spheres or to extend the procedure to four or more spheres would be impossible. We have therefore proposed an idealised model of packing,which includes the basic features of relative size and relative number of spheres in a packing which gives a general framework to approach the problem of real packings with a range of size of particle.

The basic assumption is that in a random arrangement of spheres,each sphere touches its neighbour. This allows space to be divided into tetrahedral subunits formed by lines joining the centres of continguous spheres. Figure 2.2 shows this idea in two dimensions. The hypothesis is,therefore,that gaps such as those shown by the dotted lines do not exist. This abstraction of the problem allows the effects of size distribution to be considered,apart from the less easily defined effects of the perturbation of the structure caused by the inevitable, irregular gaps. That is,elementary tetragonal subunits in the packing are calculated but their subsequent assembly to from a packing structure is ignored. The model is limited to the case in which there is an interaction between the spheres,that is a diameter ratio of 1:6.46. Above that value it is possible for a small sphere to fit in between 3 large spheres without touching them. The model was originally proposed by WISE (5) and is presented in detail in DODDS (6).

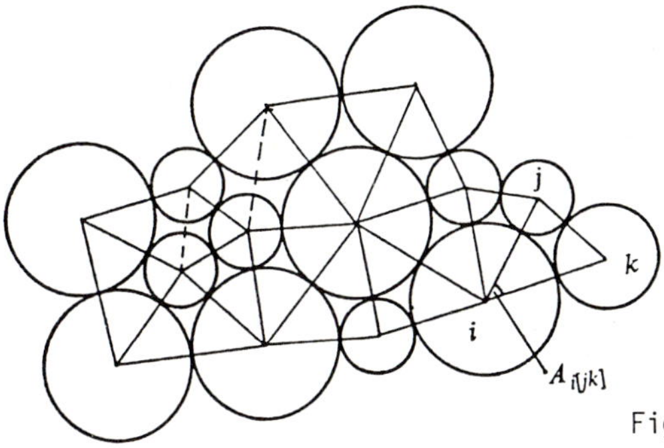

Fig. 2.2 : Two-dimensional gapless packing

In an n component packing the number N of subunit types is given by the number of different groups of 4 spheres. That is

$$N = n[(n+1)\ (n+2)\ (n+3)]/4! \qquad (2.1)$$

Thus in a binary mixture of sizes 1 and 2 we will have the following 5 different types of subunit

1111 1112 1122 1222 2222

The relative proportions of these different subunits will be given by the expansion of the expression

$$(\Sigma k_i)^4 \qquad (2.2)$$

or in the case of a binary mixture

$$k_1^4 \quad k_1^3k_2 \quad k_1^2k_1^2 \quad k_1k_2^3 \quad k_2^4$$

The probability of a sphere taking part in such a unit is given by k_i where

$$k_i = x_i/\overline{A}_i \qquad (2.3)$$

x_i being the relative number of the sphere type i in the packing. $\overline{A}_i$ being the average solid angle subtented at spheres of size i in the packing.

However as $\overline{A}_i$ is not known a priori,we must make a trial and error calculation using 2.2 and 2.3 until convergence is obtained. This is found to be quite rapid and only requires 2 or 3 iterations.

As a result, we can calculate the relative proportions of different types of subunit in a packing based on only the relative proportions of the different types of sphere and their relative sizes : in addition, as the geometry of the subunits is fully defined, we may calculate the characteristics of the idealised packing.

POROSITY

Table 2.2 shows the relative proportions of the 5 different types of subunit in a packing of spheres diameter 1 and 5 in relative proportions of 80 % by number small spheres.

	Number %	Volume %	ε	$\varepsilon + 13$
1111	13.3	0.04	22.04	35.04
1115	10.4	1.00	14.09	27.09
1155	30.2	9.06	13.66	26.66
1555	39.1	26.2	16.05	29.05
5555	19.0	63.7	22.04	35.04

$\bar{\varepsilon}$ = 19.63 "$\bar{\varepsilon}$" = 32.63

The Table also shows the porosity of each subunit type. The overall porosity of the packing is then given by the weighted sum of the porosity of 16 subunits. It can be seen that the values obtained are too small. In particular, the porosity of the subunit of identical spheres is 22.04 and not 36 as is obtained for a random packing of equal spheres. The values from the idealized model may be brought into line with practical values by adding a factor of 13 % to take into account the effect of the gaps. Without pretending to fully justify this correction factor, it may be taken that the additional void space due to the gaps is not distributed differently amongst the different subunit types. That is, the gaps occur at random and the additional void space can be added uniformly to each subunit type. The other advantage of this corrective factor is that it is simple, and does not over-mask the raw results from the model.

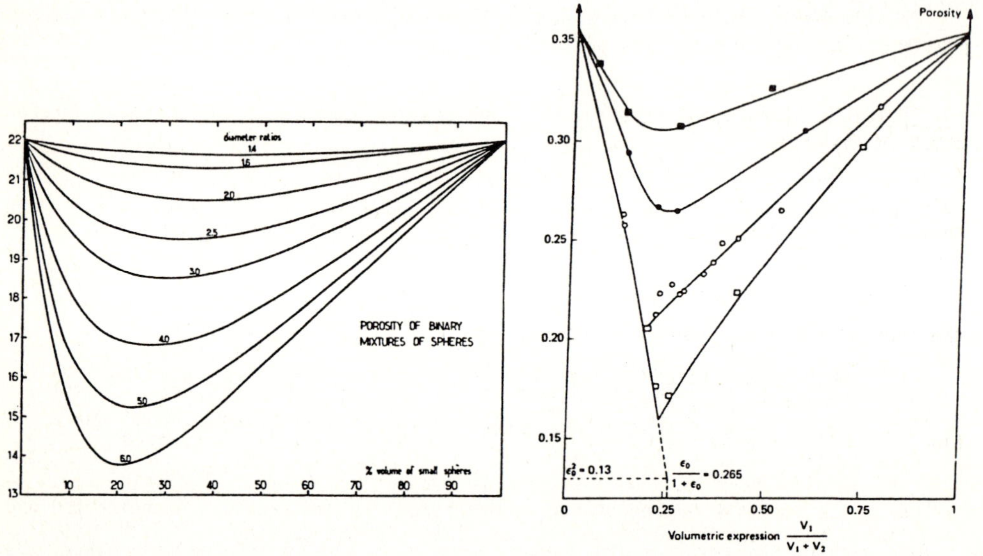

(a) results from gapless packing model

(b) experimental results (BEN AIM [7])

Fig. 2.3 : Porosity of binary mixtures of spheres

As a comparison with experimental results, Fig. 23A and Fig. 23B give diagrams of the variation in porosity in binary mixtures of spheres as a function of size ratio and mixture composition.

Figure 23A gives the results from the model ; Figure 23B gives results published by BEN AIM (7). It can be seen that the essential features are predicted by 16 model, that is a V shaped curve with a minimum tending towards the high proportion of small spheres and which deepens as the size ratio is increased.

Figures 24A and 24B give a similar comparison for ternary mixtures (8) and the comparison is with results published by STANDISH (9). Again, the essential features of the diagram are predicted, in particular that the minimum porosity is in the binary mixture with the widest size distribution, and that the addition of an intermediate component creates a trough in the centre of the diagram.

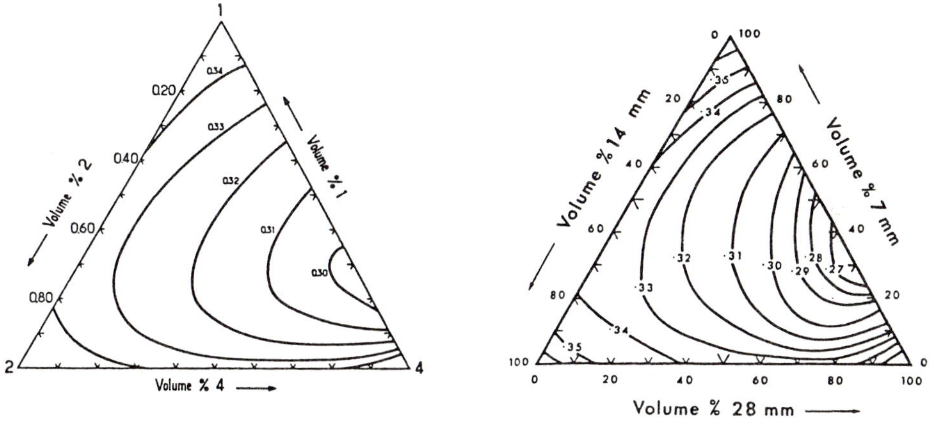

(a) results from gapless packing model

(b) experimental results (STANDISH (9))

Fig. 2.4 : Porosity of ternary mixtures of spheres

CONTACTS BETWEEN SPHERES

In addition to the distinction between real contacts, close contacts and near neighbours, it is also possible to define various types of coordination number in multicomponent packings, DODDS (6). These are the average total coordination number C

$$C = \frac{\text{number of contacts between spheres}}{\text{number of spheres in a packing}} \tag{2.4}$$

the average individual coordination number for each sphere size Z_i

$$Z_i = \frac{\text{number of contacts on spheres size i}}{\text{number of spheres of size i}} \tag{2.5}$$

finally, a further subdivision into partial individual coordination numbers p_{ij}

$$p_{ij} = \frac{\text{number of contacts of spheres j or spheres i}}{\text{number of spheres of size i}} \tag{2.6}$$

Alternatively, we may classify contact types as fractions of the total number of contacts in a packing t_{ij}. Thus in a binary mixture

$$t_{11} + t_{12} + t_{22} = 1 \tag{2.7}$$

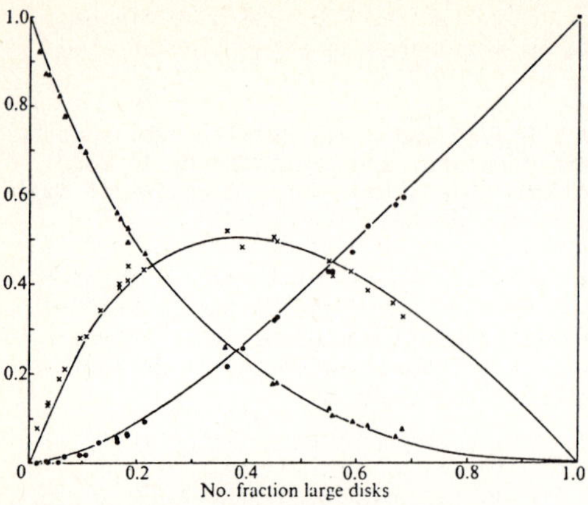

Fig. 2.5 : Comparison of coordination points from the model with computer simulation. Diameter ratio 2.5. Values of t_{ij} are shown as curves and computer simulation results are shown as points : Δ, t_{11} ; X, t_{12} ; , t_{22}.

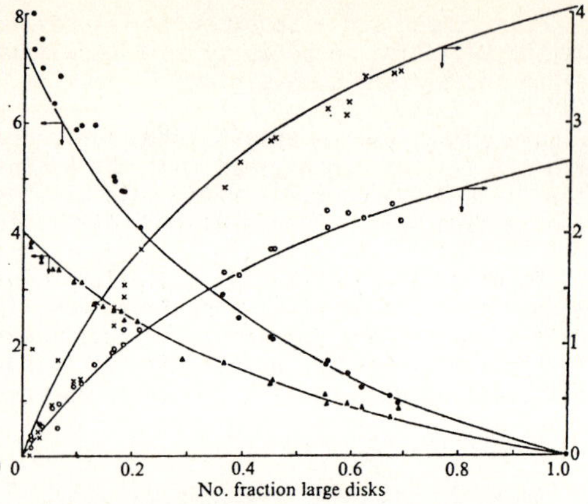

Fig. 2.6 : Comparison of coordination points from the model with the computer simulation. Diameter ratio 2.5. Values of p_{ij} from the model multiplied by the compaction factor are shown as curves, and computer simulation results are shown as points : Δ, p_{11} ; , p_{12} ; , p_{21} ; X, p_{22}.

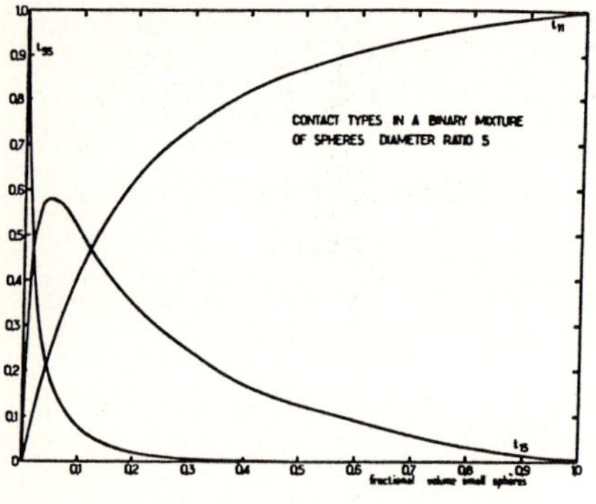

Fig. 2.7 : Contact types in a binary mixture of spheres diameter ratio 5

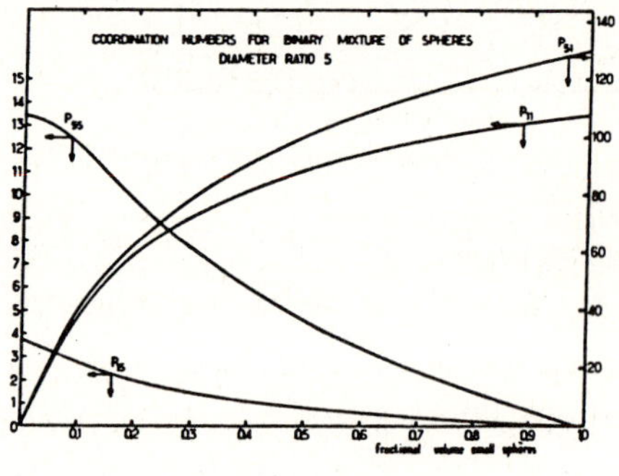

Fig. 2.8 : Coordination numbers for binary mixture of spheres diameter ratio 5

All these different values C, Z_i, p_{ij}, t_{ij} can be calculated from the idealised model as described in DODDS (6). Figure 2.7 shows the predictions for a binary mixture of spheres size ratio 1:5 as a function of mixture composition for the t_{ij} values. Fig. 2.8 shows the results in terms of Z_{ij} values. It is, however, difficult to obtain experimental results of this sort of detail to compare with the predictions. Two different comparisons have been made with 2D experimental packings of discs (10)(11) and with computer simulations (12)(13).

Figure 2.5 shows a comparison of the t_{ij} values in a binary 2D packing of discs with diameter ratio 1:2.5 with the results from a computer simulation. Figure 2.6 shows a comparison with the p_{ij} values. To make the comparison, the values from the model have been multiplied by a correction factor of 4.02/6.0 as the total average coordination number in the computer simulation was 4.02 and that in the idealized model (here 2D) was 6.0. This is equivalent to assuming that the gaps in the packing occur at random in the lattice defined by the idealized model. Recent work by

OGER et al. (13) has refined this analysis, and indicates some local ordering in real multicomponent packings due to size exclusion effects.

3 The VOID SPACE in PARTICLE PACKINGS

The most significant method for investigating the structure of the void space in porous media is by the capillary effects which result when two immiscible fluids are present in the pore space. This can be by the technique of mercury porosimetry, or more generally by capillary pressure measurements. However, other effects can be of use ; depression of the freezing point, capillary condensation, gas adsorption... Here, only the capillary pressure effects caused by displacing a wetting fluid such as water contained in a pore structure, by a non-wetting fluid such as air will be considered.

31 CAPILLARY PRESSURE CURVES

When two immiscible fluids are in contact in the interstices of a porous medium, the interface between them tries to form the smallest possible surface. This gives rise to a pressure across the interface which is shown by a curvature at every point such that

$$P = CT \tag{3.1}$$

Since the interface must be at equilibrium, it must be a surface of constant mean curvature, and may therefore be represented by two principal radii of curvature at any point

$$P = T\,(1/r_1 + 1/r_2) \tag{3.2}$$

This is called the Laplace equation : the upper limit of applicability is when the meniscus is plane and the lower limit is when the curvature is of the order of molecular dimensions and the interface has no real existence.

If liquid is contained in a circular capillary of radius v_t and has a contact angle θ then $r_1 = r_2 = r_t/\cos\theta$

$$P = \frac{2T\cos\theta}{r_t} \tag{3.3}$$

This is the familiar equation for the rise of a liquid in capillary tubes.

For irregular pore shapes, we can define the radius of an equivalent spherical meniscus which will exert the same hydrostatic stress on the liquid as the real irregular meniscus

$$r = r_1 = r_2 \tag{3.4}$$

The above phenomena leads to a standard method of measuring pore size distribution. A given saturation in a porous medium corresponds to a given value of P or mean meniscus curvature. Therefore a pore of radius r will be either empty or full, depending on whether the meniscus can be maintained in it

$$\begin{array}{ll} r > 2\,T/P_c & \text{pore empty} \\ r < 2\,T/P_c & \text{pore full} \end{array} \tag{3.5}$$

This may be explained by considering a bundle of capillary tubes with a distribution of radii and which contain water. We apply a pressure drop across the bundle and increase it, as shown in Fig.3.1, by increments, whilst waiting for an equilibrium at each increment. At first nothing will happen, until at some increment the meniscus contained in the largest tube will be too small and the tube will drain empty. When the tube is emptied and equilibrium is reached, we again increase the pressure and expel water from the next smallest tube. Continuing in this way by measuring the amount of water removed at a given pressure, we can draw a "capilla-

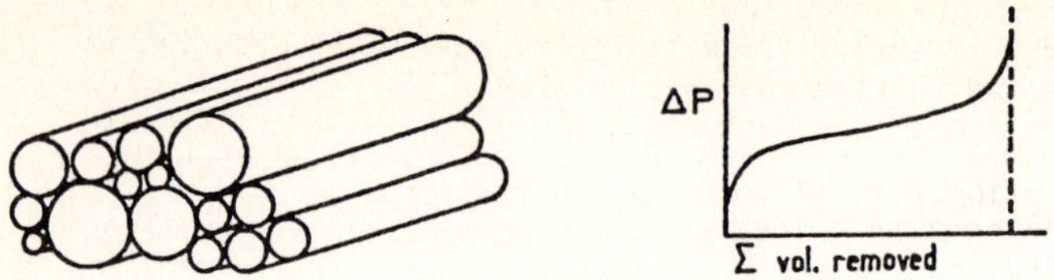

Fig. 3.1

ry pressure curve" of the tube bundle. Since pressure is related to pore size and the volume removed is related to the volume of pores, then the capillary pressure curve gives a pore size distribution.

32 CAPILLARY PRESSURE CURVES in PACKINGS of PARTICLES

Capillary pressure curves can be measured experimentally with an apparatus of the type shown schematically in Fig. 3.2. Results of these experiments, such as shown in Fig. 3.3, present several different features,which distinguish them from Fig. 3.1 for a bundle of capillary tubes. The first obvious difference is that at the end of the desaturation procedure, there remain a certain quantity of liquid trapped in the packing which cannot be removed by increasing the pressure. This residual saturation is not in hydraulic conductivity with the support plate of the apparatus,and is probably contained in pendular rings at the points of contact between the spheres. It has been shown (MORROW [14]) that in sphere packs the irreducible saturation is always about 10 % of the void volume,and is not changed by varying the wettability, the viscosity, the size distribution...

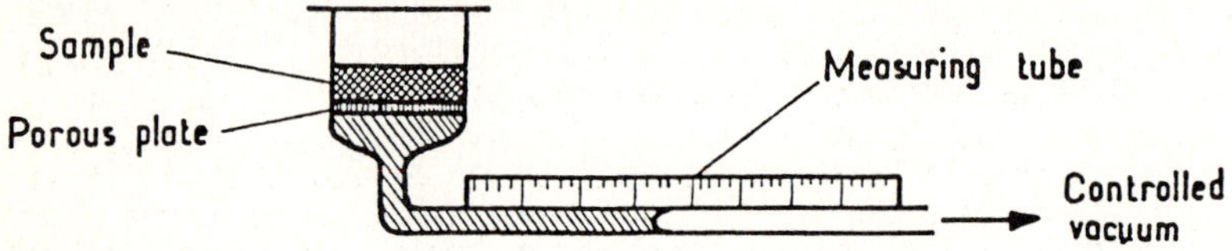

Fig. 3,2 : Capillary pressure apparatus (schematic)

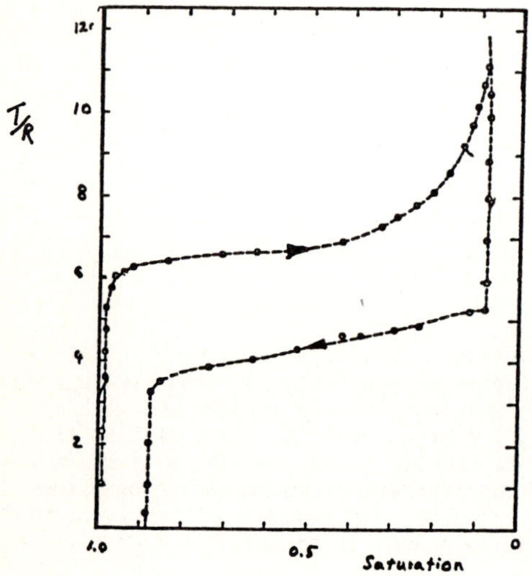

Fig. 3.3 : Capillary pressure curves in sphere packings, Haines results

A second difference between the capillary pressure curves in real particle packings and the bundle of tubes model is that there is a hysteresis in the curves. As shown in Fig. 3.3, when the pressure at the end of a desaturation experiment is progressively reduced, the packing refills at a lower capillary pressure than that at which it was emptied. In addition, the packing cannot be completely re-filled and some air remains trapped. Subsequent drainage and imbibition experiment will retrace the hysteresis loop shown in the figure.

33 HAINES THEORY

W.B. HAINES [15], in 1930, gave an explanation of these effects based on the cellular nature of the pore space in sphere packings. He considered a packing of equal spheres in close packing. Hence, in desaturation experiments, the meniscus can only penetrate fully into the packing at a pressure sufficient to empty the triangular waists between 3 touching spheres. That is a pore radius of 0.155 R

$$P_c \text{ desaturation} = \frac{2T}{0.155\,R} = 12.9\,\frac{T}{R}$$

At the end of the desaturation, when the pressure is gradually released to allow imbibition, the meniscus will sag in at the waists but not give general occupation until the voids in the centre of a unit cell can be filled. That is at

$$P_c \text{ imbibition} = \frac{2T}{0.29\,R} = 6.9\,\frac{T}{R}$$

Comparison with the Fig. 3.3 does show that these values can be considered to be the limiting case for a tight packing.

34 CORRELATION FOR THE EFFECT OF POROSITY

It is possible to take into account the effect of porosity variations on capillary pressure curves. Fig. 3.4 shows capillary pressure curves for the same sample of spheres. In one case in loose random packing with a porosity of 40 % obtained by pouring the spheres into the apparatus of Fig. 3.2. In the other case, the loose packing was vibrated to reduce the porosity to 35 %. It can be seen that the loose packing contains more liquid than the tight packing, but that the pores in the tight packing are correspondingly smaller (higher capillary pressure). The common feature of the two sets of results is that they involve the same set of spheres, and hence the same total surface area. In addition, as the area under these curves, to within the imprecision due to residual moisture, represents the work done to create the meniscus with the surface area of the spheres, then the area under the two curves is the same. If we assimilate the curves to a rectangle, we may say that

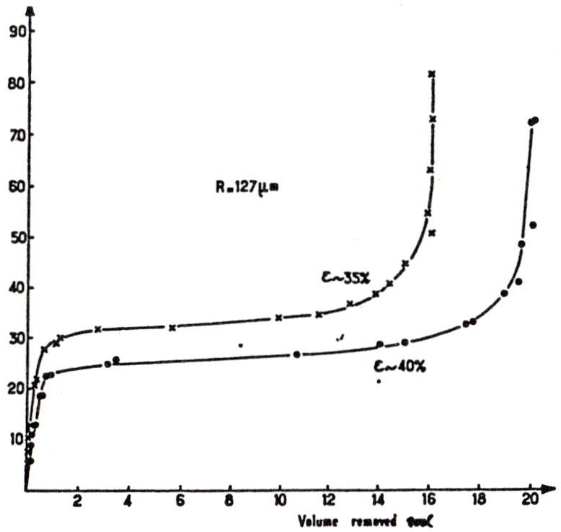

Fig. 3.4 : Capillary pressure curves for same sample in two different packing states

$$\frac{P_{c_1}}{S_o}\left(\frac{\varepsilon_1}{1-\varepsilon_1}\right) = \frac{P_{c_2}}{S_o}\left(\frac{\varepsilon_2}{1-\varepsilon_2}\right)$$

That is, by plotting capillary pressure times the ratio of void to solid fraction we can correlate the curves. Figure 3.5 shows the case of this correlation on a set of experimental capillary pressure curves.

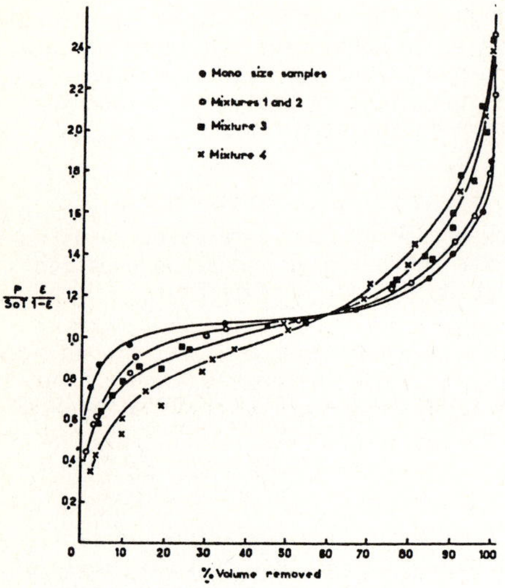

Fig. 3.5 : Correlation for capillary pressure curves

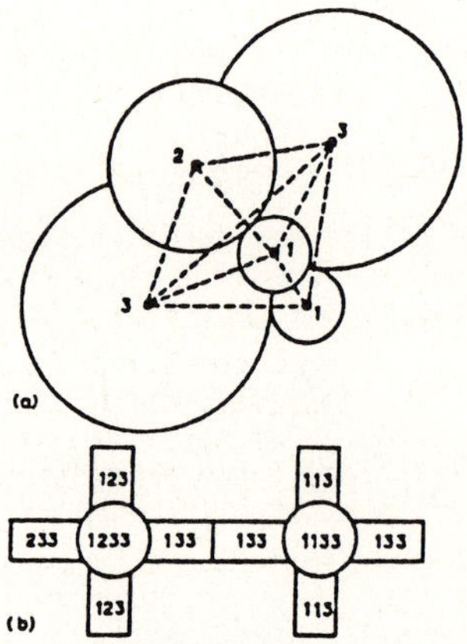

Packing of spheres sizes 11233; (b) representation of void space in packing of spheres 11233.

Fig. 3.6A and 3.6B

35 CAPILLARY PRESSURE CURVES in MULTICOMPONENT PACKINGS

Capillary pressure curves offer the advantage of giving a measure of the pore size distribution in a porous medium. It is therefore interesting to extend the treatment given by Haines to the case where a packing of spheres is composed of various different diameters. This may be done by adapting the model of gapless packing presented in the previous section. For example, Fig. 3.6A shows spheres of size 1, 1, 2, 3 3 touching together to form two tetrahedral subunits 1233 and 1133 meeting at a triangular face formed by spheres 133. The void space in these tetrahedra can be assumed to be formed of a central void with 4 connecting waists passing through each triangular face. This is approximated here as four cylindrical tubes which meet in a spherical cell. The radii and lengths of the tubes and the radius of the cell can be calculated from the sizes of the four spheres making up a given tetrahedron. Thus the pore space in Fig. 3.6A can be represented by Fig. 3.6B. These elements can be used in the network of Fig. 3.7 to give an approximation of pore space connectivity with four tubes per junction. The pores calculated from the idealised gapless packing model can be assigned randomly in this network and a Waines type desaturation procedure applied to give a capillary pressure curve. Full details of this are given in (16). One point is that the results relate to a packing with a low porosity defined by the idealised gapless packing being about 22 %. Using the porosity correlation presented previously, the results can be presented in a form comparable with experimental results. Thus Fig. 3.8 showing predictions from the network model for various mixtures of sphere sizes and some mono-size packings can be compared with Fig. 3.5.

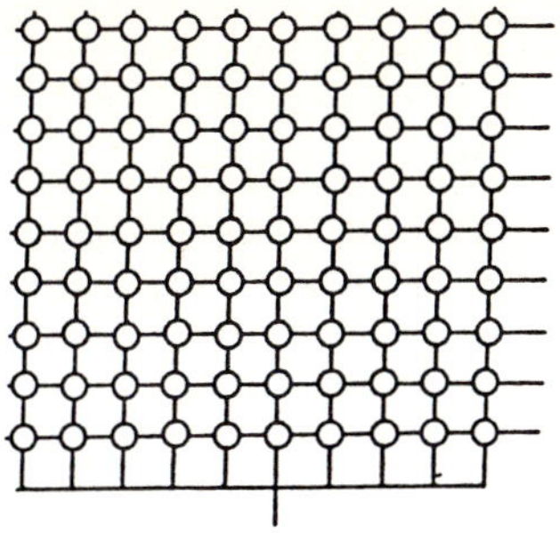

Fig. 3.7 : Four tubes per junction network model

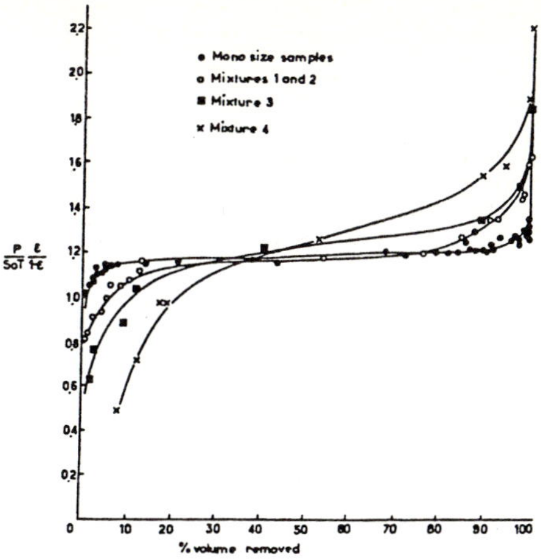

Fig. 3.8 : Capillary pressure curves from network model

It can be seen that despite this simplification of the void space, it does allow a reasonable prediction of the form of capillary pressure curves, especially considering the coarseness of the assumptions. This indicates that the model does preserve the essential features of the pore size distribution in multicomponent sphere packings and that the actual configuration of the connectivity is less important than the fact that it exists.

4 FLOW OF FLUIDS IN POROUS MEDIA

The fundamental relations for fluid flow are the Navier-Stokes equations of hydrodynamics. However, to use these general equations requires boundary conditions which are only possible to give for the simplest geometries and are out of the question for the complicated pore space in a packing of particles. For this reason flow in porous media cannot be treated by the methods of classical hydrodynamics.

In the absence of solutions to the Navier-Stokes equations, the basic relation for fluid flow in porous media is Darcys' law. This was established experimentally in 1856 by Henri Darcy, a civil engineer concerned with the water supply to Dijon. He determined that the rate of flow of fluid in a porous medium is proportional to the pressure or hydraulic head, to the cross-section of the bed normal to flow and inversely proportional to the length and the viscosity

$$Q = K \frac{A}{\mu} \frac{\Delta P}{L} \qquad (4.1) \qquad\qquad \nu = \frac{K}{\mu} \frac{\Delta P}{L} \qquad (4.2)$$

The proportionality factor K is called the permeability, has units of $(\text{length})^2$ and only depends on the structure of the porous medium. It is implied in the above that the resistance to flow is entirely due to viscous drag, and an alternative formulation used in chemical engineering serves to emphasise this

$$f \, Re = \text{constant} = \Lambda \qquad (4.3)$$

where f is the friction factor and Re the particle Reynolds number

$$f = \frac{d_p \Delta P}{\rho \nu^2 L} \qquad (4.4) \qquad\qquad Re = \frac{d_p \nu \rho}{\mu} \qquad (4.5)$$

Experimental data using this representation are given in Fig. 4.1. The change in regime at higher Reynolds number is not a transition to turbulence. Darcys law, corresponding to the horizontal part of the figure, breaks down when the distortion in the streamlines due to changes in direction of motion is great enough for inertial forces to become significant with respect to viscous forces. Real turbulence will only occur, if at all, at much higher Reynolds numbers. In what follows only the Darcy regime will be considered.

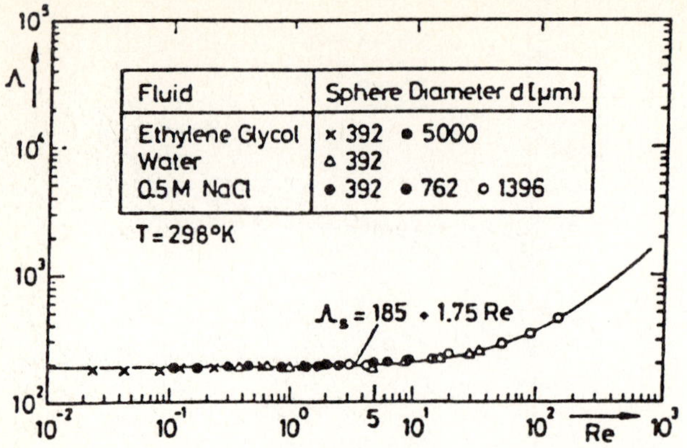

Fig. 4.1 : Flow in random packings of spheres (Kulicke and Haas IEC 23, 308, (1984))

Fluid	Sphere Diameter d [μm]
Ethylene Glycol	× 392 • 5000
Water	△ 392
0.5 M NaCl	• 392 • 762 ○ 1396

T = 298°K

$\Lambda_s = 185 + 1.75\,Re$

4.1 STRUCTURAL INTERPRETATION OF PERMEABILITY

It is known from Darcy's law that the units of permeability are (metre)^2 and the equations (4.4) and (4.5) imply that K is proportional to the square of the particle diameter. However, other features of pore space are important, and attempts have been made to predict the permeability from idealised structures of porous media. One of the most successful models, based on a description of porous media as a bundle of capillary tubes using Poiseuilles law modified in terms of a hydraulic diameter, results in the Kozeny-Carman equation.

The rate of flow in a straight capillary tube can be derived from first principles and is called Poiseuilles law

$$v^* = \frac{d_t^2}{32} \frac{\Delta P}{\mu L} \qquad (4.6)$$

For flow in tubes with a non-circular cross-section we may use the hydraulic diameter

$$m = \frac{\text{cross-sectional area}}{\text{wetted perimeter}} \qquad (4.7)$$

for a circular tube, $m = d_t/4$ and (4.6) becomes

$$v^* = \frac{m^2}{2} \frac{\Delta P}{L} \qquad (4.8)$$

The factor 2 is replaced by k_0 a shape factor which can be used to account for other geometries, for example for a triangular tube k = 1.67, for a slit k_0 = 2.65. Now if we assume that the pore space in the porous medium is as schematized in Fig. 4.2 and is composed of tubes going from one face to the other, that these tubes are longer than the length of the medium and that the fraction free cross-sectional area is equal to the porosity of the medium (ε), we can relate the apparent approach velocity (v) to the real velocity in the tubes (v*)

$$v = \varepsilon \left(\frac{L}{Le}\right)\left(\frac{L}{Le}\right) \qquad (4.9)$$

giving
$$v = \frac{m^2}{k_0 (Le/L)^2} \varepsilon \frac{\Delta P}{\mu L} \qquad (4.10)$$

going further, the hydraulic diameter of a packed bed is given by

$$m = \frac{\text{volume of void space}}{\text{surface area of solid}} = \frac{\varepsilon}{(1-\varepsilon)S_0} \qquad (4.11)$$

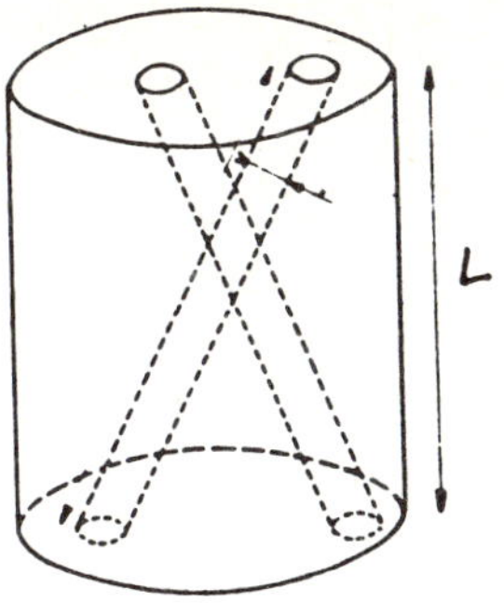

Fig. 4.2

in which for spheres $S_0 = 6/d$.

From (4.10) and (4.11), we have that the Kozeny-Carman equation for the permeability of a porous medium is

$$K = \frac{\varepsilon^3}{h(1-\varepsilon)^2 S_0^2} = \frac{\varepsilon^3}{h(1-\varepsilon)^2} \frac{d^2}{36} \tag{4.12}$$

42 DISCUSSION OF THE KOZENY-CARMAN EQUATION

In the derivation given above, it can be seen that the Kozeny constant h is made up of two terms, a shape factor and the tortuosity

$$h = k_0 (Le/L)^2 = k_0 \tau^2$$

If we assume that the average deviation of a streamline flowing in a packed bed is at an angle of 45° we obtain $Le/L = \sqrt{2}$. If we further assume a shape factor of around 2.5, this leads to a Kozeny constant h = 5.0. It is found in practice that the Kozeny constant is approximatly 5.0 in many different systems (17) and that equation (4.12) gives good results for many different materials. The equation is often used to determine specific surface or mean particle diameter from permeability experiments. However, there are important conditions which must be observed for the equation to be valid.

- No pores must be sealed off as this would affect porosity available to flow but not the absolute porosity.

- The particles, or the pores, must be randomly distributed and be reasonably uniform.

- The porosity must be neither extremely high nor extremely low.

The Kozeny-Carman equation therefore requires a reasonably uniform and random pore structure, since the large pores contribute to permeability and the small pores most to the specific surface. The porosity correlation used for capillary pressure curves in the previous section, and derived independently, is essentially a Kozeny-Carman correlation. This showed that the hydraulic diameter expression brings all capillary pressure curves or pore size distributions to the same mean value, but leaves the distribution about this mean. These differences can then only be accounted for by variations in the h factor, and neither a variation in the shape factor nor in the tortuosity can be considered to be a valid way of taking the pore size distribution into account.

Nevertheless, Fig. 4.3 and Fig. 4.4 show that the equation can give reasonable results for ternary mixtures of particles. Figure 4.3 has been calculated using the

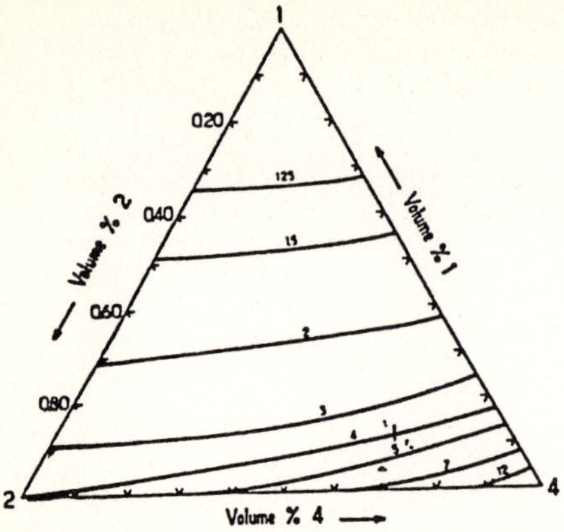

Fig. 4.3 : Permeability of ternary sphere packings by Kozeny-Carman and porosity from gapless packing model

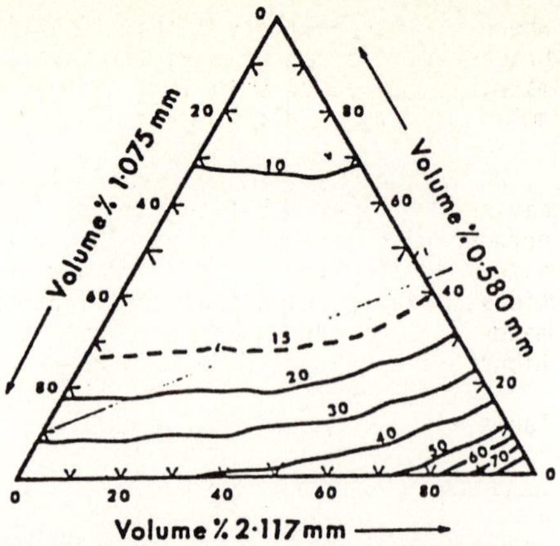

Fig. 4.4 : Experimental permeabilities for ternary mixtures of coke STANDISH (4)

ternary porosity values calculated from the model and presented in Fig. 2.4 together with mean specific surface values calculated from the mixture composition (8). Figure 4.4 gives experimental results from the literature (9).

43 PERMEABILITY of NETWORK MODEL of PORE SPACE

Instead of assuming, as in Fig. 4.2, that each pore goes from one side of the porous medium to the other, it is more reasonable to assume that they are connected in a network as shown in Fig. 4.5. This representation includes the essential features of the pore space in particle packings, that is pore size distribution and pore connectivity.

It can be shown (18) that the application of Poiseuilles law to the flow in each tube and Kirchoffs circuit law to the network that an equivalent of Darcys law can be given as

$$\psi = \left[(C)_t \; (L_i/d_i^4)(C) \right]^{-1} (P)$$

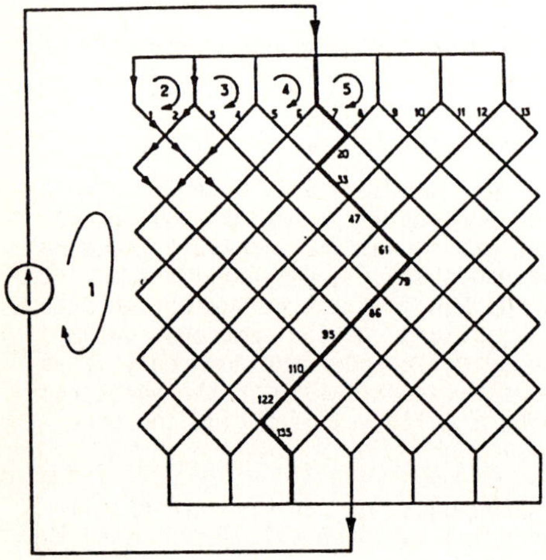

Fig. 4.5 : Network of 143 tubes and 73 loops showing one path through the network

where the permeability can be identified as the matrix operator shown in square brackets. This can be seen to contain the pore size distribution in the diagonal matrix $[L_i/d_i^4]$ and also show the way these pores are joined up in the connection matrix (C) and its transpose $(C)_t$.

We have calculated permeabilities of a network of the type shown in the figure having 143 tubes substituted with 3 different tube size distributions together, one network with one tube size only. These distributions were gaussian, all having the same mean diameter and different widths σ = 0.12, 0.23, 0.35. To simplify calculations, it was assumed that the length of each tube (L_i) was proportional to its diameter (d_i) and that therefore the resistance to flow R_i of a given tube was proportional to $1/d_i^3$. The results are given in Table 4.1.

Table 4.1 : Permeability of networks of tubes

	A	B	C	D
σ	0	0.12	0.23	0.35
$\langle d_i \rangle$	1.0	0.99	0.97	0.96
$\langle R_i \rangle$	1.0	1.1	1.4	3.4
K	1	0.94	0.86	0.72
K_{mf}	1	0.94	0.85	0.77

It can be seen that, in disagreement with the Kozeny-Carman equation, the permeability does not only depend on the mean tube size but also on the width of the distribution of tube sizes, even though the tortuosity and the shape factors are the same in all four cases. It may be concluded that the configuration of the network must be included in the averaging procedure. Obviously, a few very small tubes in a network or a porous medium are sufficient to change the mean tube size, but flow finds other paths through the network and reduces their effect on the permeability. A better averaging procedure, which includes the structure of the network, is given by the mean field theory (KIRKPATRICK [19]). For a square network as used here there leads to the expression

$$\sum_{i=1}^{1=n} \left(\frac{R_i - R_m}{R_i + R_m} \right) p_i = 0 \qquad (4.14)$$

In this p_i is the number fraction of resistances R_i and R_m is the mean resistance, which if used to relace all the resistances R_i will give the same overall resistance to the network. Values of R_m calculated from expression (4.14) are given in Table 4.1 as permeability values K_{mf}. It can be seen that there is a good correspondence with the values calculated by Kirchoffs laws. This analysis serves to emphasize that the success of the Kozeny-Carman equation is based on the essential similarity of the pore space of most particle packings, allowing the effects due to pore size distribution and connectivity to be accounted for by a simple numerical factor. Porous media with "non-standard" structure, such as consolidated rocks, flocculated filter cakes, beds of orientated fibres can require excessive values of the h factor since they have a very different structure.

44 DISPERSION DUE TO FLUID FLOW

The main conclusion from the previous section is that the structure of a porous medium is an important feature. The only really practical way of studying the structure of pore space is by residence time distribution experiments, where dispersion due to flow is followed by injecting tracers.

The phenomena causing dispersion of a tracer in flow through a porous medium are not well established, but the following four effects all contribute.

- Molecular diffusion due to concentration gradients.
- Taylor diffusion due to the interplay of velocity profiles and concentration gradients.

- Backmixing effects in the voids in the pore space.
- Dispersion due to the repeated division, recombination and re-division of the flow streams.

In the Darcy regime we may expect the last of these effects to be predominant, and furthermore this is the fundamental contribution of the structure of the porous medium to dispersion, and the other effects listed above may be considered to be superimposed on it.

The solution of the flow in a network by Kirchoffs laws, as presented in section 43, allows a calculation of these effects. We assume that at each node the probability of a division of the flow stream is given by the respective flows in each tube at the node. Furthermore, knowing the size of each tube in the network and the flow in them, we may calculate the delay time in each tube. Then, by making an inventory of all possible paths through the network from top to bottom, we can determine the time of passage in each path and the amount of fluid taking that path. For example, one path through the network of Fig. 4.3 is shown to be by the tube sequence 7 20 33 47 61 74 86 95 110 172 135. Fluid taking this path will emerge at the outlet after a time of

$$t_7 + t_{20} + t_{33} \cdots t_{135} = \sum t_i \tag{4.15}$$

and the amount of fluid following this path will be

$$p_7 \times p_{20} \times p_{33} \quad \cdots \times p_{135} = \Pi p_i \tag{4.16}$$

By applying (4.15) and (4.16) to all the 9816 possible paths through the network we can piece together the step response to flow dispersion. The results for the 4 networks of Table 4.1 are given in Fig. 4.6.

It can be seen that in the case where all the tubes are the same size (network A) there is no dispersion. All the paths through the network are identical, and a step input of tracer will give a perfect step response. However, as the width of the distribution is increased, the width of the step response is correspondingly increased.

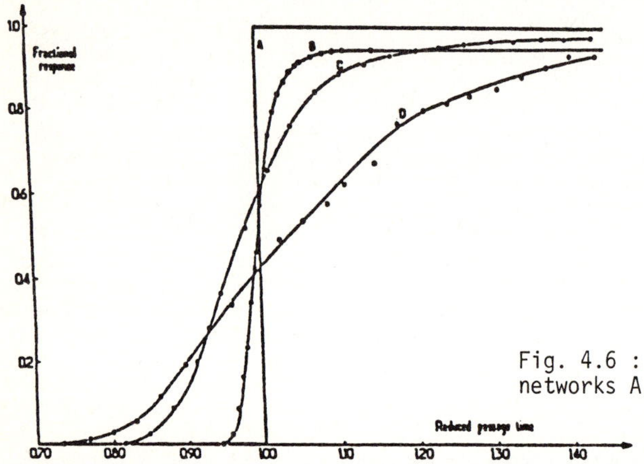

Fig. 4.6 : Residence time distribution for networks A, B, C, D

REFERENCES

1 T. Allen: Particle Size Measurement, Chapman Hall, 1981, ISBN 0412 15410 2.
2 P.P. Haughey, G.S.G. Beveridge: Canad. J. Chem. Eng. 47, 130 (1969).
3 J.D. Bernal: Proc. Roy. Soc. A280, 299-322 (1964).
4 J.L. Finney: Proc. Roy. Soc. A319, 495-507 (1970).
5 M.E. Wise: Philips Res. Rep. 18, 109 (1963).
6 J.A. Dodds: J. Coll. Interface Sci. 77, 317-27 (1980).
7 R. Ben Aim, P. Le Goff, P. Le Lec: Powder Tech. 5, 51 (1971/72).

8 M. Leitzelement, C.S. Lo, J.A. Dodds: Powder Tech. 41, 159-164 (1985).
9 N. Standish, D.G. Mellor: Powder Tech. 27, 61-68 (1980).
10 J.A. Dodds: Nature, 256, 187 (1975).
11 D. Bideau: Doctorat Thesis, Rennes University (1983).
12 J.A. Dodds, H. Kuno: Nature, 266, 614 (1975).
13 L. Oger, J.P. Troudec, D. Bideau, J.A. Dodds, M.J. Powell: to be published.
14 N.R. Morrow: Ind. Eng. Chem. 62, 32 (1970).
15 W.B. Haines: J. Agr. Sci. 20, 97 (1930).
16 J.A. Dodds, P.L. Lloyd: Powder Tech. 5, 69-76 (1971/72).
17 P.C. Carman: Flow of Gases through Porous Media, Butterworths (1956).
18 M. Leitzelement, P. Maj, J.A. Dodds, J.L. Greffe: Ch. 9. Solid-liquid Separation ed. J. Gregory, Ellis Horwood, ISBN 0853126844 (1984).
19 S. Kirkpatrick: Phys. Rev. Let., 27, 1722-25 (1971).

Dense Packings of Hard Grains: Effects of Grain Size Distribution

D. Bideau, J.P. Troadec, J. Lemaitre, and L. Oger
Groupe de Physique Cristalline, UA au C.N.R.S. N° 040804, Université de Rennes I, Campus de Beaulieu, F-35042 Rennes Cedex, France

A packing of grains, spherical or not, is often described by its mean properties :

- geometrical properties : - porosity : ϕ = Void volume/Total volume
  - coordinance z : mean number of *real* contacts/grain, etc...
- physical or physicochemical properties : - electrical conductivity (in "pore space" as well as in "grain space")
  - permeability, etc...

Then, the problem is to connect the studied mean physical property to one or several geometrical parameters. For example, Archie's law gives the conductivity σ of a porous medium (porosity ϕ) whose pore space is filled with a conducting fluid with conductivity σ_f :

$$\frac{\sigma}{\sigma_f} = a\phi^m$$

In the case of a packing of grains with comparable sizes and shapes, such a law seems to be justified, although local and global inhomogeneities, as arching effects for example, can modify that behaviour.

In a packing with a grain size distribution, local fluctuations may be very important. Is a mean parameter useful in that case, particularly in binary mixtures ? We shall study that problem, but consider only packings built under gravity.

I The simplest case : a different grain in a packing of equal spheres

I.1 Wall effect and maximum disorder

If we introduce a different sphere in a packing of equal spheres we modify the structure of the packing - at least locally -. For 2D disk packing [1, 2], only one "impurity" - disk with a different size - can lead to a long distance disorder. The figure 1 shows a 2D packing built from a triangular seed of two equal disks and the impurity. From a seed of three equal disks, the same algorithm gives an ordered packing.

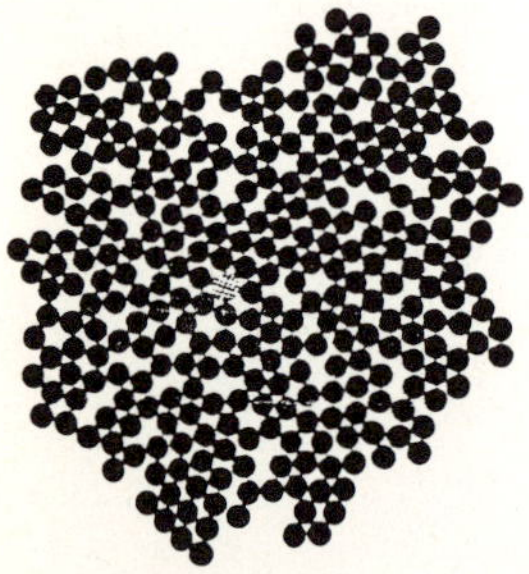

Fig. 1 2D disk packing with a central impurity

The observed disorder is generated by exclusion effects which, in 3D packings and in the cluster of the near neighbours of a grain, can be defined in 3 points :

a) the distance d between the centres of two grains with radii R_i and R_j cannot be lower than $R_i + R_j$ (hard grains).
b) it is not possible, except for particular grain sizes, to close exactly the *perimeter* around a grain on a surface (figure 2).

Fig. 2 2D stacking fault : it is not possible to close the perimeter around the central disk

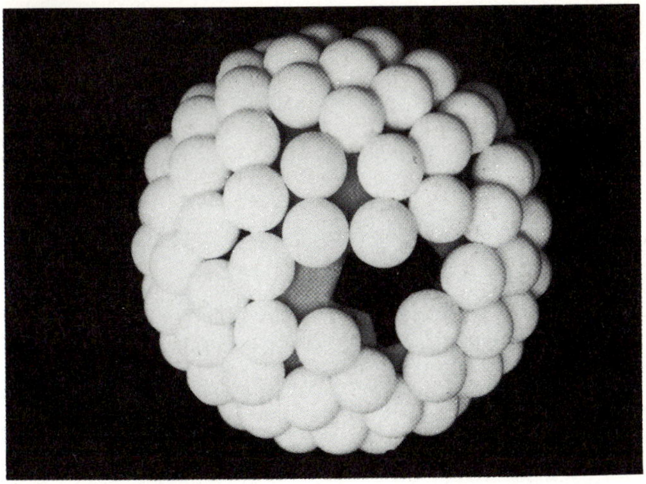

Fig. 3 3D stacking fault : it is not possible to close the surface around the central sphere

c) it is not possible, also except for particular grain sizes, to close exactly the *surface* around a grain (figure 3).

These exclusion effects are responsible for the "wall effect" described by BEN AIM [3], which is the perturbation created by a wall in a packing - the wall can be the surface of a grain -. The figure 4 gives, for example, the variations of the porosity of a packing of equal spheres versus the distance d from a plane wall. BEN AIM defines a packing with *maximum disorder* as a packing in which the perturbation involved by a wall is limited to the near neighbours of the wall. That concept is very useful to determine some quantitative results on packings.

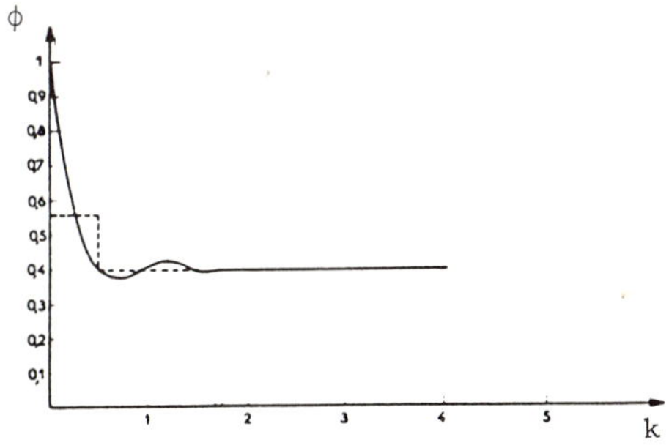

Fig. 4 Wall effect of a plane wall on the local porosity of equal sphere packings (from BEN AIM [3])
----- maximum disorder model

I.2 Evolution of the geometrical parameters

We consider a dense packing of spheres with radius R, inside which one sphere with radius R' is introduced. We shall study two parameters :

- the coordinance z' of the introduced grain ;
- its "hindrance coefficient" e', which we define as $e' = \Delta V'/W'$: $\Delta V'$ = apparent volume = volume increase of the packing, due to the introduction of the grain with radius R', W' = volume of that grain = $4/3\ \pi R'^3$.

The values of these parameters can be very different according to the value of the ratio $k = R'/R$. In the limit example of packings with tetrahedral units [4], we can define two critical ratios related to the size of the cavity within the tetrahedron :

$k'_c = \frac{1}{6.46}$: the ball R' passes just through the "windows" of the cavity.

$k_c = \frac{1}{4.45}$: the ball R' is in contact with the four spheres of the tetrahedron.

For $k \leqq k'_c$: the ball R' falls at the bottom of the box containing the packing ! z' is not defined and $\Delta V' = 0$, $e' = 0$.

For $k'_c < k < k_c$: the ball R' is free within the cavity ; $z' = 3$ (under gravity) and $\Delta V' = 0$, $e' = 0$.

For $k = k_c$: the ball R' is just blocked up within the cavity ; $z' = 4$, $\Delta V' = 0$, $e' = 0$.

For $k > k_c$: the ball R' is a wall for the other spheres.

In this latter case, z' et e' can be determined by the existing models which only take the hindrance of the grain into account. These models are based on a complete pavement of the space by tetrahedral units [4, 5, 6] or by Voronoī type cells [7], and they don't know the exclusion effects (except Dodds' model which defines random gaps). Nevertheless, starting from the simple case of one impurity in a packing of equal spheres, some calculations have been made which take the wall effects into consideration. Particularly, the case of a large sphere in a packing of small spheres, i.e. $k >> k_c$, has been studied [3] ; all the results are qualitatively comparable. From numerical simulations of spheres with radius R on the surface of a sphere of radius R', we have proposed [8] the expression :

$$z' = c/(1 - \frac{\sqrt{k^2+2k}}{1+k})$$

with c = cte. This expression gives a very good fit of numerical results for $k \geqq k_c$ (figure 5).

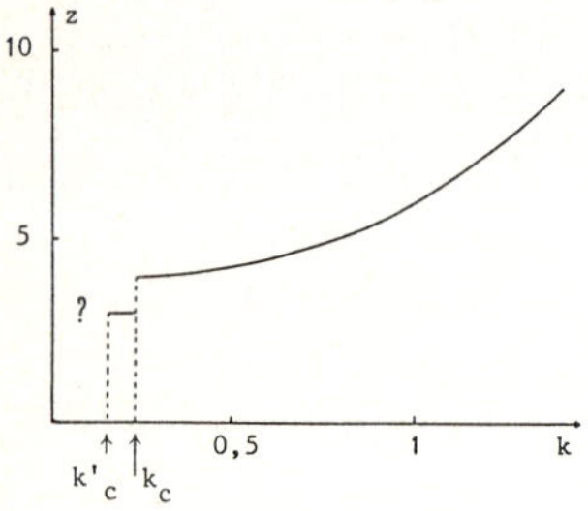

Fig. 5 Coordinance z' of a sphere with radius R' introduced in a packing of equal sphere with radius R versus $k = R'/R$

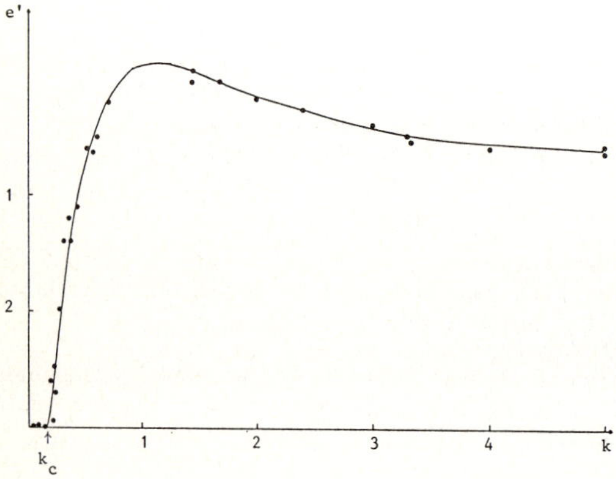

Fig. 6 Hindrance coefficient of a sphere with radius R' introduced in a packing of equal sphere with radius R, versus $k = R'/R$ (from BEN AIM)

The proposed curve for $e' = f(k)$ (figure 6) is drawn from experimental results by BEN AIM [3]. That curve shows the existence of the critical ratio k_c.
It is clear that, even in that simple model, the values of k_c or k'_c are not well defined, because the size distribution of the cavities in an equal-spheres packing.
We want to underline one fact : according to the curve of figure 6, the hindrance coefficient e' can be very different for small or large grains. In the literature, the apparent volume Δv_i of a grain R_i in a packing with grain size distribution is often given as $\frac{4}{3}\pi R_i^3/(1-\phi)$, where ϕ is the *mean* porosity of the packing, that is to say that e' is the same ($e' = 1/(1-\phi)$) for all the grains in the packing ! This is only true for packing of equal spheres.

II Binary mixtures

Here, it is not possible to define k_c or k'_c, but only a transition zone, because there is a larger distribution of the size of cavities and windows. Indeed, one can present schematically a phase diagram for the binary mixture, as a function of R_1/R_2 (where R_1 is the radius of small grains) and $v_1 = n_1R_1^3/(n_1R_1^3 + n_2R_2^3)$, the volume fraction of small grains.
On figure 7, when passing the "critical zone" along the arrow direction, we have successively :

A. A domain with two separated phases (under gravity) : the small grains in weak concentration occupy exclusively the pore space of the large grains packing : there is a *global segregation.*

B. A domain more and less large, where the small grains occupy preferentially the cavities of the large grain space : there is a strong *local segregation.*

C. Then, we have a local segregation less and less important. That local segregation in C domain varies with the "history" of the packing, and has been made evident in 2D [9] as in 3D packings [10, 11].

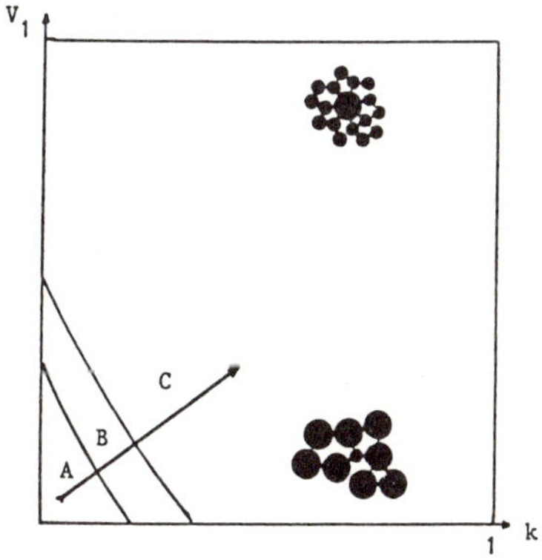

Fig. 7 Phase diagram in a binary mixture

These effects set the problem of the randomness of a packing. From Dodds model, we have proposed [12] a definition of that randomness in two points. First, the gaps of the real packing ("cut bonds") with regard to the ideal gapless packing are randomly distributed. Secondly, there is no order in the relative position of the sites (grains). This last condition has a sense only in packings with differentiable grains : A-B "alloys" or packings with different grain sizes. The local or global segregations are at variance with the second condition.
These segregations appear too in packings with large distribution of the grain sizes, and are used in models of packings with strong compacity - packings with discrete size distribution - [13, 14].
According to the above phase diagram , what are the *geometrical* parameters in a *physical* binary mixture ?

II.1 Coordinance

In some conditions, it is easy to know the mean coordinance z of a packing. Thus, a packing built sphere after sphere under gravity has a mean coordinance 6 : each grain is in equilibrium with 3 contacts, and each contact is common to 2 grains. But generally, a real packing is built collectively, and the above result can be wrong.

If one grain i is in a cavity ($k < k_c$ for that grain), it has a number of contacts which depends only on its equilibrium under gravity : $z_i = z_o$. In the same way, any grain in a packing is in equilibrium : it has a minimum of z_o contacts, and its total coordinance is function of its size. For the mean coordinance z_i of the species i in a binary mixture, we propose the expression

$$z_i = z_o + a\, f\,(n_i,\, k)$$

where $k = R_1/R_2$, and a, like z_o, depends on the shape and the surface state of the grain, and on the "history" of the packing. In the case of a random packing, the function $f(n_i,k)$ can be determined for example from Dodds model.

It can be useful to know the number of 1-1, 1-2 and 2-2 contacts in a binary mixture. It can be calculated from Dodds model in the random hypothesis, i.e. far from the "transition zone" of the above phase diagram . But local segregation can modify these numbers near the "transition zone".

II.2 Porosity

The figure 8 gives BEN AIM results [3] : porosity versus volumetric fraction of small grains for different values of $k = R_1/R_2$. The left part of that curve shows the above phase separation : the porosity decreases because of the global segregation : the small grains fill in preference the pore space of the large grains packing. The right part corresponds to the variation of the wall effect, and can be understood from the model of one sphere in a packing of equal spheres [3]. Here, the existence of a critical zone is still shown, but the mean porosity curve masks the different contributions of each species to the porosity, shown by the curve $e' = f(k)$ in chapter I.

II.3 Hindrance coefficient

That parameter has been really studied only in the case of chapter I. If one admits that the grain i is put inside a packing of spheres for which one can define a mean radius, the variation of e_i for the species i should be comparable to those

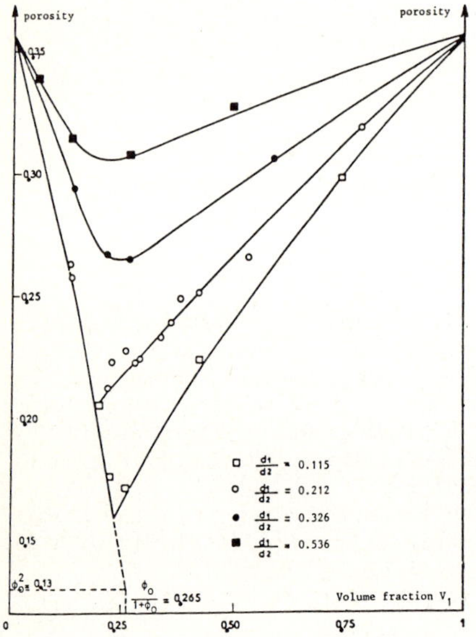

Fig. 8 Porosity in a binary mixture, versus volumetric fraction of the small spheres (from BEN AIM)

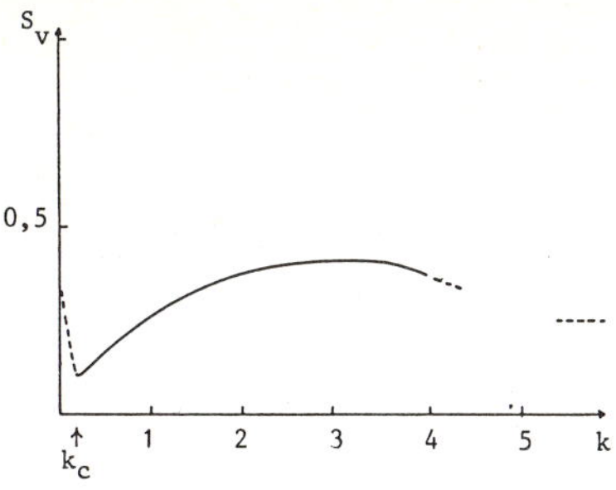

Fig. 9 Variation of the volumetric percolation threshold of a binary mixture

of the figure 5. But, particularly in the case of a packing with local segregation, that "mean radius" does not have a real sense, because each species does not have a comparable influence on the packing structure ; for example, we think that the maximum of the hindrance coefficient in binary mixtures could be for $R_2 = R_1/k_c$, rather than for $R_2 \sim R_1$.

III One application : conductivity in the grain space

Suppose that we put into a box two sphere species, with different sizes, one conducting and the other insulating. What is the minimum volume of conducting spheres for which the system is conducting, or what is the volumetric percolation threshold of the system ? The figure 9 gives the variations of S_V (= volume of conducting spheres/total volume of spheres) versus $k = R_c/R_i$ where R_c is the radius of the conducting spheres and R_i the radius of the insulating spheres. The part in dark line has been studied experimentally for spherical or polyhedral grains [11,16]. For $k<k_c$, the small grains fill the pore space of the large sphere packing ; we have a sort of "porous conducting fluid" with a porosity increasing with k. For $k >> 1/k_c$, we have the well-known problem of conducting spheres in an insulating continuum medium.

None of the parts of that curve can be fitted by the "invariant" of SCHER and ZALLEN [7], V_c/V_T # 0.17, where V_c is the conducting volume and V_T the total volume of the packing. This is because that "mean" invariant is not adapted to our problem. On the other hand, for $k > k_c$, the variations of S_V are very comparable with those of the hindrance coefficient described in chapter I, which is a more local parameter : it is the ratio of the apparent volume of the grain in the packing and its conducting volume.

Bibliography

[1] M. RUBINSTEIN and D. R. NELSON Phys. Rev. B 26, 6254 (1982)
[2] D. BIDEAU, J. P. TROADEC et L. OGER C. R. Acad. Sc. Paris 297, 319 (1983)
[3] R. BEN AIM Thèse d'Etat, Nancy 1970 (and authors cited by BEN AIM)
[4] J. A. DODDS J. Coll. and Interf. Sc. 77, N° 2, 317 (1980)
[5] M. E. WISE Philips Res. Rep. 7, 321 (1952)
[6] M. J. HOGENDIJK Philips Res. Rep. 18, 109 (1963)
[7] D. E. G. WILLIAMS Phil. Mag. B50, 363 (1984)
[8] L. OGER, D. BIDEAU, J. P. TORADEC J. Phys. Lett. Submitted to publication
[9] D. BIDEAU Thèse Rennes 1983
[10] D. E. G. WILLIAMS Private communication 1982
[11] L. OGER Thèse Rennes 1983
[12] L. OGER, J. P. TROADEC, D. BIDEAU, J. A. DODDS and M. J. POWELL Powder Technology Submitted to publication
[13] A. CAQUOT Mémoires de la Société des Ingénieurs Civils de France 562 (1937)
[14] R. OMNES J. Phys. 46, 139 (1985)
[15] L. OGER, J. P. TROADEC, D. BIDEAU, J. A. DODDS and M. J. POWELL Powder Technology submitted for publication
[16] M. AMMI, D. BIDEAU, J. P. TROADEC, F. ROPITAL and G. THOMAS Sol. St. Comm. to be published
[17] H. SCHER and R. ZALLEN J. Chem. Phys. 53, 3759 (1970)

Part IV **Percolation**

Shapes, Surfaces, and Interfaces in Percolation Clusters

Antonio Coniglio
Dipartimento di Fisica, Universita di Napoli, Mostra d'Oltremare, Pad. 19, I-80125 Napoli, Italy and
Center for Polymer Studies and Department of Physics, Boston University, Boston, MA 02215, USA

Percolation theory is reviewed. Intuitive arguments are given to derive scaling and hyperscaling relations. Above six dimensions the breakdown of hyperscaling is related to the interpenetration of the critical large clusters, and to the appearence at p_c of an infinite number of infinite clusters of zero density with fractal dimension $d_f = 4$. The structure of the percolating cluster made of links and blobs is characterized by an infinite set of exponents related to the anomalous voltage distribution in a random resistor network at p_c. The surface structure of critical clusters below p_c, which is relevant to the study of random superconducting networks, is also discussed. In particular, an exact result is presented which shows that in any dimension the interface of two critical clusters diverge as $(p_c - p)^{-1}$ as the percolation threshold is approached.

I. Introduction

Since this talk is addressed mainly to the non-expert, I shall first review briefly the standard bond percolation problem; for reviews and extensive references see Refs. [1,2]. Consider a d-dimensional hypercubic lattice of linear dimension L. Suppose that each bond has a probability p of being present or active. For small values of p, small clusters made of sites connected by nearest-neighbor bonds are formed. Each cluster is characterized by its size or mass s, the number of sites in the cluster. For large values of p in addition to small clusters we expect a macroscopic cluster that connects the opposite boundaries. This spanning cluster becomes infinite as the system size becomes infinite. For an infinite system there exists a percolation threshold p_c below which only finite clusters are present. The active bonds may represent the chemical bonds among reacting monomers, in which case the percolation describes the sol-gel transition [3].

Alternatively the active bonds may represent the capillaries of a porous media, in which case percolation is a useful model to study the flow of a fluid in a random media [4]. Fluid can flow only above p_c when there exists a path of active (open) bonds connecting the opposite boundary of the system.

If we associate an electrical conductance σ_a to an active bond and σ_B to the nonactive bonds we obtain a random composite material that describes a random resistor network if $\sigma_B = 0$ and a random superconducting network for $\sigma_a = \infty$ and σ_B finite. In the first case the conductivity tends to zero as p_c is approached from above. In the second case the conductivity diverges as p_c is approached from below.

In order to describe the percolation transition, one defines an order parameter $P_\infty(p)$ as the density of sites in the infinite cluster, which goes to zero as $p \to p_c^+$ and is zero for $p < p_c$. The mean cluster size $S(p)$ of the finite clusters that diverge at p_c. The average number of clusters $K(p)$ that exhibit a weak singularity. These quantities are all defined per site. Finally one can define the connectedness length $\xi(p)$, which is the average radius of all finite clusters for the typical linear dimension of a critical cluster. As p_c is approached from below $\xi(p)$ diverges in analogy with the critical opalescence phenomenon near the liquid-gas phase transition. Most of the above quantities can be related to the average number of clusters of s particles $n(s,p)$. Near the percolation threshold the critical behavior is characterized by critical exponents

$$K(p)|_{sing} = \sum n(s,p)|_{sing} \sim |p - p_c|^{2-\alpha} \tag{1a}$$

$$P_\infty(p) = 1 - \sum sn(s,p) \sim (p - p_c)^\beta \tag{1b}$$

$$S(p) = \sum s^2 n(s,p) \sim |p - p_c|^{-\gamma} \tag{1c}$$

$$p\xi(p) \sim |p - p_c|^{-\nu}, \tag{1d}$$

where the sum is over all finite clusters and in (1a) only the singular part has been considered.

II. Scaling

The critical exponents defined in (1) are not all independent. Scaling relations can be derived among them as for ordinary second-order phase transitions. These scaling laws are intimately related to the property of the incipient infinite cluster of being self-similar to all length scales. For such self-similar objects one can define a fractal dimensionality d_f [9] via the following relation: $M \sim R^{d_f}$, where M is the number of sites (or "mass") of the infinite cluster contained in a hypersphere of radius R. To obtain scaling laws, following Kadanoff's original idea, we perform the following three steps: (1) divide the system into cells of linear dimension b, (2) coarse grain by some suitable rule, (3) rescale the lengths by a factor b. The result is a renormalized system where the size of the large clusters s has been reduced by a factor b^{d_f} and all lengths by a factor b

$$L' = L/b, \qquad \xi' = \xi/b, \qquad s' = s/b^{d_f}. \tag{2}$$

Assuming that the large clusters **do not interpenetrate**, the sum over the large clusters must be the same before and after rescaling, i.e.,

$$\sum N(\bar{s}, \xi) = \sum N(\bar{s'}, \xi'), \tag{3}$$

where $N(s, \xi)/L^d = \bar{n}(s, \xi) \equiv n(s, p)$ is the number of clusters of s particles per unit volume. The sum in (3) is over the large clusters up to size s and can be approximated by an integral. Using dimensional analysis from (3) we have

$$sN(s, \xi) = s'N(s', \xi'). \tag{4}$$

Dividing by the volume L^d, from (2) we obtain

$$\bar{n}(s, \xi) = b^{-d-d_f}\bar{n}(sb^{-d_f}, \xi b^{-1}). \tag{5}$$

Choosing $b = s^{1/d_f}$ from (5) we have

$$n(s, p) = s^{-\tau} f((p - p_c)s^{\sigma}), \tag{6}$$

with

$$\tau = \frac{d}{d_f} + 1 \tag{7a}$$

$$\sigma = \frac{1}{\nu d_f}. \tag{7b}$$

Equation (6) exhibits the scaling form postulated by Stauffer. From (1a), (1c), and (7) we have the following relations

$$2 - \alpha = \frac{\tau - 1}{\sigma} \tag{8a}$$

$$\beta = \frac{\tau - 2}{\sigma} \tag{8b}$$

$$-\gamma = \frac{\tau - 3}{\sigma}, \tag{8c}$$

which give the scaling relation

$$\alpha + 2\beta + \gamma = 2. \tag{9}$$

From (7a) and (8a) we obtain the hyperscaling relation

$$2 - \alpha = \nu d, \tag{10}$$

and from (7), (8b), and (8c)

$$d - \beta/\nu = d_f \tag{11a}$$

$$\frac{1}{\nu}(\beta + \gamma) = d_f . \tag{11b}$$

Flory [10] first solved the gelation problem by neglecting loops. This approximation is equivalent to solving the percolation problem on the Bethe lattice or Cayley tree [11]. The critical exponents in this approximation are $-\alpha = \beta = \gamma = 1$, $\nu = \sigma = 1/2$, and $\tau = 5/2$. These exponents satisfy the scaling relation (9) and hyperscaling relation (10) only for $d = d_c = 6$. Currently we know from field theory that this is the upper critical dimensionality above which the Bethe lattice exponents are valid [12]. Therefore above $d_c = 6$ hyperscaling breaks down. Moreover, while Eqs. (11a,b) are consistent with the other hyperscaling relation $2\beta + \gamma = d\nu$, which is verified numerically below d_c, it leads above d_c to two different values for the fractal dimensionality $\bar{d}_f = d - 2$ and $d_f = 4$.

Here we want to give a geometrical interpretation of hyperscaling and why it breaks down above d_c. Critical behavior comes only from the large clusters of linear dimension ξ. If N_ξ is the number of such clusters in a volume of the order of ξ^d we have

$$K(p)|_{crit} = \frac{N_\xi}{\xi_d} \sim \varepsilon^{2-\alpha}. \tag{12}$$

Hyperscaling (10) is therefore equivalent to assuming that N_ξ is of the order of unity. Above $d_c = 6$ the Bethe lattice solution gives $n(s,p) \sim s^{-5/2} e^{-(p-p_c)^2 s}$ for large s. Therefore $N_\xi = \xi^d \sum n(s,p) \simeq \xi^{d-6}$ where $\xi \sim (p - p_c)^{-1/2}$. From (12) it then follows $(2 - \alpha) = 6 \cdot \nu$ which replaces the hyperscaling relation.

What was wrong in the derivation of Eq. (7) which leads to hyperscaling for all d? We note that one assumption was that the large clusters do not interpenetrate. Namely that there is only one cluster in a region of the order ξ^d. For $d > 6$ we have shown that this is not correct. In fact $N_\xi \sim \xi^{d-6}$. Therefore for $b \sim \xi$ Eq. (3) must be modified as

$$\sum N(\bar{s},\xi) = b^{d-6} \sum N(\bar{s'},\xi'),$$

which still leads to Eq. (6) but with different values for τ , namely

$$\tau = \frac{6}{d_f} + 1, \qquad \sigma = \frac{1}{\nu d_f}, \qquad d \geq 6.$$

From (8b), using the above values of σ and τ, we find $6 - \beta/\nu = d_f$ instead of (11a). Since $\beta/\nu = 2$, $d_f = 4$ for $d \geq 6$. Aharony et al [13] first noticed that breakdown of (11a) is due to breakdown of hyperscaling.

In conclusion, for $d > 6$ in a volume of the order ξ^d there are $N_\xi \sim \xi^{d-6}$ critical clusters of mass M given by $M \sim \xi^4$. The total mass M_{tot} is therefore given by $M_{tot} = N_\xi \cdot \xi^4 = \xi^{d-2}$, (below $d_c = 6$, $N_\xi \sim 1$, $M = M_{tot} \sim \xi^{d_f}$). The centers of these clusters are separated by a distance ξ_1, given by $N_\xi \xi_1^d \sim \xi^d$, from which $\xi_1 \sim \xi^{6/d}$.

Another length can also be introduced if we are interetsted in the total mass within a hypersphere of radius R centered at the center of the cluster. We expect for **short distances** that there is no interpenetration with the other clusters and therefore the total mass coincides with the mass of one single cluster with fractal dimension $d_f = 4$, therefore $M_{tot} \sim R^4$ with density $\rho_{tot} \sim R^{d-4}$. After a distance of the order of $\xi_2 \sim \xi^{2/(d-4)}$, $\rho_{tot} \sim \xi^2 \sim P_\infty$, the total density crossover to an homogeneous regime where the density does not change with R. This last result is based on the assumption that within a distance of the order ξ ρ_{tot} has the same behavior above and below p_c.

Similar ideas on hyperscaling have been developed independently also by Alexander et al [14] using the analogy of the percolation problem with a solution of randomly-branched chains. See also Ref. [13] for other approaches.

Note that since for $d > 6$, $N_\xi \to \infty$ as p_c is approached, we expect at p_c an infinite number of infinite cluster of zero density with fractal dimension 4. For $d \leq 6$ the above argument suggests that there is only one ∞ cluster. This result can be connected to a theorem due to Shulman and Newman that states that in percolation the number of infinite clusters, if different from zero, can be either 1 or infinity, with zero density in the latter case.

III. Cluster Structure

A very important question that has received a lot of attention concerns the structure of the incipient infinite cluster at p_c. Consider a cell of linear dimension L. The bonds in the percolating cluster as noted by Stanley [15] may be partitioned in backbone bonds that sustain shear stress or contribute to the electrical resistance. All the others are dangling bonds. The backbone bonds may also be divided into links ("red" bonds), which are singly-connected, and the remaining multiply-connected bonds ("blobs"). The question of whether the blobs could be neglected, as in the Skal Shklovskii-de Gennes model [16,17], or the links could be neglected as in the Sierpinsky gasket model [18] is non-trivial. Now we know on the basis of exact results that both links and blobs are critical. The number of links diverge as $L^{1/\nu}$ in any dimension [19]. As d approaches 6, the blobs become less important until they become irrelevant for $d \geq 6$. Numerical evidence also confirms the validity of this picture [20,21].

Although the links for $d < 6$ are much less in number than the multiply-connected bonds, they can be detected experimentally, in fact only the links determine the critical behavior of a dilute Ising model at p_c, leading to a universal crossover exponent 1 in any d. For the dilute Heisenberg system instead the crossover exponent is related to the resistivity exponent, which, because of the blobs, is larger than 1. This theory has therefore solved a longstanding controversy between the experimental data and the ε-expansion that predicted the crossover exponent = 1 for both Ising and Heisenberg dilute models.

More recently the validity of the ε-expansion for the Heisenberg case has been questioned [22]. As a consequence, another result from a new ε-expansion has been obtained [23], giving a new crossover exponent larger than 1 in agreement with the above theory. For a more detailed discussion for the cluster structure and its connection with dilute ferromagnets, see Ref. [19,22].

A hierarchical model that reproduces very well the critical exponents of the backbone has recently been introduced in the context of the anomalous voltage distribution [24].

IV. Further Characterization of the Backbone Bonds

For a more detailed analysis one can partition the backbone bonds in different categories in the following way. Associate a unit electrical resistance to each bond and apply a unit voltage at the opposite boundary of a box connected by the percolating cluster, each bond then can be characterized by the fraction $\alpha = \frac{I}{I_{tot}}$ of the total current flow through it. In this way each bond is labeled by a number $0 \leq \alpha \leq 1$, or by a corresponding color code. The higher the value of α the hotter the bonds. The hottest bonds are the cutting bonds or "red" bonds, for which $\alpha = 1$. The coldest are those that carry zero current. If $N(\alpha)$ is the number of bonds corresponding to the value α, we can define the following family of lengths ℓ_K and relative exponent $\tilde{\varsigma}(K)$ through

$$\ell_K = \sum \alpha^K N(\alpha) \sim L^{\tilde{\varsigma}(K)}. \qquad (13)$$

ℓ_0 is the number of bonds in the backbone, ℓ_2 is the resistance, ℓ_4 is related to the magnitude of the noise in a random resistor network (see Rammal, this Conference) and ℓ_∞ coincides with the number of links. In general ℓ_K is a length measure of the backbone in which each bond is weighted by a factor α^K, and is related to the anomalous voltage distribution. The prediction on the hierarchical model is that the set of exponents $\tilde{\varsigma}(K)$ are all independent and are not characterized by a "gap" exponent as usually occurs in ordinary critical phenomena. See, for example, the moments of the cluster size distribution. This prediction is verified in simulations of the two-dimensional resistor network. (For more details, see Ref. [24].) For $d = 1$ and $d \geq 6$, the $\tilde{\varsigma}(K)$ all coincides with $1/\nu$, the exponent relative to the links, since the blobs are irrelevant, and the constant gap exponent holds. On the other hand, the maximum dispersion in $\tilde{\varsigma}(K)$ occurs in $d = 2$ where the blobs are relatively most important. In conclusion, an infinite set of independent exponents is necessary to fully characterize the structure of the backbone. This may also explain the failure of many attempts related to the geometrical properties of the percolating cluster.

V. Surfaces and Interfaces below p_c

The study of the structure of the surfaces and interfaces of the large clusters below p_c has not received as much attention as the study of the internal structure of the incipient infinite cluster. This problem is relevant to the study of the dielectric constant of random composite materials, the viscosity of a gel, the conductivity of a random superconducting network, and the relative termite diffusion model.

For simplicity, let us consider a random superconducting network in which superconducting bonds are present with probability p and normal bonds carrying a unit resistance with probability $1 - p$. For small values of p we have finite superconducting clusters in a background of normal resistor. As $p \to p_c$ the superconductivity Σ diverges as $\Sigma \sim (p_c - p)^{-s}$. For a finite cell of linear dimension L just below p_c the typical configurations are characterized by two very large clusters almost touching each one attached to one of two opposite faces. Inside these clusters there are islands of normal resistors. If a unit voltage is applied between the opposite face of the hypercube there is no current flowing through the bonds in the island. We call these "dead" bonds, in analogy with the dead ends of the percolating cluster. The remaining normal bonds **connect** one superconducting cluster to the other. Similarly to the backbone bonds the "connecting" bonds are the only bonds that contribute to the resistance. The connecting bonds can be distinguished in single "connecting" bonds or "bridges," which have the property that if **one** is replaced by a superconducting bond a percolating superconducting cluster is formed, joining the opposite faces of the cell, and the remaining multiple "connecting" bonds. Similarly to the backbone links it can be proved [25] that the fractal dimensionality of the bridges is $1/\nu$. More precisely, the following relations for any lattice in any dimension can be proved

$$(1 - p)\frac{dp_{ij}}{dp} = B_{ij}, \tag{14}$$

where p_{ij} is the pair connectedness function (the probability that sites i and j belong to the same cluster) and B_{ij} is the average number of bridges between i and j. These are defined as non-active bonds, such that if one is made active i and j become connected. This result can be generalized also to the case when i and j are sets of sites. When the set of sites coincide with the opposite faces of a cell of size L, (14) becomes

$$(1-p)\frac{dR}{dp} = B,$$

where R is the probability of getting across the cell and B is the average number of bridges in the cell. Since [26]

$$\left.\frac{dR}{dp}\right|_{p=p_c} \simeq L^{1/\nu},$$

it follows that for $p = p_c$ in a cell of size L the average number of bridges $B \sim L^{1/\nu}$. Namely the fractal dimensionality of the bridge bonds is $1/\nu$. For $p \leq p_c$, choosing $L \sim \xi(p)$ it follows that $B \sim (p-p_c)^{-1}$.

Finally we mention that it can also be shown that the number of bridges coincides with the "excess" perimeter introduced by Stauffer [1]. More details will appear elsewhere [25].

Similarly to the backbone bonds, a further characterization of the normal bonds can be done by assigning to each bond a number $\alpha = \frac{I}{I_{max}}$ where I is the current flowing through the bond and I_{max} the maximum current. In this case I_{max} is the current flowing through one of the bridge bonds and is equal to I since the voltage drop between the two superconducting clusters is unity and the resistance of a normal bond is also unity. If $N'(\alpha)$ is the average number of bonds corresponding to the value α, we can again define a family of length ℓ'_K and relative exponents $\tilde{Z}(K)$ through

$$\ell'_K = \sum \alpha^K N(\alpha) \sim L^{\tilde{Z}(K)}.$$

Note that the value α associted with a bond coincides with the voltage drop V along that bond. Therefore $\tilde{Z}(2) = \tilde{Z}_s$, the exponent describing the divergence of the conductance-averaged overall nonspanning configurations of G ($G \sim L^{\tilde{Z}_s}$), which is related to the conductivity exponent $\tilde{s}$ ($\sum \equiv G/L^{d-2} \sim L^{\tilde{Z}_s - d+2}$) via $\tilde{s} = -(d-2) + \tilde{Z}_s$. For $K = \infty$, $\tilde{Z}(\infty) = 1/\nu$. In general, by duality $\tilde{Z}(K) = \tilde{\varsigma}(K)$ [27]. Since $\tilde{Z}(2) \geq \tilde{Z}(\infty)$, we have the inequality $\tilde{s} + d - 2 \geq 1/\nu$, which numerically is well verified.

A geometrical interpretation of the exponent $\tilde{Z}_s$ can be obtained by noting that if all the normal bonds are substituted with $L^{\tilde{Z}_s}$ effective bridges in parallel joining the two superconducting clusters one would obtain the conductance $G \sim L^{\tilde{Z}_s}$ of the system.

The above considerations suggest that just below p_c the system can be imagined as a superlattice made of large critical clusters whose centers are separated by a distance of the order ξ. The surfaces of these clusters almost touch, and are connected by bridges made of single bonds and other paths made of more than one bond.

VI. Anomalous Diffusion: The Ant and the Termite Models

The Ant Model. Random resistor networks can also be studied from the point of view of diffusion using the famous de Gennes model of the ant in a labyrinth. This approach has received much attention in the past few years [28,29,30]. In this model an ant is supposed to be released at random on the infinite cluster where it is confined and performs a random walk. On an undilute lattice ($p = 1$) the ant obeys the usual Fickian law for long time t:

$$< r^2 > \sim Dt, \tag{16}$$

where $< r^2 >$ is the mean square displacement and D is the diffusion constant. For $1 \geq p > p_c$ (16) still holds with D, depending on p. As p approaches p_c, the diffusion constant goes to zero. Using the Einstein relation between conductivity Σ and diffusion constant D, it is possible to predict $D \sim \epsilon^{\mu - \beta}$ where μ is the conductivity exponent [29]. Right at p_c, Eq. (11) is replaced by

$$< r^2 > \sim t^{2/d_w}, \tag{17}$$

with

$$d_w = 2 + (\mu - \beta)/\nu. \tag{18}$$

Since $d_w > 2$ from (17) follows that the ant moves very slowly compared to the usual random walk on an undilute lattice, due to the many dead ends and tortuous paths on the incipient infinite cluster.

The Termite Model. De Gennes [32] suggested a diffusion approach to the random superconducting network introducing what he called the termite model. In this model the termite goes very fast on the superconducting cluster and slows down when on the normal bonds. This model has successively been elaborated [33] and further studied by scaling approach and Monte Carlo methods [34,35].

The approach in Ref. [33] is based on the new concept in percolation of the unscreened perimeter. The termite, when on a superconducting cluster, covers it uniformly in a very short time. Afterwards it goes out of the cluster from any perimeter site. If it goes out from the perimeter of the internal lakes, eventually it will go back to the cluster. The same will occur with high probability if it comes out of a perimeter site which is deeply inside one of the fiords. However, if it comes out of one of the bridges it will leave the cluster and move to the next superconducting cluster. In general,at a given perimeter site i the termite has a probability p_i of leaving the cluster and being captured by the nearest superconducting cluster. One can replace the entire external perimeter with an effective unscreened perimeter such that when the termite goes out from one of its sites it moves to the next superconducting cluster. We expect from the previous discussion that the unscreened perimeter scales as the effective number of bridges. Therefore if the mass of the unscreened perimeter M_u of a critical cluster behaves as $M_u \sim \xi^{d_u}$ we expect $d_u = \tilde{Z}_s$. Following Ref. [34] the diffusion constant D can be written

$$D = \frac{< R^2 >}{\tau}, \tag{19}$$

where $< R^2 >$ is the radius of the cluster averaged over all configurations and is given by $< R^2 > \sim \xi^2 P_\infty$ [1] and τ is the time for the termite to leave the cluster and move into the next cluster. Since the termite can get out only when on the unscreened perimeter and has the same probability of being anywhere in the cluster, it is reasonable to assume

$$\tau = \frac{M_{tot}}{M_u}, \tag{20}$$

where $M_{tot} \sim \xi^{d_f}$ is the total mass of the cluster. Since $M_u \sim \xi^{d_u}$ and $< R^2 > \sim \xi^{2+d_f-d}$, from (19) and (20) follows $D \sim \xi^{2-d+d_u}$. From the Einstein relation $D \sim \Sigma \sim \xi^{\tilde{s}}$ and $\tilde{s} = 2 - d + Z_s$ it follows $d_u = \tilde{Z}_s$ as expected. To find $\tilde{s}$ one needs a closed expression for d_u. A mean field type of argument gives [33]

$$d_u = \frac{1}{2}(d + d_f) - 1 \qquad d_f \geq 2 \tag{21a}$$

$$d_u = d/d_f + d_f - 2 \qquad d_f \leq 2. \tag{21b}$$

Equation (21b) is taken over for $d_f \leq 2$ $(d < 1.2)$ because the superconducting cluster with fractal dimensionality $d_f \leq 2$ penetrates the nearest critical superconducting cluster more deeply than a random walk that has fractal dimensionality equal to 2. From (21) one obtains an expression for the conductivity exponent

$$\tilde{s} = 1 - (d - d_f)/2 \qquad d_f \geq 2 \tag{22a}$$

$$\tilde{s} = \frac{d}{d_f} + d_f - d \qquad d_f \leq 2, \tag{22b}$$

which, although not expected to be exact, reproduces well the numerical predictions. The first expression (22a) was conjectured without justification by Kertesz [36] for all d.

An interesting question is whether there is an anomalous diffusion for the termite as for the ant. Just below p_c the long time behavior satisfies the usual Fickian law. However for time t of the order τ one finds [37]

$$< r^2 > \sim t^{2/d_W}, \tag{23}$$

with $d_W = 2 - (s + \beta)/\nu$, which is analogous to the ant problem. However for time $t << \tau$ the termite does not follow the same law. In fact it immediately covers the all cluster and is confined within the cluster for the time τ. For more details see Ref. [37].

This talk is far from being exhaustive. Other topics not mentioned here, like numerical simulations in percolation, elasticity, noise and breaking processes in random media will be treated in other lectures at this Conference.

I would like to thank my collaborators in this field: A. Bunde, L. de Arcangelis, D. Hong, S. Redner, and H. E. Stanley.

[1] D. Stauffer, Phys. Rep. **54**, 1 (1977).

[2] J. W. Essam, Rep. Prog. Phys. **43**, 833 (1980).

[3] For review on percolation and gelation see, e.g., D. Stauffer, A. Coniglio and M. Adam, Adv. Poly. Sci. **44**, 103 (1982).

[4] S. R. Broadbent and J. M. Hammersley, Proc. Camb. Philos. Soc. **53**, 624 (1957).

[5] S. Kirkpatrick, Rev. Mod. Phys. **45**, 576 (1973).

[6] J. P. Straley, J. Phys. C **9**, 783 (1976); see also review article in Ref. [7].

[7] "Percolation Structure and Processes," eds. G. Deutscher, R. Zallen, and J. Adler, Annals of the Israel Physical Society, Vol. 5 (Hilger: Bristol, 1983).

[8] See, e.g., H. E. Stanley, **Phase Transitions and Critical Phenomena** (New York: Oxford University Press, 1971).

[9] B. B. Mandelbrot, **Form, Chance and Dimension** (San Francisco: Freeman, 1977).

[10] P. J. Flory, J. Am. Chem. Soc. **63**, 3083 (1961).

[11] M. E. Fisher and J. Essam, J. Math. Phys. **2**, 609 (1961).

[12] A. B. Harris, T. C. Lubensky, W. Holcomb and C. Dasgupta, Phys. Rev. Lett. **35**, 327 (1975).

[13] A. Aharony, Y. Gefen and A. Kapitulnik, J. Phys. A **17**, L197 (1984).

[14] S. Alexander, G. S. Grest, H. Nakanishi and T. A. Witten, Jr., J. Phys. A **17**, L185 (1984).

[15] H. E. Stanley, J. Phys. A **10**, L211 (1977).

[16] A. S. Skal and B. I. Shklovskii, Sov. Phys. Semicond. **8**, 1029 (1975).

[17] P. G. de Gennes, La Recherche **7**, 919 (1976).

[18] Y. Gefen, A. Aharony and S. Alexander, Phys. Rev. Lett. **50**, 77 (1983).

[19] A. Coniglio, Phys. Rev. Lett. **46**, 250 (1981); A. Coniglio, J. Phys. A **15**, 3824 (1982).

[20] R. Pike and H. E. Stanley, J. Phys. A **14**, L169 (1981).

[21] H. J. Herrmann and H. E. Stanley, Phys. Rev. Lett. **53**, 1121 (1984).

[22] A. Coniglio, in **Proceedings of Erice School on Ferromagnetic Transitions** (Springer-Verlag, 1983).

[23] A. B. Harris, S. Kim and T. C. Lubensky, Phys. Rev. Lett. **53**, 743 (1984).

[24] L. de Arcangelis, S. Redner, and A. Coniglio, Phys. Rev. B **31**, 4725 (1985).

[25] A. Coniglio (to be published).

[26] P. J. Reynolds, H. E. Stanley and W. Klein, Phys. Rev. B **21**, 1223 (1980).

[27] While at this Conference, S. Redner and I showed on some finite self dual lattices the following result

$$\left\langle \frac{\sum n(v) V^K}{\left[\sum n(v) V^2\right]^K} \right\rangle_p = \left\langle \sum n'(v) V^K \right\rangle_{1-p},$$

where $n(v)$ is the voltage distribution in the random resistor network problem and $n'(v)$ is the voltage distribution in the random superconducting network. The first average is made over all bond configurations for fixed p and the second for bond probability $1-p$. Although we did not prove this relation for any self dual lattices we believe that this should be true and therefore valid for the square lattice. The above relation gives for the square lattice $\tilde{Z}(K) = \tilde{\varsigma}(K)$, which checks for $K = 2$ and $K = \infty$.

[28] C. Mitescu and J. Roussenq in Ref. [7].

[29] Y. Gefen, A. Aharony and S. Alexander, Phys. Rev. Lett. **53**, 77 (1983).

[30] D. Ben Avraham and S. Havlin, J. Phys. A **15**, L691 (1982).

[31] R. B. Pandey, D. Stauffer, A. Margolina and J. G. Zabolitzky, J. Stat. Phys. **34**, 427 (1984).

[32] P. G. de Gennes, J. Phys. (Paris) Colloq. **41**, C3 (1980).

[33] A. Coniglio and H. E. Stanley, Phys. Rev. Lett. **52**, 1068 (1984).

[34] A. Bunde, A. Coniglio, D. Hong and H. E. Stanley, J. Phys. A **18**, L137 (1985).

[35] J. Adler, A. Aharony and D. Stauffer, J. Phys. A **18**, L129 (1985).

[36] J. Kertesz, J. Phys. A **16**, L471 (1983).

[37] D. Hong et al (1985).

Numerical Simulation of Percolation and Other Gelation Models

H.J. Herrmann
Service de Physique Théorique, CEN, Saclay, F-91191 Gif-sur-Yvette Cedex, France

For the description of gelation and porous media several percolation-type models have been proposed. To test how well these models describe the experimental results it is usually necessary to carry out numerical simulation of them. I will in this talk resume some of the numerical techniques to extract information out of percolation configurations and discuss the results obtained for some models for gelation.

1 Modelling of the sol-gel transition

The basic mechanism of gelation, common in all experimental realizations, is the following : one starts with many particles or small clusters of particles distributed in space (sol phase). Then a coagulation mechanism, which is often irreversible, begins. Close-by clusters (or particles) start forming connections, merging in this way to single clusters. Consequently, the average cluster size increases in time. At one point, the gel point or critical point, for the first time a cluster so big will be formed that it spans the whole system. This infinite cluster is called the gel and characterizes the gel phase.

The phase transition between sol and gel phase can be modelized in various ways. The easiest model is bond percolation [1] in which one supposes that the original particles are fixed in space all the time and form bonds to nearest neighbors completely randomly and independently. Percolation can be considered on a continuum or on a lattice, this will only change the location of the critical point but not the critical exponents that universally characterise the critical behaviour.

Two kinetic effects, that turn out to be important, are neglected in the model of percolation : the mobility of the clusters and the spatial and temporal dependence of the formation of bonds due to the detailed growth mechanism. To independently study these two kinetic effects, two alternative models are treated, each one taking into account one of the effects : (*i*) The kinetic aggregation model [2] considers many simultaneously and randomly moving clusters that form a bond when they touch each other. (*ii*) The kinetic gelation model [3] , specific for the growth mechanism of radical polymerization, considers that the bonds are grown in such a way that they form chains of bonds.

2 Numerical simulation of percolation

For two-dimensional percolation many things are known exactly, so that it is a good testing ground for numerical simulations. In three dimensions one has to rely mainly on simulations. To carry out a simulation, a percolation configuration is cast in a finite (d-dimensional) box. The larger the box the better, 23×10^9 sites in d = 2 [4] is at present the record size. The percolation configuration at p is very simple to get : for each bond a random number between 0 and 1 is chosen; if it is less than p the bond is occupied, if it is greater than p the bond is left empty.

The percolation threshold p_c can be obtained in various ways, for example by increasing p slowly until a spanning cluster appears and then taking the average over many samples. We want to describe another method for getting p_c which is very precise and at the same time yields additional information about the critical point. One calculates for different values of p the cluster-size distribution function n_s, i.e. the number of clusters that one has of s sites. We know that a cluster consists of all sites that are connected by occupied bonds. To obtain n_s on the computer one uses best what is called the Hoshen-Kopelman algorithm [5] which has the advantage that one needs to store in memory only a (d-1)-dimensional hyperplane of the system. Then one calculates the second moment χ of n_s :

$$\chi = (\sum_s s^2 n_s - s_\infty^2) / L^d \qquad (1)$$

L^d is the volume of the box and s_∞ the size of the largest cluster in the box. One knows that χ behaves close to p_c as

$$\chi = C_\pm |p - p_c|^{-\gamma} \qquad (2)$$

where C_+ is the amplitude for $p > p_c$ and C_- the amplitude for $p < p_c$. As the exponent γ is the same above and below p_c one expects in a log-log plot χ vs. $|p - p_c|$ two straight and parallel lines, one for $p > p_c$ and one for $p < p_c$. So p_c can be determined by just this condition, namely that the two lines should be straight and parallel [1)]

The function χ as a function of p is bell-shaped with a maximum at $p_c(L)p_c(L)$ is an approximate value for p_c which depends on the linear size of the box as

$$p_c(L) = p_c(1 + \text{const.}\ L^{-\frac{1}{\nu}}) \quad . \qquad (3)$$

Simulating in boxes of different sizes L one can obtain through Eq.(3) the exponent ν of the correlation length ($\nu = 4/3$ in $d = 2$ and $\nu \approx 0.88$ in $d = 3$). The fractal dimension of the infinite cluster at p_c then readily follows from [6] :

$$d_f = \frac{1}{2}(d + \frac{\gamma}{\nu}) \qquad (4)$$

d_f can of course also be obtained directly through the slope of a log-log plot of the mass of the largest cluster against the size L.

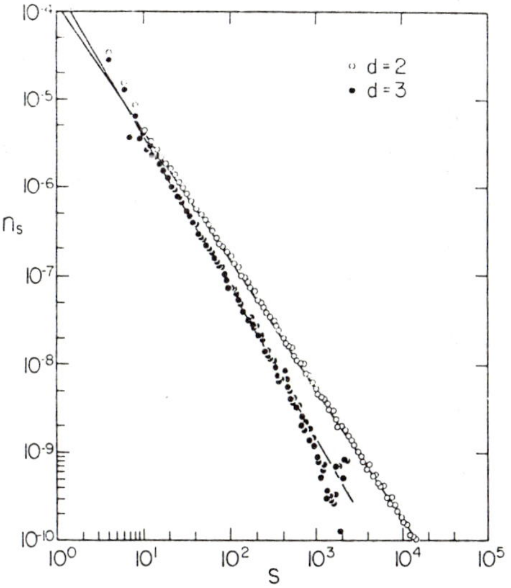

Fig. 1. Blob-size distribution function for d = 2, 3 for L = 600 (d= 2) and L = 60 (d = 3).

---

1) The slope of the straight lines is the exponent $-\gamma$ and the distance in χ-direction between the two lines yields the ratio $R = C_-/C_+$ which is like γ a universal quantity ($\gamma = 43/18$, $R \approx 200$ in $d = 2$ and $\gamma \approx 1.9$, $R \approx 10$ in $d = 3$)

The structure of the infinite cluster at p_c is fundamentally given by the backbone (see talk by A. Coniglio). A numerical simulation of the backbone implies separating all dangling ends from the cluster. One effective method is given by the "burning algorithm" [7] . The fractal dimension d_f^{BB} of the backbone is then readily obtained (d_f^{BB} = 1.62 in d = 2, d_f^{BB} = 1.74 in d = 3). The backbone consists of blobs and links (and is therefore very badly modelled by Sierpinski gaskets). This blob picture can be verified through simulation. A second "burning" method even yields the blob-size distribution m_s shown in Figure 1 in a log-log plot against s for two and three dimensions [8]. Over four orders of magnitude one verifies in Figure 1 the slope $(\nu\, d_f^{BB})^{-1} + 1$.

3 Numerical simulation of kinetic gelation

The model of kinetic gelation [3] is specifically tailored to describe radical polymerization. One assigns to each lattice site a functionality f, e.g. chooses f=4 with probability x and f=2 with probability 1-x. Then one selects randomly a small concentration c_I of sites called radicals. FInally one starts to occupy bonds : to put a bond one randomly chooses one radical and moves it to any of its nearest neighbors. The bond over which the radical moved is occupied and the functionality of both sites adjacent to this bond is decreased by one.A radical cannot be moved if $f \leq 0$ for any of the two sites.In this way it is possible that a radical gets trapped. If two radicals happen to be on the same site they annihilate each other and disappear.

Each radical produces a chain of occupied bonds. In sites with $f \geq 4$ two chains can crosslink. If we define p as the total fraction of occupied bonds there will be,as in percolation,a critical value p_c when for the first time an infinite network of crosslinked chains appears. A cluster is defined to consist of all sites within the same chain or within chains crosslinked to each other. Using these notions of a cluster and of p one can apply all the simulational techniques described in the previous section for percolation also to the model of kinetic gelation.

Some results that one obtains strikingly differ from percolation : the cluster-size distribution n_s, shown in Figure 2 for d = 3, oscillates in s [9], while it is a monotonically decaying function for percolation. At p_c and for $c_I \to 0$ these oscillations are only damped as s^{-1-d/d_f} which means that in this limit the scaling behaviour is qualitatively different from percolation. The usually universal ratio R is not only much smaller than for percolation but also seems to decrease

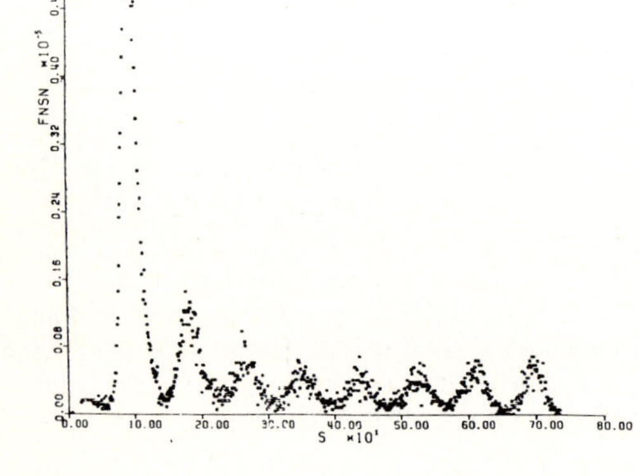

Fig. 2. Cluster size-distribution n_s as a function of s for L = 20, c_I = 0.0003, statistics : 5000

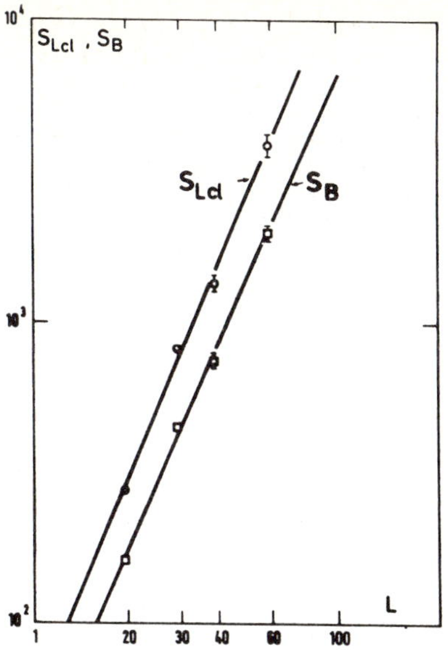

Fig. 3. Mass of the largest cluster (o) and backbone (□) plotted double logarithmically against system size L. The slope yields the fractal dimension.

slowly with c_I. In d = 3 one has R = 5.3 for $c_I = 3.10^{-2}$ and R = 2.5 for $c_I = 3.10^{-5}$ [9]. The exponents γ and ν are for d = 3, within their numerical error bars, the same as for percolation. In d=2, however, values different from percolation have been found ($\nu \approx 1.6$). Only the ratio γ/ν and thus the fractal dimension (see Eq(4)) appears to be the same for percolation and kinetic gelation in all dimensions [10].

The structure of the infinite cluster (gel) at p_c is the most recent discovery of kinetic gelation. In d = 3 the backbone has a fractal dimension which within the error bars is equal to the fractal dimension of the largest cluster itself ($d_f^{BB} \approx 2.3$). The data giving this result are shown in Figure 3 [9] . Geometrically speaking, this means that contrary to percolation the dead ends only constitute a fraction of the total mass, the dominant part of the gel consists of bundles of loops all contributing to the elastic or electric response of the gel. Consequently it would not be surprising if kinetic gelation has conductivity and elasticity exponents or a spectral dimension different from that of percolation. Work in this direction is in progress.

4 Numerical simulation of kinetic aggregation

The model of kinetic aggregation [2] is based on the movement of clusters. One starts putting particles randomly in a box until they fill up a volume fraction ρ_0. Then all particles simultaneously perform random walks. Whenever two particles become nearest neighbors they form a bond and are irreversibly attached to each other. Particles linked in this way belong to one cluster. Clusters also move randomly with a speed v that depends on their number m of particles like

$$v \propto m^{\alpha} \quad , \qquad (5)$$

where α is a parameter of the model.

Numerically, the clusters that are formed are found to be fractal. This means, that as the clusters get larger their internal density ρ_i (i.e. number of particles of the cluster per spanned volume) decreases to zero. Consequently there will be a critical time t_c, given by $\rho_i \approx \rho_0$, , when an infinite, spanning cluster, formed by the coagulation of individual, fractal clusters, appears for the first time.

The case $\rho_0 \ll \rho_i$ (flocculation regime) has been extensively studied. The case $\rho_0 \approx \rho_i$ (gelation regime) which is of interest to describe the sol-gel transition has only recently been treated [11]. It was found that when $\alpha \leq \alpha_c$ ($\alpha_c = 0.61 \pm 0.12$ in $d=2$) the gel time t_c is infinite, i.e. a gel is never formed in a finite time. Surprisingly, this is just the realistic case where smaller clusters move faster than larger ones. Although $t_c = \infty$, t_c still behaves like a critical point in the sense that scaling laws in the cluster size distribution are found to be asymptotically valid when one approaches t_c. Also a fractal dimension of the infinite cluster can be obtained ($d_f \approx 1.75$ in $d = 2$) distinct from percolation or the flocculation regime. The case $\alpha > \alpha_c$ is currently being studied. Numerical data are not yet accurate enough to give reliable predictions.

It seems that the pure mobility model of kinetic aggregation still needs an ingredient to realistically describe gelation, since it yields infinite gel times for the realistic case that larger clusters move slower than smaller ones. The important influence of mobility on the critical behaviour of gelation becomes, however, already evident at this stage not only because t_c is shifted to infinity, but also because of the new exponents, like the fractal dimension of the infinite cluster at t_c, that were found.

We conclude that several models exist for the sol-gel transition. They give different results for the critical behaviour and neither of them in general describes all the relevant effects of gelation. In some experimental situations, however, one of the models described might reproduce the reality in a good approximation, for instance if the speed of growth is much larger than the mobility of the clusters or the other way around. To obtain a complete picture of gelation still more work is needed, and this will require some non-neglectable efforts in computer simulations. Much numerical work in this sense is actually in progress.

References

1 P.G. de Gennes: J. Phys. Lett. L37, 1 (1976)
D. Stauffer: J. Chem. Soc. Farad. Trans. 2 72, 1354 (1976)

2 P. Meakin: Phys. Rev. Lett. 51, 1119 (1983)
M. Kolb, R. Botet and R. Jullien, Phys. Rev. Lett. 51, 1123 (1983)

3 P. Manneville and L. de Seze: in Numerical Methods in the Study of Critical Phenomena, eds. I. Della Dora, J. Demongeot and B. Lacolle (Springer Berlin, 1981)
H.J. Herrmann, D.P. Landau and D. Stauffer, Phys. Rev. Lett. 49, 412 (1982)

4 D.C. Rappaport, J. Phys. A

5 J. Hoshen and R. Kopelman: Phys. Rev. B14, 3428 (1976)

6 D. Stauffer: "Introduction to percolation theory" (Taylor and Francis, Bristol, 1985)

7 H.J. Herrmann, D.C. Hong and H.E. Stanley: J. Phys. A 17, L261 (1984)

8 H.J. Herrmann and H.E. Stanley: Phys. Rev. Lett. 53, 1121 (1984)

9 A. Chhabra, D. Matthews-Morgan, D.P. Landau and H.J. Herrmann: preprint

10 D.C. Hong, H.E. Stanley and N. Jan: Phys. Rev. Lett. 53, 509 (1984)

11 M. Kolb and H.J. Herrmann: J. Phys. A

Formation of Polymer Networks: Treatment of Stochastic and Spatial Correlations Using Mean-Field Approximation

Karel Dušek
Institute of Macromolecular Chemistry, Czechoslovak Academy of Sciences, 162 06 Prague 6, Czechoslovakia

1. Introduction

The elucidation of formation-structure-(properties) relationships requires a theory that satisfies the experiment with sufficient precision over the whole range of conversion of reactive groups into bonds. The available theories can be grouped in two categories :

(1) Models not associated directly with dimensionality of space - clusters are generated
  (a) statistically from (monomer) units,
  (b) kinetically (kinetic or coagulation equations) preserving the bonds formed irreversibly in clusters.

(2) Simulation of the cluster growth in n-dimensional space (percolation).

So far, almost exclusively the mean-field approach has been more or less successfully able to explain and predict the variations in the network structure determined by chemical factors and chain properties, except of the critical region dominated by fluctuations. The percolation techniques have so far poorly described the position of the critical point and the pregel and postgel structure. The neglect of conformational rearrangements within clusters occurring between formation of successive bonds and inadequate observation of the reaction mechanism were the main reasons. Only since recently, has percolation been approaching more closely the reality of the experiment.

In this contribution, the mean-field approach to the treatment of (a) long range stochastic correlations due to substitution effect and initiated reaction mechanism, (b) spatial correlations manifested by cyclization and (c) gel structure are discussed. The applicability or inapplicability of the existing mean-field methods is illustrated by experiments.

2. Statistical vs Kinetic Generation of Clusters

The statistical models of Flory and Stockmayer generalized by Gordon [1,2] generate clusters from monomer units differing in the number of reacted functional groups which participate in bonds of given type. This distribution of units is the only time (conversion)-dependent variable. If in a cluster correlations exist related to the history of cluster growth, this information is lost. The kinetic generation preserves the structure, once irreversibly formed, intact.

3. Substitution Effect.

The term substitution effect means that the reactivity of a group in a monomer unit depends on the state of neighbouring groups. If this effect is localized in the same monomer unit, the effect is called first-shell substitution effect (fsse). The distribution of units required for statistical generation of clusters is obtained by solving a set of differential equations. For example, for a f-functional step polyaddition (polycondensation) [3] the fraction p_i of units bearing i reacted functionalities given by

$$\frac{dp_i}{dt} = -(f-i)p_i \sum_j k_{ij}(f-j)p_j + (f-i+1)p_{i-1} \sum_j k_{i-1,j}(f-j)p_j \qquad (1)$$

where the rate constant k_{ij} refers to the reaction between groups in X_i and X_j.

The kinetic theory (Smoluchowski equation) schematically expressed as

$$\frac{dc_{x,\ell}}{dt} = -A + B \qquad (2)$$

where $c_{x,\ell}$ is the fraction of units in clusters composed of x units and bearing ℓ unreacted functional groups and A expresses the transformation and B the formation term for clusters x,ℓ. If the kernel of Eq. (2) $K_{ij} \propto ij$ (i and j are numbers of unreacted groups) and the rate constants k_{ij} = const., the statistical and kinetics solutions are identical [4-6].

If $k_{ij} \neq$ const., ℓ becomes a vector $\vec{\ell}$ and the A and B terms of $dc_{x,\ell}$ are split into several terms :

$$A = \sum_i \ell_i \, c_{x,\vec{\ell}} \sum_j k_{ij} \ell_j c_{x,\vec{\ell}} \qquad (3)$$

Then, the results of the statistical and kinetic treatments become irreducible and the cluster distributions different. The existing stochastic correlations are destroyed by splitting the clusters into units and reassembling them - a procedure used in the statistical theory.

Using the kinetic approach (Eqs. (2) and (3)), the distribution of clusters can be obtained only for $f = 2$ [7] ; for $f > 2$, only the moments of the distribution can be calculated. Monte-Carlo simulation is a relatively simple way of calculating the distribution and averages [3]. Significant differences between the results of both approaches have been obtained only for stronger or special values of fsse. [3,8].

4. Initiated Reactions

The statistical and kinetic treatments of initiated reactions are irreducible as well [9], which shows that the initiation mechanism also introduces long-range stochastic correlations. It can be shown on the example of linear living polymerization of a monomer M induced by initiator I

$$\begin{aligned} I + M &\xrightarrow{k_I} P_1 \\ P_1 + M &\xrightarrow{k_P} P_2 \\ &\vdots \end{aligned} \qquad (4)$$

that the statistical generation from units always gives the most probable distribution if P_1 is excluded, whereas the kinetic generation gives a different distribution depending on k_I/k_P which converges to the Poisson distribution for $k_I/k_P \gg 1$. The use of higher-order Markovian statistics by generating the chains from polyads brings the statistically generated distribution closer to the correct distribution which, however, for long chains always converges to the most probable distribution [9]. As a consequence, the gelation threshold for polymerization of a M-M monomer, even with independent reactivities of M groups, considerably differs from that calculated rigorously using the kinetic method.

5. Simplification for Systems Generated from Units Containing Groups of Independent Reactivity

The solution of the kinetic equations for polyfunctional reactions differing from the random Flory-Stockmayer process is sometimes difficult or impossible. However, the connections between groups of independent reactivity in monomers do not translate any information about the history of the process [10,11], so that they can be cut and reformed again at random. If in some monomers fsse and/or initiated mechanisms are operative, the following procedure can be used [9-11]

(a) connections between groups of independent reactivity are cut and points of cut labelled,
(b) clusters in the new system are generated kinetically,
(c) the labelled points of cuts are recombined to form the original bonds.

The advantage of the procedure can be found in the fact that the generation in (b) yields usually clusters of finite size. The treatment of postetherification initiated by the diamine-diepoxide polyaddition has shown that a purely statistical generation introduces a large error [12].

6. Spatial Correlations : Cyclization

The ring closure is a typical consequence of spatial correlations. Studies on linear and branched systems have revealed that the extent of cyclization is primarily determined by the reaction mechanism and conformational properties of chains connecting the reacting groups. The "spanning-tree" approximation [13-16] has been mostly used. In this mean-field treatment, the problem has been reduced to finding a new distribution of units, p_{ijk}, instead of p_i in which i,j and k mean, respectively, the number of functionalities engaged in branching, closing cycles and not yet reacted. The p_{ijk} s are determined by differential equations of type of Eq.(1) containing, however, the cyclization terms. These cyclization terms are obtained by counting the probabilities that a selected unreacted functionality can come into contact with other unreacted functionalities in the same cluster, which are given by the conformational statistics of the connecting paths. The treatment is, therefore, reduced to the level of statistical treatment of fsse. The contact probability between two groups takes no account of the effect of already existing cycles or excluded volume. Therefore, a good agreement with experiment for the fraction of bonds wasted in cycles in the pregel stage and at the gel point has been obtained in case the cyclization was weak [16,17].

A method based on the statistical assemblage of cyclic structures (blobs) of any cycle rank that are only single connected with other blobs [18] is more rigorous, but the approximate mean-field calculation of blob structure distribution is possible only for blobs of cycle rank one.

Cyclization has not yet been considered in the kinetic coagulation equations. Two terms, A' and B', should be added to the r.h.s. of Eq. (2) that take into account the intramolecular reaction in clusters $c_{x,\ell} \rightarrow c_{x,\ell-2}$; $A' \propto K^{(c)}_{x,\ell} \ell c_{x,\ell}$ and the kernel is determined by the probability of ring closure in a cluster. The kernel is expected to be an increasing function of x reaching a constant value at high x.

7. Post-Gel Stage

It has been shown that in the simulation of f-functional condensation in finite systems [3,20] the largest cluster continues to grow until all functional groups have reacted only if intramolecular reactions are allowed. The term A' is assumed to be proportional to $\ell^2 c^2_{x,\ell}$ but for a system of large size these terms converge to zero except of that for the largest cluster - the gel. Respecting this situation, A in Eq. (2) should be modified such that $A = 1c_{x,\ell} (\Sigma\ \ell c_{x,\ell} + c_g w_g)$, where c_g is the number of unreacted functionalities in the gel per unit and w_g is the fraction of units in the gel. The sum $\Sigma\, xc_{x,\ell} = 1 - w_g < 1$, and the modified equation describes the distribution of clusters in the sol : The missing term $d(\zeta/n)dt \propto (c_g w_g)^2$ determines the cycle rank ζ which is equal to the number of bonds to be cut in order to convert the cluster into a tree. The kinetic method gives thus the cycle rank which determines the equilibrium elastic modulus of phantom networks.

In the statistical generation, the key variable is the extinction probability, v, defined for a unit on a finite branch (sub-tree) by the following condition : if the ingoing bond has a finite continuation, all the outgoing bonds must also have finite continuation. The monomer units are now distinguished by the number of bonds with finite and infinite continuation,and it is assumed that these probabilities are independent. Thus, for a unit engaged in i bonds,the fraction of units having j bonds with infinite and i-j bonds with finite continuation, $p_{j;i-j}$ is given by binomial expansion $p_{j;i-j} = \binom{i}{j} v^{i-j}(1 - v)^j$. This weighting gives fractions of units with j=0 (sol), j=1 (dangling chains), j=2 (backbone units of elastically active network chains (EANC), $j \geq 3$ (elastically active crosslinks). Within the framework of the statistical model, this distribution is sufficient for calculating such parameters as the sol fraction, sol cluster distribution, concentration, length and degree-of-polymerization distribution of EANC's and dangling chains, etc. [20]. The number of EANC's and the sol fraction are identical with values obtained by the kinetic method,provided the stochastic correlations are not operative.

However, the gel contains also loops that do not contribute to the equilibrium elasticity,but they are counted in the cycle rank. Therefore, one has to use an effective cycle rank in which these loops are not counted. The statistical procedures for calculating the number of such loops has been offered elsewhere [14,15].

8. Comparison with Experiments

The applicability of assumptions used in the treatment of the postgel stage can be tested by the position of the gel point, sol fraction, elastic equilibrium modulus; etc. A few results given below demonstrate how the applicability of the conventional mean-field models depends on the reaction mechanism and tendency to ring formation.

In Figures 1 and 2, the theoretical and experimental sol fractions in two sets of polymer networks obtained by endlinking mechanism with a very low tendency to cyclization are compared. Because the endlinking is based on an alternating mechanism of A+B = AB type, the crosslinking density can be conveniently controlled by varying the molar ratio of the components [21,23]. In the calculation of the sol fraction for polyurethane networks, the fraction of bonds wasted in elastically inactive cycles (≈ 0.02) was taken into account. For epoxy networks, practically zero cyclization was found from the independence of the critical conversion on dilution. The no-parameter fits in Fig. 1 and 2 show a very good agreement between the model and experiment. It is to be mentioned that the experiments cannot be extended to systems with a high sol fraction. The determination of high sol fractions is generally unreliable,due to the slow diffusion of the sol from the gel and gradual degradation of highly swollen networks during prolonged extraction.

The dependence of the equilibrium modulus on the network composition also follows the predicted dependence over several orders of magnitude of the modulus, but the analysis is somewhat complicated due to several rubber elasticity models [22,23].

The network formation by chain crosslinking (vinyl-divinyl) copolymerization is not in agreement with the statistical model,even if cyclization is approximated by spanning-tree method. The experimental features of this process have been reviewed in detail elsewhere [24] and analyzed with respect to the network formation mechanism. The main features of this process can be summarized as follows :

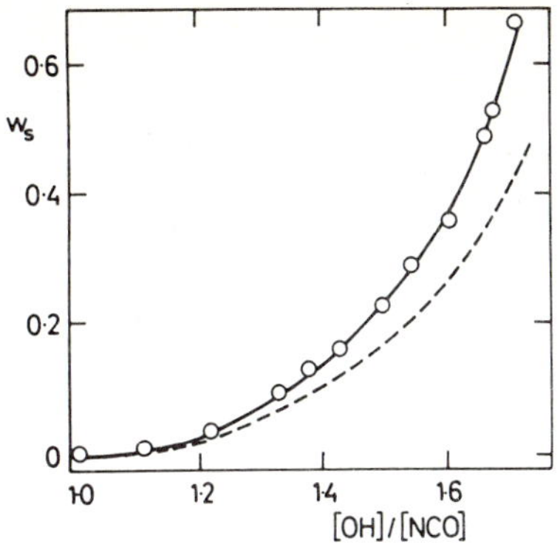

Fig. 1. Dependence of the sol fraction in networks prepared from poly(oxypropypylene) triols (MW 708) and 4,4'-diphenylmethane diisocyanate on the inital molar ratio of OH/NCO groups dependence calculated taking into account the elastically inactive cycles,--- dependence calculated assuming 100% intermolecular reaction.

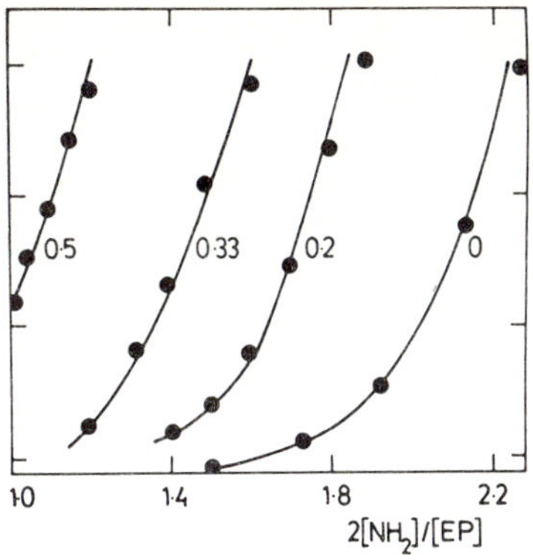

Fig. 2. Dependence of the sol fraction in networks prepared from diglycidyl ether of 4,4'-dihydroxy-2,2-propane, phenylglycidyl ether and 4,4' diaminodiphenylmethane on the initial ratio of amine hydrogens to epoxy groups. s fraction of epoxy groups of monoepoxide in the monoepoxide-diepoxide mixture, the numbers at the top of curves indicate the final conversion of epoxy groups. The curves are calculated dependences.

(a) the critical conversion is shifted to higher values by 1-3 orders of magnitude compared with the value for a ring-free system,

(b) a high fraction of the units reacted by both vinyl is found in the polymer just in the beginning of the reaction,

(c) the low conversion polymers have low intrinsic viscosity and radius of gyration,

(d) at high conversions a considerable fraction of pendent double bonds does not react.

The analysis of experimental data led to a conclusion [24,25] that network formation proceeds via microgel-like particles (clusters) which are strongly internally crosslinked. Their mass initially corresponds to the mass of the primary chain and they later grow.

It has been shown that strong cyclization as well as the chain mechanisms are responsible for this special type of network formation. In the beginning of the reaction, the free radical at the end of the growing chain can react with the monomers, but has also a choice to react with pendent double bonds on already existing chains (intermolecular reaction) or double bonds on its own chain (intramolecular reaction). In the beginning, the probability of the intermolecular reaction converges to zero because no other chains are available, while the chance for ring formation involving the groups in the same chain is finite. As a result, just the first chains are strongly internally crosslinked, and the double bonds buried inside the relatively dense core of the microgel cluster are prevented from reacting. A simplified scheme of cluster growth is shown in Fig.3.

The single-chain microgels formed in the beginning of the reaction have a dense core with non-reactive groups and a surface shell with groups capable to react. Next initiation can start again in a region where only monomers are available, or in an active region of the mono-chain cluster which includes the outer shell and a part of the surrounding volume approximately of the thickness of the radius R of a mono-chain microgel. If initiation starts in the active region, the mass of the cluster is increased by the mass of the mono-chain macrogel. Later on, the active regions of two or several microgels can overlap, and two or more clusters can be joined together.

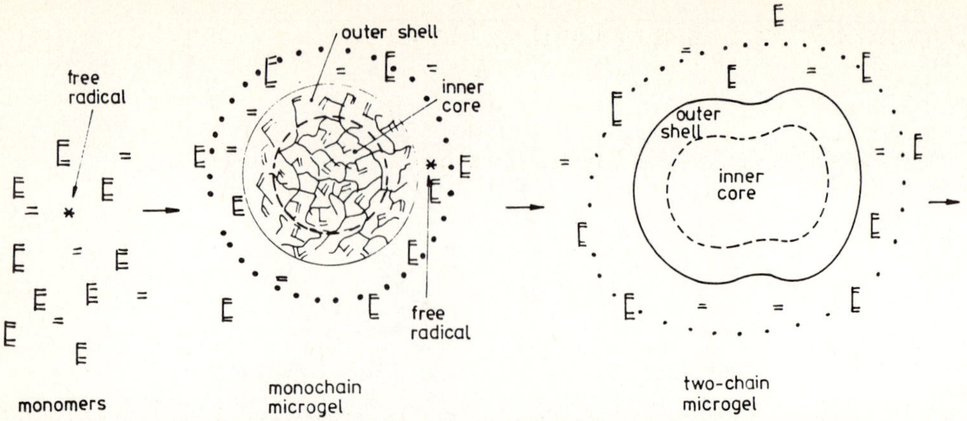

Fig.3. Sketch of network formation in vinyl-divinyl chain copolymerization.

Recent percolation models of initiated chains reactions [26,27] were able to predict cluster growth which is at least qualitatively in agreement with experiments [26,27]. However, it is expected that the process can be well described by a special type of coagulation equation.

The recent development in kinetic treatment of aggregation and gelation [28] offers new tools of elucidation of formation-structure relations for complex mechanisms of branching and crosslinking.

References

1 M. Gordon : Proc. Roy. Soc. London A268, 240 (1982).
2 M. Gordon and G.N. Malcolm : Proc. Roy. Soc. London A295, 29 (1966).
3 J. Mikes abd K. Dušek : Macromolecules 15, 93 (1982).
4 W.H. Stockmayer : J. Chem. Phys. 11, 45 (1943).
5 S.I. Kuchanov : Methods of kinetics calculations in polymer chemistry (Khimiya, Moscow 1978 - in Russian).
6 K. Dušek : Polym. Bull. 1, 523 (1979).
7 S.I. Kuchanov and E.S. Povolotskaya : Vysokomol. Soedin. A24, 2179 (1982).
8 S.I. Kuchanov and E.S. Povolotskaya : Vysokomol. Soedin. A24, 2190 (1982).
9 K. Dusek and J. Somvarsky : Polym. Bull. in press (1985).
10 V.I. Irshak and M.L. Tai : Dokl. Akad. Nauk SSSR 259, 856 (1981).
11 K. Dušek : Brit. Polym. J. 17, (1985).
12 K. Dušek : Polym. Bull. in press (1985).
13 M. Gordon and G.R. Scantlebury : J. Polym. Sci., Pt. C 16, 3933 (1968).
14 K. Dušek, M. Gordon and S.B. Ross-Murphy : Macromolecules 11, 236 (1968).
15 K. Dušek and V. Vojta : Brit. Polym. J. 9, 164 (1977).
16 R.F.T. Stepto in Developments in Polymerization (ed. R.N. Haward) (Appl. Science Publishers, London 1982).
17 L. Matejka and K. Dušek : Polym. Bull. 3, 489 (1980).
18 S.I. Kuchanov, S.V. Korolev and M.G. Slinko : Polym. J. 15, 775 (1983).
19 R.M. Ziff and G. Stell : J. Chem. Phys. 73, 3492 (1979).
20 K. Dusek : Macromolecules 17, 716 (1983).
21 M. Ilavsky and K. Dušek : Polymer 24, 891 (1983).
22 K. Dušek and M. Ilavsky : J. Polym. Sci., Polym. Phys. Ed. 21, 1323 (1983).
23 M. Ilavsky, L.M. Bogdanova and K. Dusek : J. Polym. Sci. Polym. Phys. Ed. 22, 265 (1984).
24 K. Dusek in Developments in Polymerization 3. (ed. R.N. Haward) (Appl. Science Publishers, London 1982).
25 K. Dusek, H. Galina and J. Mikes : Polym. Bull. 3, 19 (1980).
26 H.M.J. Boots and R.B. Pandey : Polym. Bull. 11, 415 (1984).
27 R. Bansil, H.J. Herrmann and D. Stauffer : Macromolecules 17, 988 (1984).
28 Kinetics of Aggregation and Gelation (eds. F. Family and D.P. Landau) (North Holland, Amsterdam 1984).

Dynamical Processes in Random Media

S. Redner
Center for Polymer Studies and Department of Physics, Boston University,
Boston, MA 02215, USA

1. Introduction

A large number of percolation problems have now been thoroughly studied [1]. Most of these problems are either static in nature, such as cluster topology and structure, or of a steady-state type, such as electrical conductivity or diffusion on percolating clusters. In this paper, I discuss some aspects of dynamical percolation problems which appear to exhibit new and rich phenomena. Various portions of this work were performed in collaboration with L. de Arcangelis, A. Coniglio, H.J. Herrmann, J. Koplik and D. Wilkinson. I also thank Schlumberger-Doll Research for their support and hospitality during a leave where portions of this work were performed.

The first problem that will be treated is the breaking of random media. This process is modelled by a random fuse network in which each bond can break irreversibly when the voltage across the bond exceeds a critical value. We wish to understand how cracks form, grow and ultimately break the network. A second problem is that of dispersion in random media. This is the spreading of a dynamically neutral tracer as it is carried along in a fluid following in a porous medium.

Both problems can be formulated as random resistor network models. Breaking occurs at the most highly stressed bonds, and in the corresponding network model, these are the bonds with the highest potential drop across them. Dispersion is controlled by the bonds with the slowest flow, i.e., the bonds with the smallest potential drop across them. These insights motivate a general study of the distribution of potential drops across the bonds of a random resistor network.

2. Voltage distribution in a random resistor network

In order to discuss the distribution of potential drops analytically, we first introduce [2] a new hierarchical model to describe the percolating backbone at threshold (Fig.1). This model is a regular self-similar fractal consisting of links and blobs. Here links are defined as those bonds which, if cut, would render the network disconnected, while the remaining bonds comprise the blobs. The hierarchical model explicitly retains the self-similar links and blobs picture; an essential topological feature of the percolating backbone.

For this model, it is relatively simple to calculate a variety of geometrical quantities as a function of the iteration index, N. Now we use the fundamental relation that the number of links N_ℓ, which equals 2^N in this model, scales exactly as $L^{1/\nu}$, where L is an effective linear dimension of the model lattice. With this connection, it is possible to obtain good numerical values for many two-dimensional percolation exponents. It is also possible to generalize these considerations to arbitrary dimensionality.

If a unit voltage is applied across the lattice, it can be readily verified for an Nth order hierarchy, that the possible values of the voltage drop across

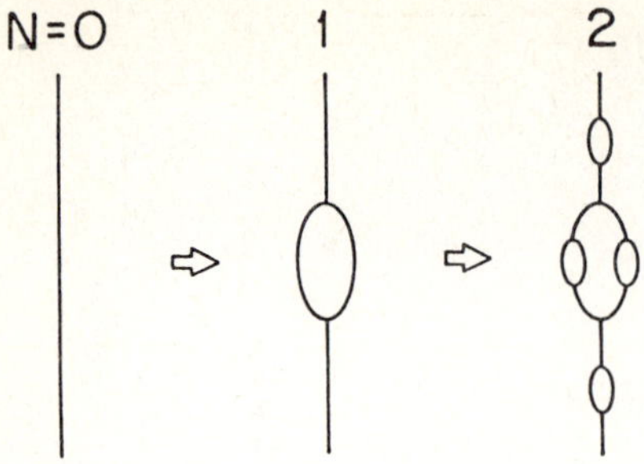

Fig.1

each bond are $V(k) \equiv 2^k/5^N$, with $k = 0,1,2,\ldots,N$. The number of bonds with voltage drop $V(k)$ is

$$n(V(k)) = 2^N \binom{N}{k} \qquad (1)$$

Since k is proportional to log V, the distribution is log binomial, hence log normal in the continuum limit.

From (1) it follows immediately that $\langle V^k \rangle \equiv \sum_V n(V)\, V^k$ and $\langle V \rangle^k$ scale differently. Thus an infinite set of exponents are required to completely specify the moments of the voltage distribution. These moments probe progressively finer details of cluster structure as k increases, ranging from the number of bonds in the backbone for $k = 0$, to the number of links for $k \to \infty$. All of these peculiar features have been observed in our numerical simulations of the voltage distribution for a square lattice resistor network at the percolation threshold.

3. Breaking of a random fuse network

Consider a regular square lattice network in which each bond is a fuse, defined as a device which behaves as a resistor if the applied voltage V is less than a critical value V_c, and which breaks irreversibly if $V > V_c$. We treat a model where each fuse has the same value of the conductivity but a different value for the "breaking point", V_c. We shall investigate how the network breaks apart as the external voltage is increased [3].

The behaviour of the model depends on the distribution of breaking points for each fuse. For simplicity, we choose a uniform distribution with average value unity and variable width w. Our detailed simulation procedure is the following : First the external voltage is set so that the breaking point of the weakest fuse is just exceeded. This weakest fuse is now broken and the new equilibrium voltages at each node of the network are then recalculated. If additional "overstressed" bonds exist, the most stressed one is now broken, and the above process is repeated. If no overstressed bonds exist, the external voltage is raised until one fuse just becomes overstressed. This bond is now broken and voltages are recalculated. This general procedure continues until the network breaks.

For $w \to 0$, the numerical simulation leads to a straight crack. This arises because the extra current that is diverted around the crack causes the bonds at the edges of the crack to become the most overstressed. Thus, a notch spontaneously grows to a straight crack which breaks the system.

For $w \to 2$, the breaking process initially resembles random percolation, as the randomness of the bond strengths is more important than the correlation effects induced by the extra voltage at the crack ends (Fig.2a). Rather suddenly, however, a new regime is reached where cracks are self sustaining and the network breaks apart (Fig.2b,c). These qualitative features can be seen more clearly by measuring the external voltage across the network and the network conductivity as a function of the number of broken bonds (Fig.3). These quantities show clearly the two general regimes of behaviour mentioned above : Independent crack initiation when relatively few bonds are broken, and spontaneous crack growth at the later stages of the breaking process.

(a) 393 bonds broken

(b) 508 bonds broken

(c) 535 bonds broken

Fig. 2a,b,c

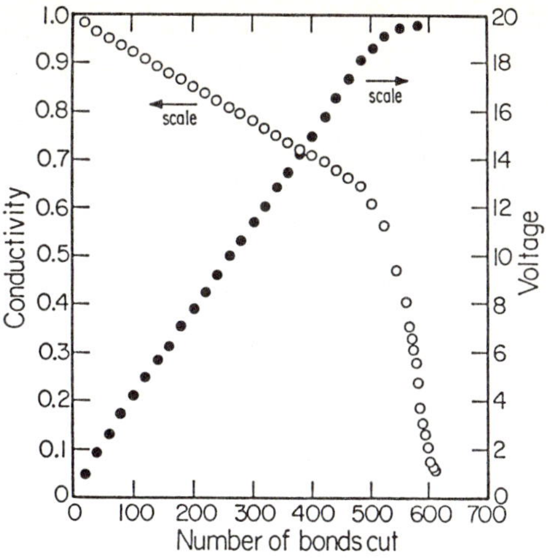

Fig. 3

At this point more theoretical interpretation is needed, and many interesting questions can be asked. For example, is the breaking process a first or a higher-order transition ? How does the value of the external breaking potential and its distribution depend on the initial distribution of randomness ? Is it possible to give some insight of the crack size distribution ? What is the time dependence of breaking processes? These questions suggest promising directions for future research.

4. Hydrodynamic dispersion

Consider injecting a spatially localized pulse of a dynamically neutral contaminant in a fluid flowing in a porous medium. Due to the divergence and convergence of streamlines, and the different velocities of various streamlines, the pulse of tracer particles will spread as it moves downstream. This is the phenomenon of dispersion [4-6]. More quantitatively, dispersion can be quantified by $\langle r^2\rangle - \langle r\rangle^2 = 2D_{/\!/}\,t$, where $\langle r\rangle$ is the average position of the contaminent pulse parallel to the flow, t is the time, and $D_{/\!/}$ is defined as the dispersion coefficient. In this section, various aspects of dispersion phenomena in random media will be discussed.

We formulate the problem in terms of a resistor network model. In a steady current flow, the mean and the mean-square transit time for a tracer to traverse the network will be calculated. There are two basic rules for the tracer motion :

(i) the time required for a tracer to pass through a bond which carries a current I is proportional to I^{-1} (convective flow), and

(ii) the probability for the tracer to enter a particular bond emanating from a node equals the fractional current entering the bond (perfect mixing).

In a mean-field type approximation, the dispersion coefficient can be found by replacing the network by a linear sequence of tubes and mixing chambers, which are separated by the original lattice spacing, ℓ . Each tube contains all the streamlines of the network which are assumed to not interact, while perfect mixing between all streamlines occurs at the mixing chamber. In a frame of reference moving at the average flow velocity, U, a given tracer particle is sometimes moving faster and sometimes slower than U. If the distribution of individual stream velocities is not pathological, the tracer executes a random walk of step length ℓ and time between steps $\tau \sim \ell/U$. This leads immediately to a dispersion coefficient $D_{/\!/} \sim \ell^2/\tau \sim U\ell$.

This mean-field picture is a useful first approximation, but it is inadequate because it neglects the possibility of tracer particles getting stuck in very slow bonds. This effect can be appreciated simply in an almost balanced Wheatstone bridge (Fig.4). Suppose the bond conductivities are such that the voltages at the midpoints of the bridge are $1/2 \pm \varepsilon$. Since the current flowing in the vertical bond is approximately 1/2, the transit time going straight down the bridge is approximately $(1/2)^{-1} + (1/2)^{-1} = 4$. However, if the tracer takes the horizontal path (with probability proportional to ε), the transit time is of order $1/\varepsilon$. In fact it is readily shown that $\langle t \rangle = 5 + O(\varepsilon)$ and $\langle t^2 \rangle \sim 1/\varepsilon$. Since $D_{/\!/} \sim \langle t^2 \rangle - \langle t \rangle^2$, the minimum value of ε must be cutoff in order to obtain a finite result. This cutoff originates in molecular diffusion ; the transit time across any bond cannot be any longer than the molecular diffusion time. In a random but well-connected medium, a uniform distribution of slow bonds down to the diffusion cutoff is expected. Averaging over such a distribution leads ultimately to $D_{/\!/} \sim U\ell \log(U\ell/D_{mol})$, where D_{mol} is the molecular diffusion coefficient.

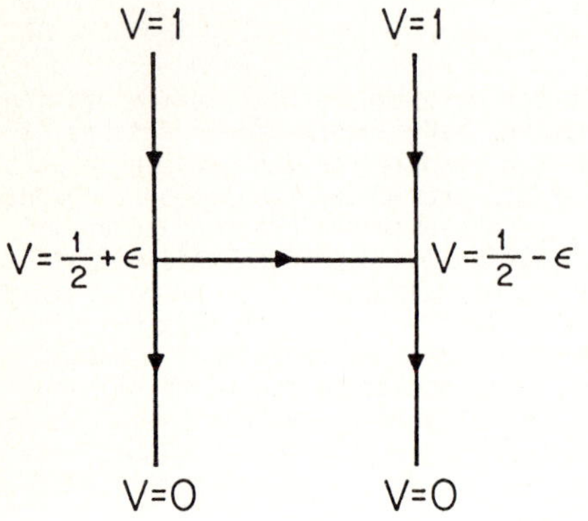

Fig.4

Thus the effects of molecular diffusion are quite important and they are required to give a sensible description of dispersion.

Turning now to dispersion on poorly connected networks, we first introduce a modified hierarchical lattice. It is obtained from figure 1 by taking the conductivity of a bond on the right edge of each bubble to be one-half the conductivity of the corresponding bond on the left edge. For this network it is possible, but somewhat tedious, to calculate the asymptotic form of the transit time distribution. A primary result of this calculation is that the dispersion coefficient is scale dependent and it may be written as $D_{/\!/} \sim UL$, where L is the effective linear dimension of the lattice as defined in section 2. As a result of this scale dependence, a localized pulse of tracer will spread out at a rate faster than $t^{1/2}$. This anomalously fast dispersion is similar, but also complementary, to the phenomenon of anomalously slow diffusion on self-similar structures [7].

Attempts to account for the effects of molecular diffusion are in progress. An important new feature for poorly connected media is that regions of stagnation can form, rather than isolated slow bonds, for which the discussion of the beginning of this section applied. Furthermore, stagnation can occur in large blobs in the backbone, and on dead ends ; blobs and dead ends play the same role in dispersion. Both types of stagnation need to be included to give a full description of dispersion on the percolating cluster. For this case, it has been suggested [8] that $D_{//}$ will vary as $U^2\xi^2/D_{ant}$, where ξ is the correlation length, and D_{ant} is the (scale-dependent) molecular diffusion coefficient on the percolating cluster. The complete understanding of this result, and the crossovers caused by the interplay of convective and diffusive effects, should be fruitful areas for new results.

Flicker (1/f) Noise in Percolation Networks

R. Rammal
Centre de Recherches sur les Très Basses Températures, CNRS, B.P. 166 X,
F-38042 Grenoble Cédex, France

1. Introduction

Statistical self-similarity is emerging as an important concept underlying the behavior of disordered systems. In percolation clusters, for example, the fractal dimension has been identified first /1/. Much later, the notion of spectral dimension /2/ was introduced in order to describe the spectrum of the Laplacian operator, which appears in a large variety of linear physical problems. Another intrinsic property is the recently introduced spreading dimension /3/. Intuitively, it is plausible that an infinite number of exponents must be used to characterize a fractal. In the following, we show that at least one physically measurable quantity, the magnitude (not the frequency dependence) of the resistance noise spectrum (1/f noise) depends on a new exponent pertaining to fractal lattices. We show that this exponent, b, can be seen as a member of an infinite family of exponents, which includes the fractal and spectral dimensions. Physically, the exponent b comes from a well-known fact in the 1/f noise problem : the macroscopic mean square resistance fluctuations are much more sensitive to local inhomogeneities than the square of the macroscopic resistance itself.

For a more detailed exposition we direct the reader to reference /4/, where some relevant references (prior to 1985) for 1/f noise problem are given.

2. Flicker (1/f) noise

Flicker (1/f) noise refers to the low-frequency spectrum of excess voltage fluctuations measured when a constant current is applied to a resistor. The spectrum $S_V(\omega) = \int e^{i\omega t} < V(t)\ V(0) > dt$ (where brackets refer to time average) almost always has a power law form $\omega^{-\alpha}$ with α close to unity. The origin of this power law has been the subject of innumerable controversies, and is not the purpose of the present discussion. We rather use two well-established properties of 1/f noise : a) 1/f noise is resistance noise. In other words, a simple application of Ohm's law suggests that if there are voltage fluctuations δV in the presence of a constant current, they are caused by resistance fluctuations, $\delta V = I\delta R$. This naive picture is confirmed by the fact that i) the noise spectrum is proportional to I^2, ii) the resistance fluctuation spectrum which can be inferred from 1/f noise experiments can also be directly measured with no applied current from higher order equilibrium correlation functions /5/. b) At low frequencies, resistance fluctuations are nearly gaussian /6/ and are correlated over microscopic scales only. This has been verified experimentally in many systems /7/ and most mechanisms suggested for 1/f noise (except diffusion /8/) are consistent with this hypothesis.

3. Resistance fluctuations model /4/

Given the previous considerations, we consider the following model for 1/f noise on resistor networks of arbitrary shape. Only purely resistive networks will be considered here. It is easy to generalize all the results for : networks made of complex impedances or non-linear elements (see below) and for a continuous medium.

The resistance of each branch (α) in the network has a small fluctuating part so that the associated resistance is assumed to be : $r_\alpha + \delta r_\alpha$, where $\delta r_\alpha'$ s are time-dependent uncorrelated random variables with mean zero and covariance

$$< \delta r_\alpha(\omega)\delta r_\beta(-\omega) > = \rho_\alpha^2(\omega)\,\delta_{\alpha\beta} \qquad (1)$$

These fluctuations in the resistance could be produced by an arbitrary noise mechanisms. For identical r_α' s, $\rho_\alpha^2 = \rho^2$ is assumed to be independent of (α). Given a constant current source configuration, we have to calculate the fluctuating part of the resistance R measured in a one-port configuration, the correlation between the measured resistances in a two-port configuration, etc. As long as each branch resistance fluctuates independently with the same spectrum, the explicit frequency dependence can be discarded. Actually, the spectrum is of the form $\rho^2(\omega) \sim \omega^{-\alpha}$.

The magnitude of the relative noise $\mathcal{S}_R$ for the resistor network is defined by

$$\mathcal{S}_R \equiv S_R/R^2 = <\delta R\delta R> \quad /R^2 \qquad (2)$$

where R and δR are respectively the overall resistance (one-port case) and its time fluctuation. Note that similar quantities can be defined for two-port configurations. S_R so defined is the relevant quantity for the magnitude of 1/f noise, measured under constant external current.

4. Computational methods and general results

As defined by Eq. 2, we need the expression of R as a function of all r_α in order to calculate S_R. However, it is clear that such a direct approach leads to formidable calculations and cannot be used in practice. Simplified procedures can be used however : composition rules and sensitivity calculations.Here we shall give the main results.

4.1 Composition rules

Series resistances : $\mathcal{S}_R = \sum_\ell (R_\ell/R)^2 \mathcal{S}_{R_\ell}$, $R = \sum_\ell R_\ell$ (3)

Parallel resistances : $\mathcal{S}_R = \sum_\ell (R/R_\ell)^2 \mathcal{S}_{R_\ell}$, $R^{-1} = \sum_\ell R_\ell^{-1}$ (4)

4.2. General method

For a network of arbitrary shape, one can show /4/ :

$$\delta R = \sum_\alpha \delta R_\alpha (i_\alpha/I)^2 \qquad (5)$$

where δR is the variation of $R = V/I$, measured in a one-port configuration, resulting from a small variations (δR_α) in the value of the resistance inside the network. Here i_α denotes the current through the branch α in the unperturbed original network. This result is a particular case of the following one (with obvious notations) :

$$I.\delta V - V.\delta I = \sum_\alpha (i_\alpha \delta v_\alpha - v_\alpha \,.\, \delta i_\alpha) \qquad (6)$$

where $v_\alpha = i_\alpha R_\alpha$ is the potential drop across the branch α.

Eq. 5 leads immediately to a general expression for S_R. In particular, for uncorrelated fluctuations (Eq. 1) and identical resistances $R_\alpha = r$, one obtains the simple result:

$$\mathcal{S}_R = \frac{\rho^2}{r^2} (\sum_\alpha i_\alpha^4)/(\sum_\alpha i_\alpha^2)^2 \qquad (7)$$

where $R = \sum R_\alpha i_\alpha^2/I^2$ has been used.

The result Eq. 7 can easily be generalized to higher order correlation functions. Higher order cumulants of the resistance fluctuations then have a simple expression in terms of the steady state currents and of the cumulants of the elementary resistance fluctuations (which are assumed all identical). For example (recall $< \delta R > = 0$) :

$$\langle\delta R^4\rangle - 3 \langle \delta R^2 \rangle^2 = (\sum_\alpha i_\alpha^8)(\langle \delta r^4\rangle - 3 \langle \delta r^2 \rangle^2) \tag{8}$$

we thus define the quantities (n integer) :

$$G_{2n} \equiv \sum_\alpha i_\alpha^{2n} \tag{9}$$

which for an arbitrary geometry of the network relate microscopic and macroscopic correlation functions of the resistance fluctuations.

4.3. Lower and upper bounds for the noise

It is easy to show that for any current pattern $\{i_\alpha\}$, $N_b \sum_\alpha i_\alpha^4 \geqslant (\sum_\alpha i_\alpha^2)^2$ leading to

$$\mathcal{S}_R \geqslant s/N_b \tag{10}$$

where N_b denotes the number of conducting bonds in the network, and $s \equiv \frac{\rho^2(\omega)}{r^2}$. Similarly, an upper bound for the noise can be obtained as

$$\mathcal{S}_R \leqslant s \cdot \frac{r}{R} \tag{11}$$

5. Self-similar networks - new hierarchy of exponents

For an Euclidean network made of $d.L^d$ identical resistances, arranged in a d dimensional hypercubic lattice, the measured resistance between two parallel electrodes is given by $R = r/L^{d-2}$ (L is measured in units of the lattice spacing). Assuming uncorrelated resistance fluctuations, we obtain

$$\mathcal{S}_R = s \cdot L^{-d}$$

This simple result shows that for Euclidean networks $_R$ scales as L^{-d}, i.e. the inverse of the volume. A natural question therefore arises : what happens for fractal networks ? Is there a new exponent controlling the size dependence of $\mathcal{S}_R$?

This question is motivated by the anomalous size-dependence of the resistance R(L) in the case of fractal lattices /2/ :

$$R(L) \sim L^{-\beta_L} \quad (L \gg 1) \tag{12}$$

where $\beta_L = \frac{\bar{d}}{\tilde{d}}(\tilde{d} - 2)$ is an exponent controlling the transport properties on the considered lattice. Here, $\bar{d}$ and $\tilde{d}$ denote respectively the fractal and spectral dimensions of the structure.

Another motivation comes from the recent progress in our understanding of random resistor networks. There,the concept of self-similarity has also been shown to be very useful /10/.

Using scaling arguments, it has been shown /4/ that $_R$ obeys the following scaling form :

$$\mathcal{S}_R(\lambda L) = \lambda^{-b}\mathcal{S}_R(L) \quad , \text{ i.e. } \mathcal{S}_R(L) \sim L^{-b} \quad (L \gg 1) \tag{13}$$

where λ denotes a scaling factor, and b is a characteristic exponent of the lattice. For instance $b = \bar{d}$ for linear structures such as the von Koch curve ($\bar{d} = \ln 4/\ln 3$, $\beta_L = -1$), the Peano curve ($\bar{d} = 2$, $\beta_L = -2$) and the random walk trajectory ($\bar{d} = 2$, $\beta_L = -1$). For these structures, the currents $\{i_\alpha\}$ are equal and Eq. 10 becomes an equality.

The scaling form, Eq. 13 for $\mathcal{S}_R$ has been checked on a large number of fractal structures /4/ where b has been calculated using renormalization group methods : Sierpinski gaskets, Branching von Koch structure, Phi lattices, checkerboards, etc. For instance, in the case of the known 2D sierpinski gasket ; $b = \ln(25/11)/\ln 2 = 1.1844$, whereas $\beta_L = \ln(3/5)/\ln 2$ and $\bar{d} = \ln 3/\ln 2$. In general, b appears as a new exponent not related to previously introduced exponents. Lower and upper bounds for b can also be derived. For instance :

$$- \beta_L \leqslant b \leqslant \bar{d} \tag{14}$$

More generally, the new exponent b, associated with the noise magnitude, appears as the third in a series of new exponents associated with the various "moments" of currents distribution. In fact, the first two in this series are given by $\bar{d}$ and β_L :

$$\sum_\alpha 1 \sim L^{\bar{d}} \quad \text{and} \quad R(L) = r \sum_\alpha i_\alpha^2 \sim L^{-\beta_L} \tag{15}$$

as recalled above. The next one is given by S_R which is proportional to $\sum_\alpha i_\alpha^4$.

Clearly, all of the above quantities are special cases of the more general G_{2n} defined in Eq. 9. We have seen that the G'_{2n} have a natural physical meaning even for $n > 2$. It is legitimate to associate an exponent x_n to G_{2n} which describes its power law behavior : $G_{2n} \sim L^{-x_n}$ for large L. As for b, these exponents can be bounded /4/ :

$$x_n \geqslant x_{n-1} \qquad \text{(all n)} \tag{16}$$

$$x_n \leqslant \frac{n}{n-1} x_{n-1} + \frac{\bar{d}}{n-1}, \quad (n > 1) \tag{17}$$

and

$$x_n \leqslant n\beta_L + (n-1)\bar{d} \tag{18}$$

Obviously, $x_o = - \bar{d}$, $x_1 = \beta_L$, $x_2 = b + 2\beta_L$.

Note that the intrinsic exponents /11/ are given by the combinations $x_n/\bar{d}$ and not by $\{x_n\}$: $x_n/\bar{d}$ are independent of the embedding Euclidean space (for a different purpose /12/, the exponents x_n were calculated numerically for n = 0, 0.5, 1, 1.5 and 2 on percolating networks : $\mathcal{Y}_{2n}/\nu$ in the notation of Ref. 12, is $-x_n$ is ours).

6. Application to percolation networks

The fractal geometry of percolation clusters is at the basis of our present understanding of percolation problems. At $p < p_c$, the random resistor network is divided into isolated finite clusters which are self-similar and have a fractal dimension $d_p = d - \beta_p/\nu_p$ and a spectral dimension $\tilde{d} = 2(d\nu_p - \beta_p)/(t - \beta_p + 2\nu_p)$. The spreading dimension $\hat{d}$ is known from numerical simulations /3/ at all d ($1 \leqslant d \leqslant \infty$). At $p > p_c$, the infinite cluster remains self-similar at short length scales up to the correlation length $\xi_p \sim (\Delta p)^{-\nu_p}$, $\Delta p = p - p_c$. The branches carrying the current belong to the backbone. Therefore, the exponent b satisfies

$$- \beta_L \leqslant b \leqslant \bar{d}_B \tag{19}$$

where $\bar{d}_B$ denotes the fractal dimension of the backbone.

Close to p = 1, a finite network (L^d sites) is slightly perturbed and its behavior is given by the Euclidean regime : $\mathcal{S}_R = s\, L^{-d}$ at p = 1. Starting from p = 1, $\mathcal{S}_R$ increases when p is lowered : the number of conducting bonds N_b decreases (Eq. 10). $\mathcal{S}_R$ increases at least as fast as $1/B(p)$, where B(p) denotes the volume of the backbone. This Euclidean regime is maintained for values of p far from p_c, where $\xi_p \ll L$. For value of p very close to p_c, an Euclidean-to-fractal crossover is expected to occur. Both the resistance R(p,L) and the noise $\mathcal{S}_R(p,L)$ exhibits actually this crossover :

$$R(p,L) = L^{-\beta_L} f(L/\xi_p) \tag{20}$$

$$\mathcal{S}(p,L) = L^{-b} g(L/\xi_p) \tag{21}$$

where f and g denote scaling functions, with : $f(u \ll 1) \simeq 1$, $f(u \gg 1) \simeq u^{2-d+\beta_L}$ and $g(u \ll 1) \simeq 1$, $g(u \gg 1) \sim u^y$. Therefore, one deduces $y = b-d$ and $\mathcal{S}_R(p;L) \sim L^{-d}(\Delta p)^{-\nu_p(d-b)}$ at $\xi_p \ll L$.

This analysis implies the divergence of $s(p)/s \equiv \mathcal{S}_R(p,L)/\mathcal{S}_R(p=1,L)$ as $(\Delta p)^{-\kappa}$ where $\kappa = \nu_p(d-b)$. The exponent κ is actually a positive number : $b \leqslant \bar{d}_B \leqslant d$ (see Eq. 19).

The exponents κ and b have been calculated, using different approaches. In the following, we will summarize the actual situation.

6.1. Monte Carlo simulations

Performed /13/ for bond percolation on square lattices ($p_c = .5$). The results, shown on figure 1, agree very well with the previous analysis. s(p) diverges at $p \sim p_c$, and the value of b so obtained (from the data at p_c) is : $b=1.16\pm0.02$, giving ($\nu_p = 4.3$ at $d = 2$) : $\kappa = 1.12 \pm 0.02$. The value of b lies within the theoretical bounds (Eq. 19) with : $-\beta_L = 0.973 \pm 0.005$ and $\bar{d}_B = 1.62 \pm 0.02$ in two dimensions.

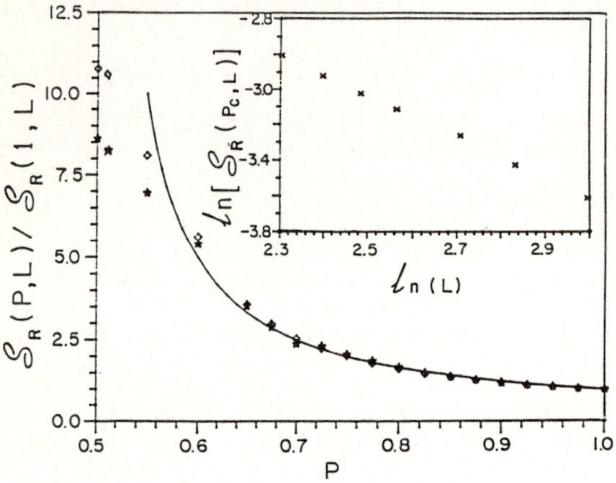

Figure 1. Monte Carlo results for the relative noise $s(p) = \mathcal{S}_R(p,L)/\mathcal{S}_R(1,L)$ for bond-percolation problem on a two-dimensional square LxL lattice ($p_c = .5$). Different symbols correspond to different L : ★ L = 15, ◇ : L = 20. The solid line represents the effective-medium theory prediction $s_m(p) = (2p-1)^{-1}$. The insert shows the power-law behavior of $\mathcal{S}_R(p,L)$ versus L at threshold (after ref. 13).

The exponent κ appears as an increasing function of d since it starts from $\kappa = 0$ at $d = 1$, is equal to 1.12 at $d = 2$ and reaches the value $\kappa = 2$ at $d \geqslant 6$ (see Ref. 9). Using the same numerical data, the first members of the hierarchy x_n introduced above have been calculated :

$$x_0 = -\bar{d}_B = -1.65 \pm 0.02$$

$$x_1 = \beta_L = -0.978 \pm 0.01$$

$$x_2 = b + 2\beta_L = -0.81 \pm 0.02$$

$$x_3 = -0.77 \pm 0.03$$

and

$$x_4 = -0.74 \pm 0.02$$

These values are consistent with Eqs. 16-18. Furthermore, the asymptotic result : $x_\infty = -1/\nu_p = -3/4$ is actually in perfect agreement with the numerical results.

6.2. Effective-medium theory (EMT)

Such a theory, previously developed for the conductivity has been extended for the noise /9/. We find that the EMT value for the relative noise $s_m(p)$, normalized at p = 1 (bond percolation) is given by

$$s_m(p)/s = \frac{z-2}{z}(p-p_c)^{-1} \quad , p > p_c \qquad (22)$$

where $p_c = 2/z$ and z is the lattice coordination number. For the particular case of the square lattice Eq. 22 implies $s_m(p)/s = (2p-1)^{-1}$ in very good agreement with the numerical results ($z = 4$). For both site and bond percolation model, the EMT leads to $s_m(p) \sim (p-p_c)^{-1}$, i.e. $\kappa = 1$ in any dimension.

6.3. Nodes-links-blobs picture

In this picture of the backbone /14/, the conducting "backbone" of the infinite cluster is imagined to consist of a network of quasi 1D string segments ("links"), tying together a set of "nodes" whose typical separation is ξ_p. Each string is supposed to consist of several sequences of singly-connected bonds, in series with thicker regions, or "blobs", where there are two or more conducting bonds in parallel. Ignoring the resistance of the blobs, the resistance between two points at distance L on the backbone is given by sum over $L_1 \sim L^{1/\nu_p}$ singly-connected bonds. Let us consider G_{2n} given by Eq. 9, where $i_\alpha \leqslant I = 1$ for each α. It is clear that for $n \to \infty$, only the singly-connected bonds can contribute to G_{2n}. This gives trivially : $x_\infty = -1/\nu_p$ as was anticipated. In this spirit, and assuming the validity of this approximation, one obtains : $x_n = -1/\nu_p$ for all $n \geqslant 1$. This result underestimates the true value of $x_1 = \beta_L$ at d = 2 and 3. However, using this value only for x_2, one deduces a most accurate estimation for b and κ :

$$-b = 2\beta_L + 1/\nu_p \qquad (23)$$

At d = 2, this gives b = 1.196 ± 0.01 to be compared with b = 1.16 ± 0.02 obtained by Monte Carlo simulations. At d = 1, Eq. (23) gives the correct result b = 1 as well as at d = 6 where b = 2 (see Ref. 4).

Actually, Eq. 23 must be viewed as an upper bound for the exponent b. Then, for all d, b is constrained by :

$$-\beta_L \leqslant b \leqslant -2\beta_L - 1/\nu_p \qquad (24)$$

6.4. Real space renormalization

In order to go beyond the EMT approximation, position space renormalization group transformation can be used. We quote just the results so obtained

Table 1. Upper and lower bounds for the noise exponent b at different dimensions d

d	$-\beta_L$	$-2\beta_L - 1/\nu$
1	1	1
2	.97	1.19
3	1.16	1.18
4	1.59	1.66
5	1.69	1.82
$\geqslant 6$	2	2

a) Differential PSRG /8,13/ :

The ratio κ/ν at $d = 1 + \varepsilon$ dimensions is given by

$$\kappa/\nu = (d-1)I(a)/a \tag{25}$$

where $I(a) = \int_0^1 dx\,[\ell n(1 + ax)]/x$ and $a = p/(1-p)$ taken at $p = p_c$. This leads to $\kappa/\nu = 0.822$ at $d = 2$ and 1.911 at $d = 3$ to be compared with the previous results.

b) Cell transformation :

Using different iterated cells, the following values for b were obtained :

Diamond hierarchical lattice /16/ :

$b = 1.18$ at $d = 2$ ($\nu = 1.63$)
$b = 1.09$ at $d = 3$ ($\nu = 1.22$)

in perfect agreement with Eq. 25.

Wheatstone bridge /9/ :

$\kappa = 1.70 \pm 0.02$

7. Extension to other models

The above approach can easily be extended, in different directions, in relation with possible implications in ongoing experiments on real materials. The experimental measurement of the relative noise is actually under active consideration by various groups : Marseille, Urbana, Cornell, IBM, etc. In this respect, the critical region,as well as the region of small disorder,are of great interest. The measurement of s(p) on metal-insulator mixtures /17/ and possibly Ni-Al_2O_3 cermets /18/ at low temperatures is actually possible.

7.1. Networks of non-linear elements

Anomalous noise scaling has been observed in a non-linear composite conductor /18/. In this case, the presence of tunneling links, at microscopic scales, suggests the study of the noise magnitude on such a non-linear system. Actually the noise arises to a very large extent from those links which carry an unusually large amount of current. It should then show the effects of non-linearity at lower levels of the total current than does the I-V characteristic.

The calculation of s(p) in the case where non-linear elements are included in the network can actually be performed as above. For instance, in the case of power-law conductors : $V = r\,|I|^{\alpha}\,\mathrm{sgn}\,I$, recently investigated /19/ on percolation networks, only a slight modification is needed. More precisely, in the basic equation Eq. 5, one has to replace the exponent 2 by $\alpha+1$. The same approach leads /20/ to new exponents $b(\alpha)$, $x_n(\alpha)$,..., etc., as for the linear case $\alpha = 1$ described here.

7.2. Continuum percolation

More generally, random resistor networks with a singular distribution of conductances can exhibit non-universal behavior depending on the distribution of conductances near $g = 0$. This is actually the case of lattice percolation models with a probability distribution of conductances

$$P(g) = (1-p)\delta(g) + ph(g)$$

For instance, the critical conductivity exponent $t = (d-2-\beta_L)\nu_p$ is known /21/ to increase with α for $h(g) \sim g^{-\alpha}$ near $g = 0$ $(0 \leqslant \alpha \leqslant 1)$. Similarly, the noise exponent b can increase in this case /20/.

The mapping of the "Swiss-cheese" continuum models /22/, where spherical holes are randomly placed in a uniform transport medium, onto a type of discrete random network, leads also to a singular distribution of conductances. In this case, the exponent t increases and the same is expected for the noise exponent b /20/.

References

1. S. Kirkpatrick : in "Ill Condensed Matter, Les Houches Summer School", Ed. R. Balian, R. Maynard and G. Toulouse (North-Holland), Amsterdam, 1979, p. 321.
2. S. Alexander, R. Orbach : J. Physique Lett. (Paris), 43, L625 (1982). R. Rammal, G. Toulouse: ibid, 44, L13 (1983).
3. R. Rammal, J.C. Angles d'Auriac and A. Benoit : J. Phys. A17, L491 (1984). J.C. Angles d'Auriac, A. Benoit, A.B. Harris, R. Rammal : to be published.
4. R. Rammal, C. Tannous, A.M.S. Tremblay : Phys. Rev. A31, XXX (1985).
5. A.M.S. Tremblay, M. Nelkin : Phys. Rev. B24, 2551 (1981), and references therein.
6. See for instance P.J. Restle, R.J. Hamilton, M.B. Weissman, M.S. Love : Phys. Rev. B31, 2254 (1985).
7. R.D. Black, M.B. Weissman, F.M. Fliegel : Phys. Rev. B24, 7454 (1981). J.H. Scofield, D.H. Darling, W.W. Webb : Phys. Rev. B24, 7450 (1981).
8. R. Rammal : J. Physique Lett. 45, 1007 (1984), and references therein.
9. R. Rammal : J. Physique Lett. 46, 129 (1985).
10. R. Rammal, M.A. Lemieux, A.M.S. Tremblay : Phys. Rev. Lett. 54, 1087 (1985). A.B. Harris, S. Kim, T.C. Lubensky : Phys. Rev. Lett. 53, 743 (1984).
11. R. Rammal, G. Toulouse, J. Vannimenus : J. Physique (Paris), 45, 389 (1984).
12. L. de Arcangelis, S. Redner, A. Coniglio : Phys. Rev. B (in press). We thank S. Redner for letting us know of this work prior to publication.
13. R. Rammal, C. Tannous, P. Breton, A.M.S. Tremblay : Phys. Rev. Lett. 54, XXX (1985).
14. A. Coniglio : Phys. Rev. Lett. 46, 250 (1981).
15. A.M.S. Tremblay : Private communication.
16. J.M. Luck : Subm. to J. Phys. A (1985).
17. D.A. Rudman, J.C. Garland : Bull. Am. Phys. Soc. 29, 352 (1984).
18. J.V. Mantese et al. : Sol. State Commun. 37, 353 (1981).
19. J.P. Straley, S.W. Kenkel : Phys. Rev. B 29, 6299 (1984).
20. R. Rammal : to be published (1985).
21. P.M. Kogut, J.P. Straley : J. Phys. C 12, 2151 (1979). A. Ben-Mizrahi, D. Bergman : Ibid, 14, 909 (1981).
22. B.I. Halperin, S. Feng, P.N. Sen : Subm. to Phys. Rev. Lett.

Part V Properties of Branched Polymers and Gels

Structure of Branched Polymers

W. Burchard, A. Thurn, and E. Wachenfeld
Institute of Macromolecular Chemistry, University of Freiburg,
D-7800 Freiburg i. Br., Fed. Rep. of Germany

1 Introduction

Recent theoretical studies on aggregates and polycondensates [1-4] revealed that the f-functional random polycondensation (Flory-Stockmayer theory [5,6]) and the diffusion-limited coagulation of Smoluchowski [7] are two special examples of a general Smoluchowski coagulation process. Valuable insight into the structure of clusters was obtained by computer simulations, and has led to a satisfactory understanding of simple branching processes [8]. For a polymer scientist, the situation nevertheless remains disappointing, mainly for two reasons.

The one is that no full use can be made of the laws of universality which are strictly valid only at very large degrees of polymerization. These are mostly not attainable in practice. The other is that the functionalities of the monomers have to meet special chemical conditions. For instance, the system may consist of two types of monomers, i.e. bifunctional RA_2 and f-functional RB_f units, where reaction can take place exclusively between A and B functional groups. Frequently, the one monomeric unit has two functional groups, A_1 and A_2 of different reactivity. The primary and secondary OH-groups in glycerol are a typical example. The kinetic coagulation theory has been ineffectual with these problems, and the same holds true for the percolation theory. The cascade branching theory developed in 1873 [9] and first applied to polymer science in 1962 [10] has been more successful. This theory proved to be equivalent to the equilibrium branching theory of Flory and Stockmayer.

In 1970 the cascade theory was extended to the derivation of conformational properties and the angular dependence of scattered light or scattered neutrons, where the full molecular weight distribution is taken into account. The need of considering scattering functions is essential in polymer characterization, and has received a new impetus by the work of Witten [11] who calculated by computer simulation a diffusion-limited aggregation (DLA) process. He tried to characterize the structures by the space or density correlation function

$$G(r) = \langle \rho(0)\rho(r) \rangle \tag{1}$$

Here $\rho(r)$ is the density of segments in a volume element at the position r in space. Witten postulated a rather direct correlation of G(r) to a Hausdorff dimension D, which is defined through the equation

$$N \sim R^D \tag{2}$$

where N is the number of segments within a sphere volume of radius R which, however, is not the radius of gyration. Since

$$\int_0^R G(r)4\pi r^2 dr = \int_0^R \langle\rho(0)\rho(r)\rangle 4\pi r^2 dr = N \tag{3}$$

is a function of R it appears obvious that G(r) can be described by a Hausdorff dimension. This is certainly true if G(r) obeys an exponential law of the kind

$$G(r) = Kr^{-A} \tag{4}$$

Then eq(3) leads immediately to

$$N \sim R^{3-A} = R^D \qquad \text{or} \qquad D = 3-A \tag{5}$$

The density correlation function can be derived from the particle scattering factor, since G(r) and P(q) are related through Fourier transforms

$$P(q) \sim \int_0^\infty G(r)\sin(qr)/(qr)\, dr \tag{6a}$$

$$G(r) \sim r^{-1} \int_0^\infty qP(q)\, \sin(qr)\, dq \tag{6b}$$

Since the particle scattering factors can be calculated for many models by the cascade theory, it appeared interesting to check the scaling behaviour of G(r) as postulated by Witten.

2 The Basis of Cascade Theory

The cascade theory is essentially a percolation on a Bethe lattice (no ring formation) and this confines the model to a specific universality class, which probably is incorrect in the critical region near the gel point, but it nevertheless gives satisfactory results in the realistic region of polymers.

In the following, a graph-theoretical representation is preferred to that of a shell-like Bethe lattice, as this representation allows a clearer distinction of the various shells, which now appear as generations of a rooted tree. The main idea in cascade theory consists in the derivation of a general expression for the average number of units in the n-th generation followed by a summation over all generations, which leads to the avearge degree of polymerization. The average size, and also the size distribution, can then be calculated by statistical means if certain assumptions are made. In the original Flory-Stockmayer theory, the only assumption made is that all functional groups of a monomer RA_f have equal reactivity; the extent of reaction of such a functional group may be α ; then the average number of units in the first generation is

$$\langle N(1)\rangle = f\alpha \tag{7}$$

To find the average number of units in n-th generation, select one unit from the (n-1)-th generation which with its one functionality is linked to the preceding generation. This unit can bind on average $(f-1)\alpha$ further repeating units, which now occur in the n-th generation. Thus the total number is

$$\langle N(n)\rangle = \langle N(n-1)\rangle (f-1)\alpha = \langle N(1)\rangle [(f-1)\alpha]^{n-1} \tag{8}$$

Summation over all generations results in Stockmayer's well-known equation for the weight average degree of polymerization

$$P_w = 1 + \Sigma \langle N(n)\rangle = (1+\alpha)/[1-(f-1)\alpha] \tag{9}$$

3 Conformational Averages

Radius of Gyration. Inspection of a rooted tree shows, that $\langle N(n)\rangle$ is also the number of paths of length n (from the n-th generation to the root). Assuming Gaussian statistics for every subchain in the molecule, the mean square end-to-end distance of such a path is

$$\langle r_n^2\rangle = b^2 n \tag{10}$$

where b is the effective bond length. The mean square radius of gyration of the molecule is given by

$$\langle s^2\rangle_z = (1/2P_w^2)\sum\sum\langle r_{ij}^2\rangle \tag{11}$$

which can be recast into a single sum [12]

$$\langle s^2\rangle_z = (1/P_w)\sum\langle N(n)\rangle\langle r_n^2\rangle \tag{12}$$

Insertion of eq.(8) and eq.(11) yields

$$\langle s^2\rangle_z = (b^2/2P_w)(\alpha f/[1-(f-1)\alpha]^2) \tag{13}$$

Particle Scattering Factor. Next the angular dependence of the scattered light may be considered, which is given by the Debye equation

$$P_z(q) = (1/P_w^2)\sum\sum\langle\sin(qr_{ij})/(qr_{ij})\rangle = (1/P_w^2)\sum\sum\exp(-b^2q^2|i-j|/6) \tag{14}$$

where the second equality implies the assumption of Gaussian statistics. Again eq.(14) can be written as a single sum

$$P_z(q) = (1/P_w)\sum\langle N(n)\rangle[\exp(-b^2q^2/6)]^n \tag{15}$$

which with $\Phi = \exp(-b^2q^2/6)$ and eq.(8) gives

$$P_z(q) = (1/P_w)(1+\alpha\Phi)/[1-(f-1)\alpha\Phi] \tag{16}$$

Thus, the angular dependence is formally obtained from the equation for the particle scattering factor by multiplying the link probability α by the scattering function Φ for one bond.

4. Copolymers

Transition probabilities. Evidently the factor $(f-1)\alpha$ in eq.(8) is a transition probability p which describes the probability to reach the n-th generation if a repeating unit has arrived at the (n-1)-th generation. Thus, more generally, one comes to

$$\langle N(n)\rangle = \langle N(1)\rangle p^{n-1} \tag{8a}$$

and

$$P_w = 1 + \langle N(1)\rangle(1-p)^{-1} \tag{9a}$$

P_w of a Copolymer. For a copolymer composed of the two monomer types RA_2 and RB_f two rooted trees have to be considered, where the first starts with a RA_2 unit as root, the other with a RB_f unit. The population in the n-th generation can be calculated from that in the preceding generation if (i) the average number of RA_2 and RB_f units in the (n-1)-th generation is known and (ii) if the transition probabilities are known. Let α and β be the the extents of reaction for the functional groups A and B respectively. Furthermore p may be the fraction of A-groups which have reacted with a B-group and ρ the corresponding fraction for a B-functionality that has reacted with an A-group. Then, the four transition probabilities can be written in the form of a table or as a matrix **P**

n \ n+1	A	B
A	$\alpha(1-p)$	αp
B	$(f-1)\beta\rho$	$(f-1)\beta(1-\rho)$

(18a)

$$\mathbf{P} = \begin{pmatrix} \alpha(1-p) & \alpha p \\ (f-1)\beta\rho & (f-1)\beta(1-\rho) \end{pmatrix} \tag{18b}$$

where the row denotes the functionality with which a unit in the n-th generation is linked to the preceding one and the column denotes the functionality to which this unit is linked to the (n+1)-th generation. Similarly, the number of the various repeating units occurring in the first generation can be written as a population matrix

$$\langle \mathbf{N}(1) \rangle = \begin{pmatrix} \langle N(1) \rangle_{AA} & \langle N(1) \rangle_{AB} \\ \langle N(1) \rangle_{BA} & \langle N(1) \rangle_{BB} \end{pmatrix} \tag{19}$$

where the first row refers to the A-rooted and the second to the B-rooted tree. In this matrix notation the degree of polymerization becomes

$$P_w = 1 + \mathbf{n}\,(\langle \mathbf{N}(1) \rangle (\mathbf{1} - \mathbf{P})^{-1}\, \mathbf{1}^t) \tag{20}$$

where $\mathbf{n} = (n_A, n_B)$ is the composition vector, in which the components denote the mole fractions of the RA2 and RBf units in the molecule and the superscript t denotes the transposed unit vector. The gel-point is given in this notation by the concise equation

$$|\mathbf{1} - \mathbf{P}| = 0 \tag{21}$$

since then P_w goes to infinity.

Particle-Scattering Factor. The extension for deriving $P_z(q)$ is accomplished in the same manner as described for the homopolymer, and leads to the recipe that the various link probabilities have to be multiplied with the corresponding bond-scattering function Φ_{AA}, Φ_{AB} or Φ_{BB} , where for instance

$$\Phi_{AA} = \exp(-b_{AA}{}^2 q^2/6) \tag{22}$$

and the other two Φ-functions are correspondingly defined. Hence

$$P_w P_z(q) = 1 + \mathbf{n}\,(\langle \mathbf{N}(1)\Phi \rangle (\mathbf{1} - \mathbf{P}\Phi)^{-1} . \mathbf{1}^t) \tag{23}$$

where

$$\langle \mathbf{N}(1)\Phi \rangle = \begin{pmatrix} 2\alpha(1-p)\Phi_{AA} & 2\alpha p \Phi_{AB} \\ f\beta\rho\Phi_{AB} & f(1-\rho)\beta\Phi_{BB} \end{pmatrix} \tag{24}$$

$$\mathbf{P}\Phi = \begin{pmatrix} \alpha(1-p)\,\Phi_{AA} & \alpha\, p\Phi_{AA} \\ \beta(f-1)\rho\,\Phi_{BA} & \beta(f-1)(1-\rho)\,\Phi_{BB} \end{pmatrix} \tag{25}$$

5 Density Correlation Function (DCF)

In the past, the particle-scattering factors were calculated for various models and for some branched products which have actually been measured. One striking example is discussed below. The cascade theory allows calculation of the influence of heterogeneities in branching. Here the question arises to which extent these heterogeneities can be detected from scattering measurements. The work on DLA by Witten suggests that the density correlation function G(r) may be easier and more directly interpreted than the particle scattering factor; therefore G(r) was calculated from $P_z(q)$ using eq.(6b).

In the following 5 examples are discussed. The first two are the linear and randomly branched f-functional polycondensates, i.e. the RA_2 and RA_f models of the original Flory-Stockmayer theory. The correlation functions show three typical sections. The first one at very short radii is determined by the structure of the individual bonds, where G(r) remains constant. In the second section an exponential decay is observed with an exponent of A = 1. This exponent corresponds to a Hausdorff dimension of D = 2. Finally, a rapid decrease in G(r) occurs if r is of the order of the radius of gyration of the molecules. (See Fig.1). It is worth emphazising that the scaling behaviour becomes clearly apparent not before $\varepsilon = (1 - \alpha/\alpha_c) < 10^{-4}$. The branched and linear polycondensates show no difference in behaviour at large r, but at short r the effect of the bonds becomes more pronounced for the branched chain.

The kink in this region is more distinct for the third example, an epoxy resin [13] (Fig.1). This system has the remarkable property that new functional groups A' are created at the RB_2 monomer when a B-functionality reacts with an A-group of the RA_2 monomer. Thus the RB_2 monomers change during the reaction into a monomer of the type $A_2'RB_2$. At larger r this copolymer resembles the simple RA_f model. The Hausdorff dimension again is D = 2, but for lower molecular weight an apparently lower dimension of $D \simeq 1.8$ is obtained.

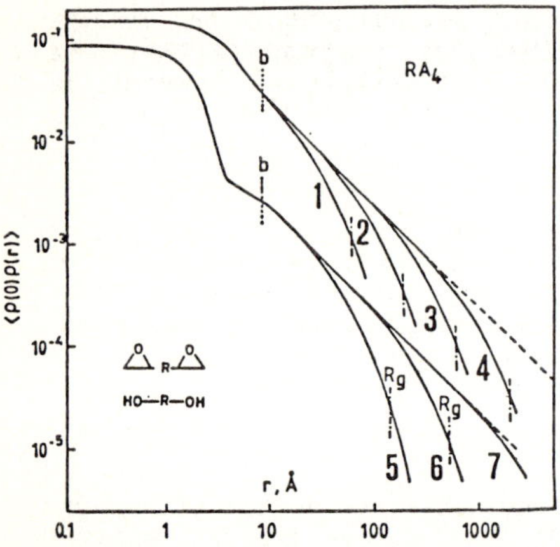

Figure 1 : Log-log plot of the density correlation function (DCF) for the random tetrafunctional homopolycondensate RA_4 and a copolymeric epoxy resin. b: effective bond length; R_g: radii of gyration. Curves 1-4 refer to $\varepsilon = 10^{-2}$, 10^{-3}, 10^{-4}, 10^{-5}, curves 5-7 to M_w = 2.5×10^5, 3.3×10^6, 4.4×10^8 respectively. $\varepsilon = (1 - \alpha/\alpha_c)$

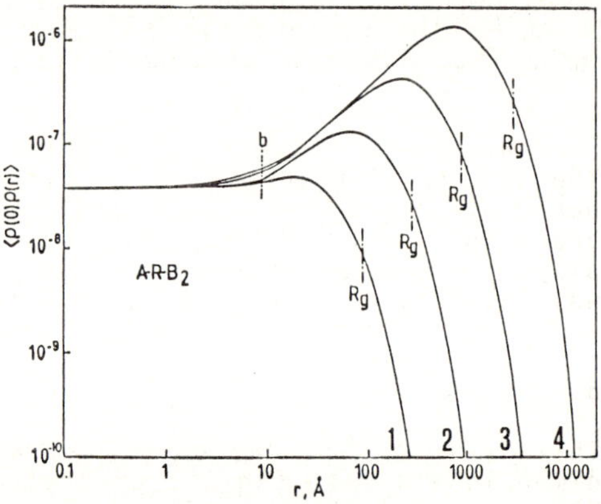

Figure 2 : Log-log plot of the DCF for the homopolymeric ARB_2 condenstae. Curves 1-4 refer to $\varepsilon = 10^{-2}$, 10^{-3}, 10^{-4}, 10^{-5}.

The fourth model represents the homopolymeric ARB_2 polycondensate. Strikingly different behaviour is obtained with an increase of the density correlation function. The corresponding Hausdorff dimension approaches for very large structures D = 4;for $\varepsilon = 10^{-5}$, i.e. $DP_w = 10^9$, only D = 3.88 was obtained.(Fig.2).

The last structure to be discussed is that of amylopectin (AP) which is the branched component in starch. It is a biological macromolecule and can specifically be degraded by special enzymes. Two of these enzymes cleave exclusively the α-(1,6) glycosidic branching bonds,and thus debranch completely the whole molecule. GPC fraction of the resulting mixture of linear chains gave a multinomial distribution,with three to four different main chain lengths of 22, 44, 66 etc. units. From these results biochemists deduced a heterogeneously branched structure for AP. The insert in Fig. 3 shows a simplified version. We carried out light scattering measurements at five different wavelengths, determined the molecular weight (350×10^6), the radius of gyration (2600 Å) and the angular dependence. The AP structure was then reconstructed by statistical means and compared with the measured angular dependence. Excellent agreement was observed.[14].

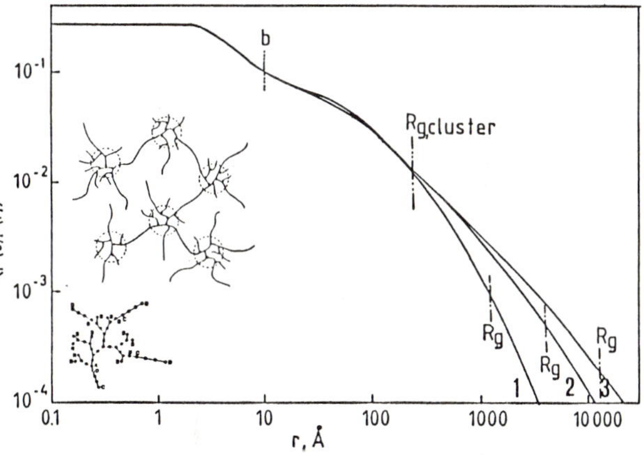

Figure 3 : Log-log plot of the DCF for the amylopectin β-limit dextrin molecule. Insert shows the heterogeneous structure. $R_{g,cluster}$, R_g radii of gyration one cluster and of the total molecule respectively. Curves 1-3 refer to $\varepsilon = 10^{-2}$, 10^{-3}, and 10^{-4}.

The result of the Fourier transform is shown in Fig. 3. A similar behaviour as in the other examples is observed; the heterogeneity in branching is not detectable. The clusters are too small as to display an increase in G(r) in the region of the cluster radius which was expected for the cluster of the ARB_2 type.

6 Discussion

This contribution may be closed with some remarks on the Hausdorff dimensions and on the DLA process of aggregates. For the RA_f and the RA_2/RB_f models Hausdorff dimensions of D = 2 are obtained from the scaling behaviour in the intermediate r-region , where a self-similarity in structure is expected. This dimension agrees with that for the radius of gyration, where

$$\langle s^2 \rangle_z \sim P_w \qquad (26)$$

is found. This result is a little unexpected, since eq.(5) is restricted to the region of validity of the scaling behaviour; the radius of gyration, however, results from an averaging over <u>all</u> distances.

The same consistency between scaling in G(r) and $\langle s^2 \rangle_z$ is observed with the ARB_2 model, but here the Hausdorff dimension is D = 4 and larger than the topologic dimension of d = 3. At first sight this finding may be considered as a physical nonsense. A closer inspection reveals that D > 3 means an increase of the segment density with increasing radius. Such an increase is physically meaningful as long as the segment density remains significantly lower than for a close sphere packing. Apparently this condition is fulfilled even for the largest molecule calculated. For even larger structures, the Gaussian chain approximation fails and excluded volume effects become essential. Angular dependences similar to that of the ARB_2 model have indeed been observed with glycogen, a highly branched energy-storage polsaccharide in the liver of animals [15].

7 References

1. R. M. Ziff, G. Stell, J. Chem. Phys. 73, 3492 (1980)
2. R. L. Drake, in: "Topics in Current Aerosol Research", eds. G. M. Hidy, J. R. Broek, Pergamon , New York 1972, Vol. 3, p. 201
3. R. J. Cohen, G. B. Benedek, J. Phys. Chem. 86, 3696 (1982)
4. J. L. Spouge, J. Phys. A.: Math. Gen. 16, 767 (1983)
5. P. J. Flory, "Principles of Polymer Chemistry", Cornell University Press, Ithaca 1953
6. W. H. Stockmayer, J. Chem. Phys. 11, 45 (1943)
7. M. Smoluchowski, Z. Phys. 17,557,585 (1916)
8. D. Stauffer, A. Coniglio, M. Adam, Adv. Polym. Sci. 44, 103 (1982)
9. H. W. Watson, Educational Times 19, 17 (1873)
10. M. Gordon, Proc. Roy. Soc. (London) A 268, 240 (1962)
11. T. A. Witten, L. M. Sander, Phys. Rev. B 27, 5686 (1983)
12. W. Burchard, Adv. Polym. Sci. 48, 1 (1983)
13. E. Wachenfeld, W. Burchard, Polymer submitted
14. W. Burchard, A. Thurn, Macromolecules, in press
15. M. Rinaudo, W. Burchard, see Ref. 12, p. 72

Size Distribution and Conformation of Branched Molecules Prepared by Cross-linking of Polymer Solutions

L. Leibler
Laboratoire de Physico-Chimie Macromoléculaire, E.S.P.C.I., 10, rue Vauquelin, F-75231 Paris Cedex 05, France
F. Schosseler
Institut Charles Sadron, C.R.M., 6, rue Boussingault, F-67083 Strasbourg Cedex, France

Irreversible gelation reactions are typical examples of kinetic growth process. At the beginning of the reaction, the system contains only small precursor units (monomers or linear polymer chains) which then react to form larger branched molecules (clusters). The system becomes more and more polydisperse, and eventually at the gel point the reaction bath contains a macroscopic cluster as well as precursor units and molecules of all intermediate sizes. It is interesting from both practical and fundamental points of view to understand the temporal evolution of cluster size distribution, as well as the morphology of large clusters.

The analogy between the gelation reaction and kinetic irreversible aggregation processes was already recognized some forty years ago by Stockmayer, who modelled the vulcanization process by a simple Smoluchowski coagulation equation [1]. This model, and most kinetic models developed afterwards, predict a scaling law for the time evolution of the size distribution function of large clusters below the gel point (e.g. [2,3]) :

$$\phi(M, t) \sim M^{-(\tau-1)} f(M/M^*(t)) \tag{1}$$

$\phi(M, t)$ denotes the weight fraction of monomers belonging to clusters with molecular weight M at reaction time t and $\phi(M, t)/M$ yields the number of such clusters. The function $\phi(M, t)$ is normalized so that the sum over all molecular weights is equal to 1 for each reaction time t. This normalization reflects the fact that the overall number of monomer units does not change during the reaction ; what varies is solely the repartition of monomers among clusters with different molecular weights M.

The important point is that the time-dependence of the cluster size distribution $\phi(M, t)$ is a universal function of a single variable $M/M^*(t)$, that does not depend on the initial distribution $\phi(M,0)$ at t = 0. The size $M^*(t)$ plays a role of a characteristic cluster size. It diverges when the gel point t_g is approached

$$M^*(t) \sim (t_g - t)^{-1/\sigma} \tag{2}$$

Therefore, $M^*(t)$ appears as the unique size scale which determines the divergence of the moments of the distribution function when $t \to t_g$ or equivalently, the spread of the cluster size distribution. The function f(x) may be interpreted as a cut-off function : $f(x) \simeq 1$ for $x \ll 1$ and $f(x) \ll 1$ for $x \gg 1$. At the gel point the number of clusters decreases according to a power law with a critical exponent τ.

In kinetic theories based on Smoluchowski coagulation equation, the exponents τ and σ are simply related, $\tau = 2 + \sigma$, and their value depends on details of the coagulation probability (ErnE) [2,3]. In the classical Flory-Stockmayer theory $\tau = 5/2$ and $\sigma = 1/2$ [1].

Cluster size distribution function has been also extensively studied in equilibrium models, and in particular in the percolation model (e.g. [4] and references given therein). These theories predict a similar scaling law (1) but the fundamental quantity is now the number of cross-links formed during the reaction rather than the time t. In equilibrium theories, the sol-gel transition is analogous to thermal critical phenomena and exponents τ and σ appear to be independant one from another. Moreover, it follows from the hyperscaling hypothesis that the radius of large clusters of molecular weight M should scale like $R \sim M^{\rho}$ with $\rho = (\tau - 1)/3$ (at 3 dimensions). Therefore, in scaling theory the values of two exponents σ and τ determine not only the divergence of the moments of the distribution function but also the confomation of branched molecules in the reaction bath.

Though very important the validity of the scaling picture was not demonstrated experimentally. An indirect test of scaling laws (1) and (2) would require a simultaneous measurement of at least three moments of the distribution function and verification of relations between critical exponents. This is difficult to achieve in in situ experiments because standard physicochemical methods of measuring the moments of molecular weight distribution functions and conformation of molecules are nonoperating in concentrated solutions. This obstacle does not exist in dilution experiments, in which reaction is stopped at some stage and the system is then diluted and characterized. We have thus used this approach to test the scaling theories. We could then measure not only simple moments, but also the whole size distribution function, and verify directly the validity of the concept of the typical size M^*.

In this paper we present some first results concerning the gelation of polystyrene solutions in cyclopentane induced by irradiation with ^{60}Co γ-rays. At t = 0 only linear chains are present in the solution. When the solution is irradiated polystyrene chains cross-link at random and ultimately a gel may be formed. The mechanism and kinetics of the cross-linking reaction are rather complex. The adsorbed γ-rays create free-radicals in the solvent which is particularly radioactive. These radicals may attack the polystyrene chains which then are able to cross-link. The radicals may be also produced directly on PS chains under influence of γ-irradiation. The advantage of this reaction lies in the fact that it can be stopped immediately at any moment by stopping the irradiation. The polymer molecules may be then easily extracted from the sol phase, and characterized with the help of various physico-chemical techniques [5,6].

In order to measure the size distribution of polymer molecules extracted from the sol phase, we need to divide these molecules into narrow fractions and to measure the number of molecules in each fraction and their average molecular weight. Here we have used the size-exclusion chromatography (gpc) to separate molecules into fractions. The molecules in each fraction were characterized by a small-angle light scattering detector and differential refractometer, coupled on-line with chromatographic columns. From these measurements a cluster size distribution $\phi(M, t)$ can be obtained, by a procedure described and discussed in detail in a forthcoming paper [7,8].

Figure 1 shows the measured cluster size distribution function for typical irradiated samples. We have chosen to use the precursor chains of relatively low molecular weight $M_w(0) \simeq 5.5 \times 10^4$ in order to be able to study the whole distribution function from the beginning of the reaction up to nearly the gel point t_g ($t/t_g \simeq 0.92$). For samples closer to the gelation threshold the method does not work, since the huge molecules present in such samples are retained by the chromatographic columns.

We have used the precursor chains with the weight average molecular weight $M_w \simeq 5.5 \times 10^4$ and with relatively narrow molecular weight distribution spreading from about 10^4 to 3×10^5.

Several important qualitative conclusions can be drawn from the figure. The maximum in the distribution function corresponds to the linear precursor chains.

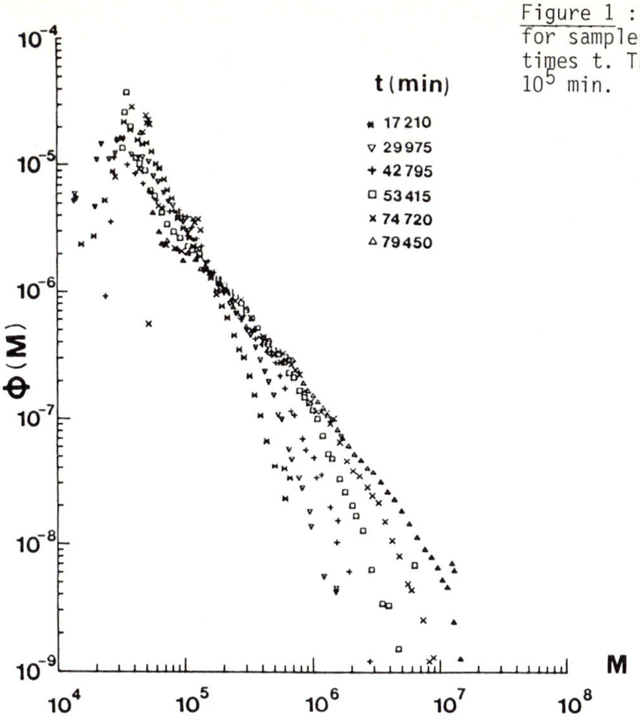

Figure 1 : Size distribution functions for samples with different irradiation times t. The gel is obtained for $t_g \simeq 10^5$ min.

The concentration of these linear chains decreases monotonically during the reaction as they cross-link to form larger branched molecules. If we look, however, at the branched molecules of a given size, their concentration first increases but then decreases when the reaction proceeds further. This agrees with the idea of randomly cross-linking clusters.

The most striking feature of the irradiated samples is their enormous polydispersity. For example,a sample with the weight average molecular weight $M_W \simeq 5.4 \times 10^5$ $(t/t_g \simeq 0.8)$ contains molecules with molecular weight ranging from about 10^4 to 2×10^7. We observe that for fixed irradiation time t the concentration of large clusters decays algebraically as a function of M. There exists a characteristic size above which this decay cuts off. The cut-off size increases very rapidly with time,and the distribution spreads more and more as the gel point is approached. This qualitative discussion suggests that the scaling picture may be valid for our gelation process.

Actually,an analysis of the scattered light spectrum confirms this notion of the typical cut-off size $M^*(t)$ and enables us to measure directly $M^*(t)$ [7 ,9]. $M^*(t)$ diverges as t approaches the gel time t_g. Unfortunately, experimental uncertainty in the position of the gel point does not allow to determine precisely the critical exponent σ. This is related to a general drawback of dilution experiments, namely the necessity of preparing many absolutely identical samples in which reaction proceeds under strictly identical conditions. In practice this is not really possible and for instance, the fluctuations of a few percents in the gel point for identically prepared samples are not unusual. This problem may be overcome by measuring simultaneously two different types of diverging quantities [10]. It is then possible to find some combination of critical exponents without knowing exactly the advancement of the reaction with respect to the gel point for each particular sample. By this method we can find the exponent τ since we measure for example the weight average molecular weight $Mw(t) = \phi(M,t) M dM$ and $M^*(t)$. If the scaling law (1) holds, $Mw(t)$ will be an algebraic function of $M^*(t)$, $Mw(t) \sim M^*(t)^{(3-\tau)}$. This is the case and we find $\tau \simeq 2.3 \pm 0.1$ [7,9].

Hence we can test the validity of the scaling law (1) using only experimentally determined values of $\phi(M,t)$, $M^*(t)$ and τ without any adjustable parameters. In fig. 2 we have plotted the quantity $G = \phi(M,t)\, M^*(t)^{\tau-1}$ which according to (1) for large clusters should be a universal function independent of time t. We find a very satisfactory collapse of our data for eight samples whose relative distance from the gel point ranges from $t/tg \simeq 0.18$ to 0.92. This demonstrates beautifully the validity of the scaling law (1) for cluster size distribution function. The test is rather severe, since $M^*(t)$ varies by a factor 80 from $1.35\ 10^5$ to $1.01\ 10^7$ for the most irradiated sample.

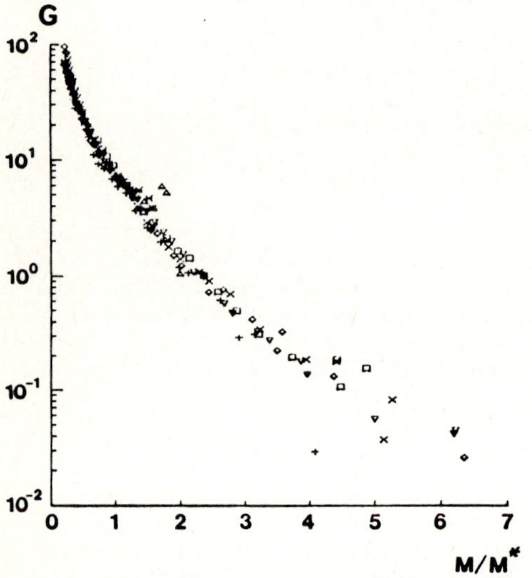

Figure 2 : The scaling function $G = (M,t)\, M^*(t)^{(\tau-1)}$

It is interesting to note that for some samples we measured the concentrated $\phi(M,t)$ for molecular weights M well above the cut-off size M^* ($M/M^* \simeq 7$). These data show that the cut-off function appearing in (1) is not a simple exponential function, as would predict for instance a percolation model [11]. The best fit seems to be a function $f(x) \simeq \exp(-x^{\zeta})$ with $\zeta \simeq 0.8 \pm 0.1$.

Measurements of the cluster size distribution provide a very useful basis for the studies of the conformation or diffusion properties of large clusters. In fact, most techniques such as viscosimetry, elastic or dynamic light scattering do not measure the properties of individual molecules but the averages over the size distribution [10]. For example, elastic light scattering yields the weight average molecular weight $Mw(t) = \int \phi(M,t) M dM$ and the z-average radius of gyration

$$\langle R^2(t) \rangle_z = \int \phi(M,t)\, M\, R^2(M)\, dM/Mw \qquad (3)$$

with R(M) being the radius of clusters of molecular weight M which is expected to scale like M^{ν}. We have measured [5] the average radius of gyration $\langle R^2(t)\rangle_z$ for diluted sol molecules in a good solvent and we have found that $\langle R^2(t)\rangle_z$ scales like $Mw(t)^{2\nu_{eff}}$ with $\nu_{eff} \simeq 0.58 \pm 0.06$. Since we have verified the scaling law for $\phi(M,t)$ we can relate ν_{eff} and ν. From (3) we obtain $\nu = \nu_{eff}(3-\tau)$ [10] and conclude that $\nu \simeq 0.41 \pm 0.05$ for the swollen branched molecules. The same value for swelling exponent ν is obtained more directly from the simultaneous measurements of the average radius of gyration

$<R^2(t)>_z$ and of the cut-off size $M^*(t)$. Eqs. (1) and (3) yield the scaling law $<R^2(t)>_z \simeq M^*(t)^{2\nu}$ which shows that the average radius is just equal to the radius of the typical cluster.

The measured value of the swelling exponent ν is well below $\nu = 0.5$ expected for randomly branched molecules swollen in a good solvent [12]. This result seems to suggest that there are many intramolecular loops that limit the swelling of our branched chains. This might be expected, since the solution is cross-linked in a nearly theta solvent at the concentration close to the overlap concentration C^*.

To conclude, our experiments on cross-linking of polymer chains in a theta solvent seem to confirm the validity of the concept of the typical cluster which determines the behaviour of the moments of the cluster size distribution and its spreading. We find that the cluster size distribution function obeys the scaling law $\phi(M, t) \simeq M^{-\tau} \exp -[M/M^*(t)]^{\zeta}$ with $\tau \simeq 2.3 \pm 0.1$ and $\zeta \simeq 0.8 \pm 0.1$. The value of the exponent τ seems to be smaller than that ($\tau = 5/2$) given by classical theories. In a good solvent the branched molecules swell less than would be expected for randomly branched chains. This may suggest the relevant role of intromolecular loops in such clusters.

For a better understanding of gelation processes, it would be very interesting to study by the methods introduced in this work the systems in which the advancement of the reaction (i.e. the number of cross-links that are formed) can be followed by a spectroscopic technique. It would be then possible to find not only the exponent τ but also σ and test the validity of different gelation models. On theoretical side, the calculations beyond the mean-field theories for cross-linking in theta conditions will be certainly very useful.

The authors would like to thank J. Bastide, H. Benoît, N. Bikales, S. Candau, M. Daoud, Z. Gallot, P.G. de Gennes, J.F. Joanny, A. Lapp, M. Rinaudo and C. Strazielle for their help and interest during this work.

[1] W.H. Stockmayer, J. Chem. Phys., 11, 45 (1943)
[2] F. Levzyaz, H.R. Tschudi, J. Phys., A 15, 1951 (1982)
[3] E.M. Hendriks, M.H. Ernst, R.M. Ziff, J. Stat. Phys., 31, 513 (1983)
[4] D. Stauffer, A. Coniglio, M. Adam, Adv. Polym. Sci., 44, 103 (1982)
[5] F. Schosseler, L. Leibler, J. Physique Lett., 45, L 501 (1984)
[6] F. Schosseler, L. Leibler, Macromolecules, 18, 398 (1985)
[7] F. Schosseler, Thesis, Université Louis-Pasteur, Strasbourg (1985)
[8] L. Leibler, F. Schosseler, H. Benoit, Z. Gallot, C. Strazielle, to be published.
[9] F. Schosseler, L. Leibler, to be published.
[10] M. Daoud, F. Family, G. Jannink, J. Physique Lett., 45, 193 (1984)
[11] D. Stauffer, Phys. Rep., 54, 1 (1979).
[12] G. Parisi, N. Sourlas, Phys. Rev. Lett., 46, 871 (1981).

Viscoelastic Effects Occurring the Sol-Gel Transition

B. Gauthier-Manuel
Laboratoire d'Hydrodynamique et de Mécanique Physique, E.S.P.C.I.,
10, rue Vauquelin, F-75231 Paris Cedex 05, France

I INTRODUCTION

The gels can be defined as a composite material made of two phases : a three-dimensional polymeric network embedded in a fluid. The main characteristic of such materials is that only a few per cent (w/w) of network is sufficient to give a weak elastic solid behavior, and to prevent the flow of the fluid phase due to the friction of solvent molecules with the network.

Most gels networks are built with polymeric chains crosslinked in order to assure the three-dimensional edifice. According to the nature of the crosslinks the gels can be divided into two main groups.

i) Physical gels (gelatine, agarose) in which the crosslinks are realised with low energy bonds (hydrogene, Van der Waals). These gels are thermally reversible. They present a sol phase (solution of molecules without crosslink) at high temperature and a gel phase at low temperature. Gelation is realised by decreasing temperature from the high temperature state; a critical temperature (under which gel phase may be obtained) can be defined.

ii) Chemical gels (acrylamide-bisacrylamide, styrene-divinylbenzene) in which the crosslinks are realised with high energy bond (covalent bonds of multifunctional molecules). This process is thermally irreversible. Gelation is realised with chemical reaction from a solution of reactive molecules, and a critical time of reaction (above which a gel phase may be obtained) is defined.

Sol-gel transition is characterized by the appearance, into the solution, of an infinite cluster which connects all the sample. This cluster results from the growth of the small clusters during the kinetic reaction of the formation of the crosslinks. It is responsible for the elastic behavior of the gel.

The evolution of the system between the sol phase and the gel phase is a function of a parameter : the conversion factor p which represents the fraction of bonds realized. The measurement of this parameter is not as easy in chemical systems, as the radicalor copolymerisation of acrylamide-bisacrylamide (the determination of p would require to stop the reaction at different times and to titrate the free bonds into the solution). In order to alleviate this problem, the study is realised as a function of the time t under the following assumption.

Near the gelation threshold a regular behavior of the law p(t) gives the same value of the exponent for the first order expansion of the power law as a function of time

$$\left(\frac{p-p_c}{p_c} \right)^{\mu} = \left(\frac{t-t_c}{t_c} \right)^{\mu} \times \dot{p}(t_c) + \dots \qquad (1)$$

this regular behavior can be proved in the case of the mean field theory [1].

The critical conditions for the appearance of the infinite cluster were first investigated by FLORY and STOCKMAYER |2| who described the sol-gel transition with a mean field theory.

Recently de GENNES |3| and STAUFFER |4| suggested an approach through which the gelation phenomena can be explained by analogy with bond percolation theories on lattices. These theories give a critical threshold p_c of the conversion factor above which an elastic mechanical behavior is observed. Different quantities as the correlation length ξ, the intrinsic viscosity $\bar{\eta}$ and the shear elastic modulus E are supposed to have a critical behavior near the gelation threshold. This behavior is characterized by a power law :

$$\xi \propto |\varepsilon|^{-\nu} \qquad (2)$$

$$\bar{\eta} \propto |\varepsilon|^{-k_I} \qquad \varepsilon < 0 \qquad (3)$$

$$E \propto \varepsilon^{\mu} \qquad \varepsilon > 0 \qquad (4)$$

where ε is the reduced distance to the threshold $(p-p_c)/p_c$.

The values of these exponents differ from one theory to another. Table 1 gives different values for these exponents.

A new kind of approach has been developed recently using kinetic pattern of aggregation of clusters |5,6| which would apply best to the present case due to the role of the activator in the progression of the reaction. This theory also gives a threshold which corresponds to the building of an infinite cluster, as in the gelation process.

The originality of this modelisation of the sol-gel transition is to take into account the mobility of the clusters and the non-equivalence of accessibility of different sites in a building cluster.

The results are, for the moment, limited to the geometrical aspect (fractal nature) of the generated clusters. In particular they do not make any prediction on the behavior of macroscopic quantities such as elasticity or viscosity. More theoretical work must be done in order to quantify the importance of the mobility effects near the gelation threshold.

The purpose of this paper is the study of the evolution of such a system just near the transition threshold, when mechanical behavior goes from a viscous liquid behavior to an elastic solid one.

We limit our discussion to the case of chemical gels obtained with copolymerisation reaction exemplified by the acrylamide-bisacrylamide system.

II STATIC MEASUREMENT

The macroscopic viscosity (at zero shear) in sol phase and the shear elastic modulus in gel phase are measured with a sphere rheometer |9| in the vicinity of the gelation threshold during the sol-gel process. This device allows us to make experiments at very low shear rate (10^{-4} s^{-1}) and to measure low amplitude forces (10^{-9} N).

Table 1

exponent	classical theory	percolation
ν	1/2	0.8
h	?	?
μ	3|7|	1,7|7|,2|8|

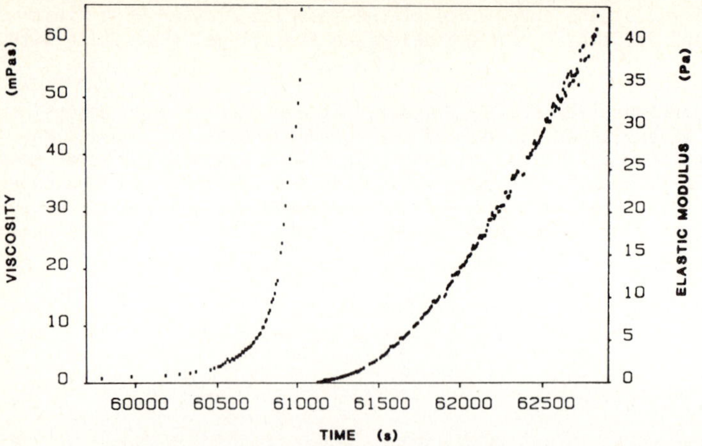

Fig. 1 : Evolution of macroscopic viscosity (x) and shear elastic modulus (■) for the gelation of the acrylamide-bisacrylamide system as a function of time. The concentration in monomer is equal to 0.01749 and the crosslink ratio (ratio between the quadrifunctional units of bisacrylamide and the bifunctional units) is equal to 0.0267

Figure 1 shows the results obtained for an acrylic gel. When chemical reaction progresses we see a particular time (T_c) defined with an accuracy of 60 s from which an elastic behavior is observed. Below this time,we measure a fast increase of the viscosity and the appearance of a viscoelastic behavior. This behavior is characterized by time-constants in the measured signal of the rheometer. This viscoelastic behavior leads to difficulties in the experiment :

i) in order to measure the zero shear viscosity,the shear rate should always be lower than the time-constant of the viscoelastic system, thus the experiment is done by decreasing the shear rate when the threshold is approached;

ii) this viscoelastic behavior produces errors on the measurement. In order to measure the asymptotic value of the viscosity (when all the time-constants are released) we should wait an infinite time to have the static value of the viscosity.

In practice we wait only for a finite time (longer than the time-constant of the system but shorter than the evolution time of the physical state of the system).

This leads to a region near the gelation threshold where no measurement is possible (which defines the limit of accuracy on the value of the gelation threshold).

In the gel phase,the same phenomenon leads to an inaccuracy in the value of the measured elastic modulus (all measurements were done by waiting for the relaxation of the signal during 7 seconds) the measured elastic modulus has a larger value than the static modulus.

The research of the critical behavior for the increase of the shear elastic modulus leads to a linear behavior of the function g_1

$$g_1(t) = | E(t) |^{1/\mu} \tag{5}$$

for a particular value of the exponent μ. The best value of μ, equal to 1.9 ± 0.1, gives a linear evolution of the function g_1 along three decades of values of elasticity (Fig. 2 right curve) |10|.

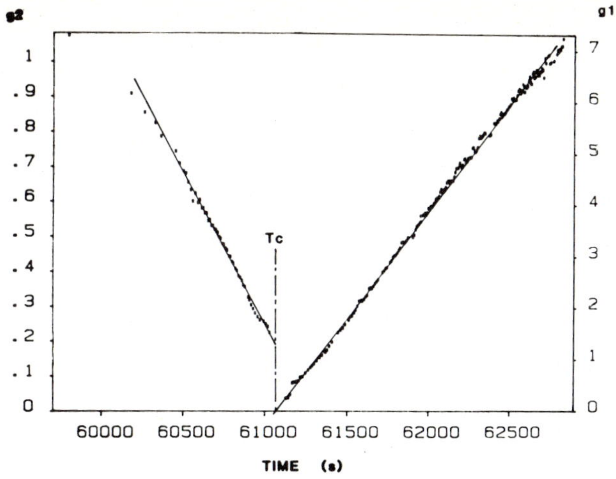

Fig. 2 : Plot of the two functions g_1 (■) and g_2 (x) for the best values of the exponents $\mu = 1.9$ and $k = 2.8$.

The value of this exponent is perfectly reproducible for all the gels realised with the same concentration in monomers. However, we have observed an increase of this exponent value when the concentration in monomer or the crosslink ratio is increased (this value is shifted from 1.9 to 2.4 when the concentration is increased of 50 % w/w).

A similar procedure for the viscous behavior gives a linear variation of the function g_2

$$g_2(t) = | \eta(t) |^{-1/k} \qquad (6)$$

for the value $k = 2.7 \pm 0.2$ (Fig. 2 left curve).

This value of the exponent k, greater than the value obtained with other polymeri systems |11|, can be compared with predictions of theories proposed by WALLES |12| and GORDON |13|. They relate the macroscopic viscosity to the average molecular mass with a scaling law of the form :

$$\log \eta = K_1 + K_2 \log Dp_w$$

If we assume a critical behavior of the average molecular mass with the exponent γ then the exponent k of the macroscopic viscosity would be equal to $\gamma x K_2$.

Table 2 gives the different values of the exponents for mean field and percolation theories. The value of k in percolation theory is larger than the value k_I predicted for the intrinsic viscosity ($k_I = 0.8$). The finite value of the macroscopic viscosity measured at the gelation threshold means probably that viscoelastic effects become too important to neglect the frequency effects : the measured quan-

Table 2

	γ	K_2	k
mean field theory	1	1	1
percolation 3D	1.7 |7|	1.2|12|	2.04
	2.3 |15|		2.76

tities are not static quantities and the effects of the experimental time-scale could be studied in order to give correct results of this experiment. This will be the subject of the next part of this work.

The validity of the values of the exponents predicted with percolation assumes an analogy between electrical and mechanical problems. This assumption is now contravened, and mechanical experiments done on weak mechanical systems give values for mechanical exponents notably greater than for electrical ones. But an argument due to ALEXANDER [14] stating that a gel is not obtained in a non-equilibrium state, (internal stresses are important when the gel is obtained) suggests that the analogy between a weak mechanical lattice having compression of bending forces and a gel is not adequate, and that scalar elasticity (equivalent to an electrical problem) may apply.

III DYNAMICAL MEASUREMENT

In order to quantify the error done on the critical "static" behavior previously measured, we can impose an experimental time-scale by the choice of a frequency for the study of this phenomenon. Experiments were done with the alternating part of the sphere rheometer |16| in the range of frequencies 0.2-20 Hz. In order to overcome the poor reproductiveness of the gelation time (2 %), we use two different frequencies on the same sample at the same time. One of these frequencies is kept at a constant value (reference frequency), and the other is changed during the different trials.

The measurement of the relative amplitude and of the phase-shift between the imposed stress and the measured strain leads to the determination of the complex viscosity $\eta^*(\omega)$. The choice of an experimental time-scale $T_{ex} = 1/\omega$ cuts the population of clusters in two parts : one part characterised by a relaxation time lower than T_{ex}. These clusters give a viscous response (they have enough time to flow entirely during the experiment).

The other part gives an elastic response (these clusters don't have enough time to relax during the experiment).

Figure 3 shows the evolution of the real part η' of the complex viscosity (dissipative part) for the two different frequencies on the same sample.

A qualitative analysis of the results obtained with the phase-shifts measurements allows us to give evidence for the variation of the apparent gelation time (at the measured frequency) as a function of the frequency |17|. The higher the frequency, the earlier the elastic behavior of the system appears.

In order to measure this effect in a quantitative way, we assume a dynamic scaling law between the displacement of the gel point and the imposed frequency ω

$$\omega = k.(t_0 - t_c)^{\delta} \tag{7}$$

where t_0 is the gelation threshold at zero frequency and t_c is the threshold at ω frequency. The exponent δ is the dynamic critical exponent which characterizes this effect. This relation implies that the real part of viscosity η' is of the form

$$\eta' \propto f\left(\omega(t_0-t)^{-\delta} \right) \tag{8}$$

the problem is to calculate the value of the exponent δ when the value of t_0 is unknowned.

The method applied is a renormalisation procedure, in order to have a universal curve of $\eta'(t)$ when the frequency is shifted. We note that the ratio

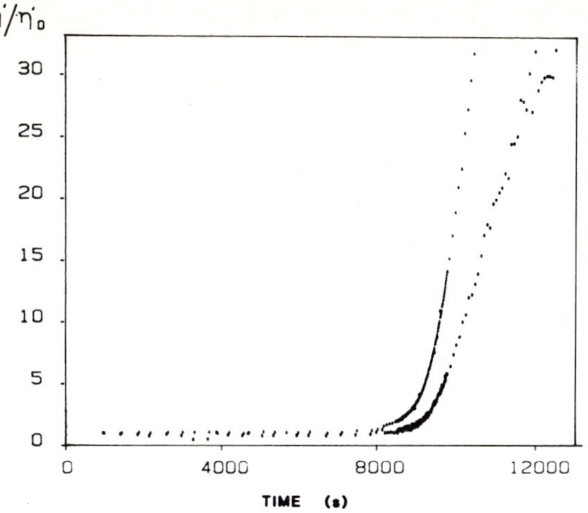

Fig. 3 : Evolution of the real part of viscosity η' a function of time (normed with the initial value η_0') at two different frequencies 0.2 Hz (▲) 10 (●) Hz for the same acrylamide-bisacrylamide system undergoing a sol-gel transition.

$$\frac{t_o - t_2}{t_o - t_1} = \left| \frac{\omega_1}{\omega_2} \right|^{-1/\delta}$$

gives the renormalisation law between the two time scales t_1 and t_2 of the two frequencies used

$$t_2' = t_2 \left| \frac{\omega_1}{\omega_2} \right|^{-1/\delta} + t_o \left| 1 - \left(\frac{\omega_1}{\omega_2} \right)^{-1/\delta} \right|$$

The similarity coefficient a

$$a = \left| \frac{\omega_1}{\omega_2} \right|^{-1/\delta}$$

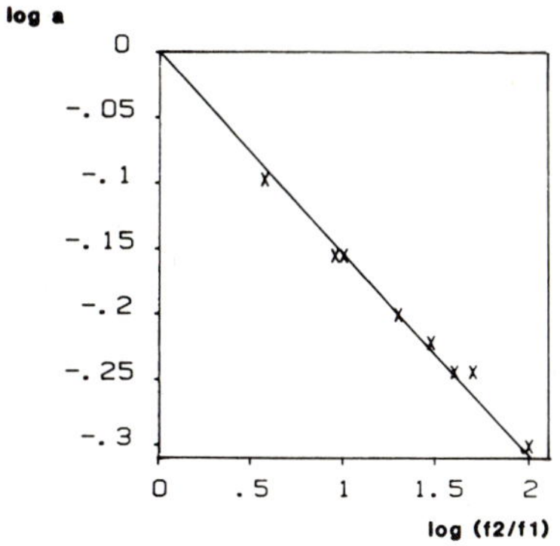

Fig. 4 : Log-Log plot of the renormalization coefficient a as a function of the ratio f_2/f_1 between the two different frequencies used for measurement.

is measured for each experiment as a function of the frequency ratio, and the log-log plot of the result is sketched on Fig. 4. We obtain a straight line with $1/\delta$ shape. We find the following result :

$$\delta = 6.5 \pm 0.2$$

This value of δ can be compared with the theoretical prediction [18] of dynamical exponent of percolating clusters.

In sol phase below the gel point we have clusters of different sizes characterized by two relaxation times :

i) an elastic time t_{el} of the deformation of the cluster (maxwell time) equals to

$$t_{el} = \frac{\bar{\eta}}{\bar{E}}$$

where $\bar{\eta}$ is the viscosity of the solution in the vicinity of the cluster and $\bar{E}$ is the elastic medulus of the cluster;

ii) the rotation time t_{rot} of the rotational diffusion of the cluster is

$$t_{rot} = \frac{r_p^2}{D}$$

where r_p is the largest anisotropic cluster scale of size p and D is the diffusion coefficient.

If we assume scaling laws of elastic modulus and viscosity

$$\bar{E} \sim E \left|\frac{\xi}{r_D}\right|^{\mu/2}$$

$$\bar{\eta} \sim \eta \left|\frac{r_p}{\xi}\right|^{k/\nu}$$

We can express the initial behavior of these two relaxation times as

$$t_{el} \sim \varepsilon^{-(k+\mu)}$$

$$t_{rot} \sim \varepsilon^{-(k+3\nu)}$$

In the case of percolation values of the different exponents, rotational effects are the most important, and the value of the dynamic exponent $(k+3\nu)$ is found to be 3, a value lower than our experimental result.

The discrepancy between this theory and experimental results suggests a deeper theoretical analysis. The fractal nature of the percolating cluster leads to a particular vibration spectrum [19]. The critical behavior of the relaxation time is modified by the "spectral" dimension, and the value of this dynamic exponent should relate to the fractal dimensions of the cluster. The problem of the analogy between electrical and mechanical behavior which we have evoked for the static behavior should also induce a different behavior of the dynamic properties of gels near the transition threshold.

IV CONCLUSION

The displacement of the gel point can be calculated for the frequency corresponding to the experimental duration of the static measurement. This gives a value equal to

$$\frac{\Delta T_s}{T_o} = 3.5\ 10^{-3}$$

This result is in good agreement with the shift measured directly on the static result (with the extrapolation of the function $g_2(t)$ at zero value) equals to 4 10^{-3}.

We see that viscoelastic effects are responsible for the finite value of the "static viscosity" measured at the gelation threshold. The static results, for the elastic behavior, are in good agreement with percolation results. This result gives evidence for the validity of the analogy between electrical and mechanical problem in the particular case of gelation. But a knowledge of the exponent value in the case of kinetic theory will be necessary in order to define the importance of the diffusion process during the sol-gel transition.

The author thanks E. GUYON and C. ALLAIN for helpful comments and stimulating discussions.

REFERENCES

1 R.M. Ziff, G. Stell, J. Chem. Phys. 73, 3492 (1980)
2 P.J. Flory, J. Am. Chem. Soc. 63, 3083, 3091, 3096 (1941)
W.H. Stockmayer, J. Chem. Phys. 11, 45 (1943)
3 P.G. de Gennes, Scaling Concepts in Polymer Physics (Cornell University Press, Ithaca, N.Y. 1979)
4 D. Stauffer, J. Chem. Soc. Faraday Trans II 72, 1364 (1976)
5 T.A. Witten, L.M. Sander, Phys. Rev. Letters 47, 1400 (1981)
6 M. Kolb, R. Botet, R. Jullien, Phys. Rev. Lett. 51, 1123 (1983)
7 D. Stauffer, A. Coniglio, M. Adam, Adv. in Polym. Sci. 44, 105 (1982)
8 C.D. Mitescu, M.J. Muslof, J. Phys. Lett. (France) 44 16 L 679 (1983)
9 B. Gauthier-Manuel, R. Meyer, P. Pieranski, J. Phys. E : Sci. Instrum. 17, 1177 (1984)
10 B. Gauthier-Manuel, E. Guyon, J. Phys. (Paris) 41, L 503 (1980)
11 M. Adam, M. Delsanti, Pure Appl. Chem. 53, L 539 (1980)
12 E.M. Walles, C.W. Macosko, Macromolecules 12, 521 (1979)
13 M. Gordon, K.R. Roberts, Polymer 20, 681 (1979)
14 S. Alexander, J. Phys. (France) 45, 1939 (1984)
15 H.J. Herrmann, D. Stauffer, D.P. Landau, J. Phys. A. 16, 6, 1221 (1983)
16 B. Gauthier-Manuel, J. Phys. E : Sci. Instrum. 17, 1183 (1984)
17 B. Gauthier-Manuel, C. Allain, E. Guyon, Comptes Rend. Acad. Sciences (Paris) 296, série II, 217 (1983)
18 J.F. Joanny, J. Phys. (Paris) 43, 467 (1982)
19 S.G. Grest, I. Webman, J. Phys. Lett. (France) 45, 1155 (1984)

Neutron Scattering Investigation of the Loss of Affineness in Deswollen Gels

Jacques Bastide
Institut Charles Sadron (CRM-EAHP), CNRS-ULP Strasbourg, 6, rue Boussingault, F-67083 Strasbourg Cedex, France

Introduction

Polymer networks swollen with a solvent of the linear chains of same nature are usually called gels. In these materials the rubber elasticity manifests itself not only in the classical way, when they are uniaxially stretched, but also tridimensionally : as first recognized by Frenkel [1], the swelling equilibrium can be considered as the result of a competition between the "osmotic" tendency of polymer dilution and an "entropic" elastic retraction of the network. Because polymer networks are commonly synthesized in presence of a solvent, i.e. in a gel state, an important part of the research activity in the rubber elasticity field has been devoted to the study of the swelling and deswelling mechanism. In this regard, it is of particular interest to know how the macroscopic volume change of the sample affects the chain conformation. This kind of information can be gained using the neutron scattering technique applied to selectively labelled networks.

Recall of Previous Experiments

First series of experiments were performed on samples containing labelled meshes [2-4]. The gels were synthesized (in a good solvent) by reaction of bifunctional primary chains on polyfunctional crosslinking nodules. Some of the chains were deuterated, and therefore their conformation could be investigated using the small angle neutron scattering technique. Practically, the radius of gyration Rg of these labelled meshes was measured as a function of the swelling degree of the sample. The gross result of these experiments was that the radius of gyration change was very weak, far less important than that of a linear dimension of the sample. In itself, such a result is not surprising : an essential feature of the rubber elasticity (both in uniaxial and tridimensional cases) is the loss of affineness of the deformation when decreasing the scale of observation. In order to quantify this loss of affineness process, and thus to estimate the elastic properties, the "classical" models of rubber elasticity assume that the junction points between chains play a particular role. In a first "family" of theories [5-7], the instantaneous positions of the junctions are supposed to be transformed affinely in the macroscopic strain. In other words, the affineness is assumed to be transmitted down to the scale of the crosslinks, and then progressively lost at shorter scales, the chains being free to rearrange their conformations between their fixed extremities. As a consequence, the radius of gyration of the elementary chains, which takes into account the correlations between all pairs of points on the chain, will be less deformed than the end-to-end distance. In a second "family" of theories [7-9], which consider so-called "phantom networks", only the average position of the junctions are supposed to be affinely transformed in the macroscopic strain. The crosslinks are assumed to fluctuate, independently of the strain, around their mean positions. As a result, the affineness is expected to be partly lost already at the scale of the mean square end-to-end distance, which is a sum of the deformed squared mean distance and an unperturbed fluctuation. Consequently, at the scale of the radius of gyration, the loss of affineness is expected to be more important than for the "junction affine" model. The theoretical expressions of the radius of gyration change within the framework of these two models have been calculated, in the case of initially gaussian chains [10-12]. Therefore, the confrontation of experimental data to the predictions of the models is, in principle, possible. Practically, since the state of the sample for which the elementary chains are gaussian is difficult to precise for networks prepared in good solvents, the conclusion of this confrontation cannot be considered as certain. It however seems to indicate that the deformation of the elementary mesh could be even smaller than that predicted by the "phantom network" model.

A New Series of Experiments: Deswelling of Gels Containing Labelled Paths

Such a result of quite low deformation at the scale of the mesh led us to the idea of an investigation of the deswelling mechanism in a larger range of scales of distances. To this purpose, we decided to study samples containing labelled paths, i.e. deuterated chains linked to the network structure by more than two points. Moreover, in order to remove some ambiguities in the comparison between the data and the predictions of the models, it appeared necessary to dispose of samples for which the state where the chains are gaussian is well defined. In these two aims, we have developed a new crosslinking procedure, consisting of a γ-irradiation of semi-dilute solutions (under vacuum and after degassing) of polystyrene (H + 1% or 2% D) in cyclopentane. Cyclopentane is a radioreactive solvent and its presence increases significantly the efficiency of the reaction (by a factor ten approximately with respect to the case of radio reticulation of a melt). Correlatively, the number of chain cuts, using this method, seems to be negligible [13]. Additionally, in the conditions of the reaction ($T \sim 25°C$), cyclopentane is nearly a θ solvent for polystyrene ($\theta \sim 20°C$). Actually, we have verified that the conformation of the deuterated chains of the solution, which behave as labelled paths in the obtained networks, can be considered as gaussian in the state of preparation.

Characterization of the samples

Unfortunately, this crosslinking procedure does not allow a direct estimation of the size of the elementary mesh. Therefore, this quantity must be characterized by an indirect method. We have adopted the following one. (i) we measured the swelling degree Qe, the elastic modulus G and the cooperative diffusion constant Dc (determined with the quasi-elastic light scattering technique [14]) of the samples swollen at equilibrium in toluene. (ii) then we selected in the literature gels synthesized by end-linking of primary chains (at ϕ_c=0.1), i.e. with well defined chemical meshes, that exhibited the same values for these three quantities. (iii) at last, we postulated that gels with same triplets Q_e,D_c,G should have roughly the same mesh.

Radii of gyration of labelled paths linked to deswollen gels

We have first considered a series of gels with the same crosslinking density called C, and corresponding, according to the estimation procedure presented above, to a mesh of molecular weight ranging between 45000 and 50000. These samples were differing by the molecular weights of the labelled paths they contained ($M \sim 1.5.10^5$ to $M \sim 1.6\ 10^6$). They were submitted to a deswelling by a factor 10 between the state of preparation and the dry state. In Figure 1 are plotted the radii of gyration of the labelled paths linked to deswollen gels as a function of their molecular weights. On the same figure are also plotted the radii of gyration of the same deuterated chains, uncrosslinked and dispersed in a polystyrene melt. These latter, which follow approximately a $M^{\frac{1}{2}}$ variation law, can, within a good approximation, be considered as equal to the radii of gyration of the paths before deswelling. One observes that the apparent slope of the variation of Rg with M (for the paths in the deswollen gels) is, in the logarithmic representation adopted here, smaller than $\frac{1}{2}$. This indicates that longer are the paths, more important is the collapse induced by the deswelling.

Qualitatively, this is easy to understand:the longer the paths, the larger is the number of junctions they cross,and therefore greater is the constraint that the network configuration exerts on their conformation. More quantitatively, the

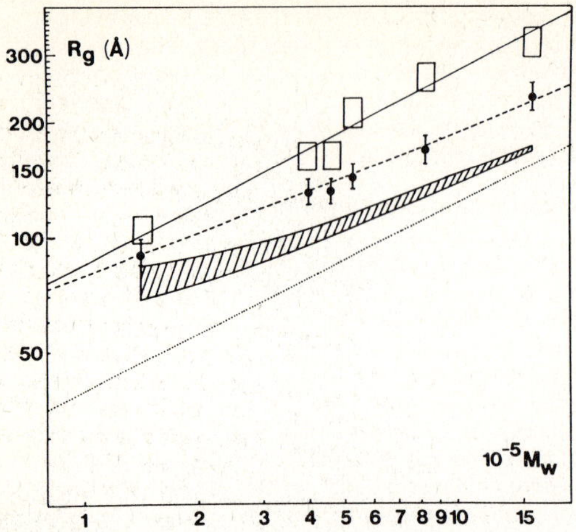

Figure 1 : Logarithmic representation of the radii of gyration of deuterated chains as a function of their molecular weights.

□ Uncrosslinked chains dispersed in a melt matrix.

Ī Labelled paths linked to gels deswollen by a factor 10. Estimation of the molecular weight of the mesh : 45000<M<75000.

Hatched zone : Calculated variation [15] of the radius of gyration as a function of the molecular weight for paths running through affinely displaced junctions (deswelling by a factor 10). The width of this zone corresponds to the uncertainty on the number of junctions per path for a given length.

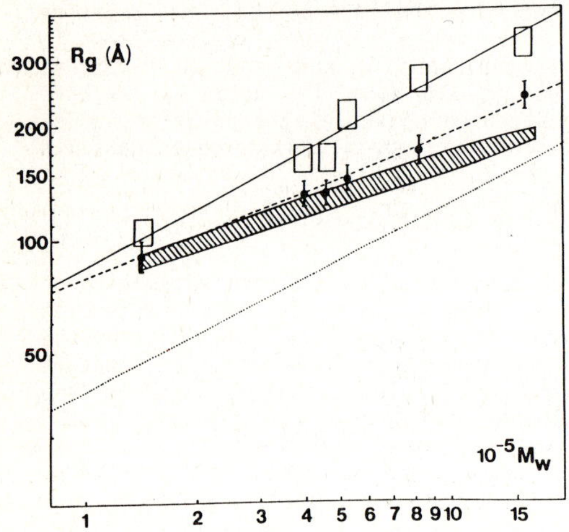

Figure 2 : The experimental results are the same as for Fig. 1.

Hatched zone : Calculated variation [16] of the radius of gyration as a function of the molecular weight, for paths linked to a "phantom network" (deswollen by a factor ten).

results can be compared to the classical deformation models. On Figure 1, we have plotted a hachured zone which represents the calculated values [15] of the radii of gyration for labelled paths of increasing mass, running through affinily displaced junctions. The width of this area corresponds to the uncertainty on the mesh of the network, and consequently on the number of junctions per path. One observes that this calculated variation does not fit our results. As in the previous series of experiments, it seems that the affine deformation assumption, for the junction positions, leads to an over-estimation of the deformation. The same type of representation has been adopted on Figure 2, where the experimental data are now compared to the predictions of the "phantom network" model. These latter are figured, as in the preceding case, by a hachured zone [16]. The experimental points are now closer to the calculated variation. Nevertheless, it still seems to overestimate slightly the deformation. It is important to notice that such a result is in contradiction with all the available theories, since it is generally assumed that the deformation of a real network should be intermediate between that corresponding to the "junction affine" and the "phantom" models [7]. Therefore, this new experimental result, as the previous ones, seems to be in disagreement with the classical assumptions. This conclusion may however be considered with some carefulness, since the mesh dimension was not determined directly . As a matter of fact, the choice of a larger value for this quantity would lead to a better agreement. This is the reason why we have proceeded with a new set of experiments which provides more precise information.

Study of the intensity scattered by the labelled paths, in the intermediate regime of scattering vector amplitude q

For these experiments, we have prepared a new series of samples, containing three types of labelled paths (of molecular weight, $M \sim 4\ 10^5$, $8\ 10^5$ and $2.8\ 10^6$). For each type, we have considered the four crosslinking densities A,B,C,E. The crosslinking density C is the same as before ($M_{mesh} \sim 45000$ to 75000). Networks of type E appeared to be equivalent to gels of well-defined chemical meshes ranging between M=35000 and M=50000 (more probably $M \sim 35000$). We have foregone estimating the meshes of samples of types A and B, because of the lack of data concerning comparable networks. Again, these gels have been submitted to a deswelling by a factor 10 between their state of preparation and the dry state. We want now to compare the scattering functions S(q) of the labelled paths linked to deswollen gels and that of uncrosslinked chains dispersed in a melt, these latter figuring that of the paths before deswelling. For this purpose, we have adopted the Kratky-Porod representation, which consists in plotting $q^2S(q)$, versus q. It is well known that this representation leads to plateau-shaped curves for gaussian unperturbed chains, in the intermediate regime of q ($1/Rg < q < 1/b$, b statistical unit length). This is actually what can be observed on the example given in Figure 3, (small crosses for the uncrosslinked chains). The height of the plateau is proportional to the mass of the chain per square unit length. Therefore, if the collapse of the labelled paths was homogeneous at any scale of distance, the scattering function would be simply shifted parallel in the upper direction, with respect to that of the unperturbed chains.

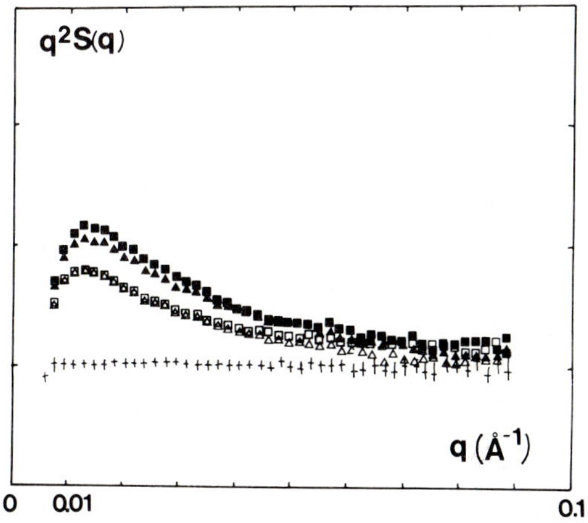

Figure 3 : Kratky-Porod representation for the scattering functions of deuterated chains with molecular weight $M=2.8\ 10^6$

\+ uncrosslinked chains, dispersed in a melt matrix

The hump-shaped curves correspond to labelled paths of same molecular weight, linked to deswollen networks of different crosslinking densities. The crosslinking density can be characterized by an estimated molecular weight of the mesh M(mesh).

■ 35000<M(mesh)<50000 (more probably $M \sim 35000$)

▲ 45000<M(mesh)<75000 (more probably $M \sim 45000$)

□, △ M(mesh)>75000

Instead of that, hump-shaped curves are obtained, which indicate the existence of a loss of affineness, more especially important as the scale of observation is local (i.e. q is large) (see some examples in Figure 3). As mentioned before, this is a characteristic feature of rubber elasticity, qualitatively predicted by all the models. More quantitatively, we have to compare the obtained curves to those can be calculated from the "junction affine and "phantom network" assumptions. One example of this confrontation is given in figures 4 and 5 for the curves of the longer path ($M \sim 2.8\ 10^6$) linked to a network of type E.

One observes that no agreement can be obtained between the theoretical and experimental curves, even for values of the meshes much larger than estimated. A result of same kind was obtained also for the other types of paths and crosslinking densities. The nature of the loss of affiness process seems therefore to differ significantly from the one assumed by the classical

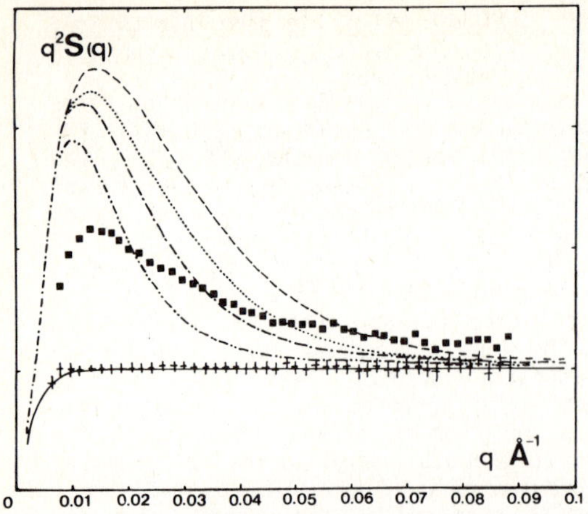

Figure 4 : Kratky-Porod representation for scattering functions of deuterated chains ($M \simeq 2.8\ 10^6$)

■ labelled paths linked to a gel deswollen by a factor 10 (estimated mesh, 35000<M(mesh)<50000)

\+ uncrosslinked chains, dispersed in a melt matrix

— calculated scattering function (Debye function) for unperturbed chains of same radius of gyration. The fit procedure needs an arbitrary constant in order to match the experimental and calculated plateaus.

Other curves : Calculated values of the intensity scattered by labelled paths running through affinelydisplaced junctions, for a macroscopic deswelling equal to 10. The curves differ by the molecular weight of the mesh assumed for the network. The arbitrary constant for the fit was set so that the curves were equal to the Debye function, for a deswelling equal to 1.

M(mesh) :----- 35000 ; 50000
M(mesh) :-·-·- 75000 ; ··-··- 150000

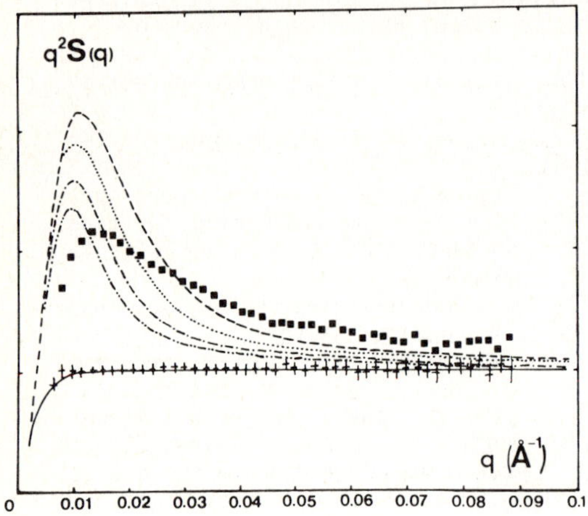

Figure 5 : Please refer to the caption of Fig. 4, except for the calculated hump - shaped curves. They are now obtained within the framework of the "phantom network" model [16], again for different values of the mesh dimension.

··-··- M(mesh)=100000

approaches. It must be stressed that, in the present case, the disagreement cannot be attributed to a wrong estimation of the mesh dimension, since no better agreement could be obtained by acting on this variable. However, as shown in Figure 6, it is possible to give a first order description of the experimental curve, within the framework of the "junction affine" model, by introducing in the calculation an apparent deswelling Q_{app} much smaller than 10 (equal, in the present case, to 4). Comparable results were obtained for the two other types of paths : (M=8 10 , Q_{app}=2.8 ; M=4 10 , Q_{app}=2.2). Therefore, everything happens as if, for a given length of the path, the deformation of the junctions to which the chains are linked, was affine in a deformation smaller than the macroscopic one and depending on the length of the path.

Conclusion

Clearly, this conclusion must not be taken literally. As a matter of fact, it cannot be admitted that the displacement of a given junction could depend on the length of the **labelled** path which is linked to it. However, this gross description indicates that the distance between two junctions is probably affinely deformed only when the shortest path, that connects them along the network, is significantly larger than the mesh. Such a behaviour would qualitatively be in agreement with the existence of the non-affine reorganization of junctions positions that has been proposed to occur in loose networks [17].

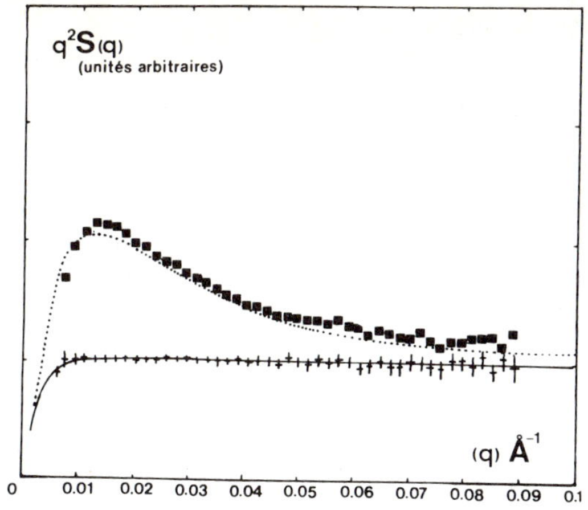

Figure 6 : Please refer to the caption of fig. 4, except for the dotted curve, which is now calculated within the framework of the "junction affine" model, but with a deswelling ratio $Q_{app} = 4$ much smaller than the experimental one (equal to 10). The molecular weight of the mesh is assumed to be equal to 35000.

References

1. J. Frenkel, Acta Physicochimica USSR 9 235 (1938)
2. R. Duplessix, Thèse d'Etat, Université Louis Pasteur Strasbourg (1973)
3. H. Benoît et al., J.Polym.Sci. 14, 2119 (1976)
4. J. Bastide, R. Duplessix, C. Picot, S.J. Candau Macromolecules 17, 83 (1984)
5. W. Kuhn, Kolloid Zeitschrift 76 258 (1936)
6. F.T. Wall, P.J. Flory, J.Chem.Phys. 19 1435 (1951)
7. P.J. Flory, Proc.R.Soc. London Ser. A 351 1666 (1976)
8. H. James, E. guth, J.Chem.Phys. 15 669 (1947)
9. R.T. Deam, S.F. Edwards, Phil.Trans.R.Soc. London, Ser. A 280 1296 (1976)
10. H. Benoît, unpublished calculations (1964)
11. R. Ullman, J.Chem.Phys. 71 436 (1979)
12. D.S. Pearson Macromolecules 10 696 (1977)
13. F. Schosseler, L. Leibler to be published
14. J.P. Munch, S. Candau, G. Hild, J.Herz, J.Phys. (Paris) 38 971 (1977)
15. R. Ullman Macromolecules 15 1395 (1982)
16. M. Warner, S.F. Edwards J.Phys.A Math.Gen. 11 1649 (1978)
17. J. Bastide, C. Picot, S. Candau, J.Macromol.Sci.Phys., B19 13 (1981)

Scaling NMR Properties in Polymeric Gels

J.P. Cohen-Addad
Laboratoire de Spectrométrie Physique associé au CNRS, Université Scientifique et Médicale de Grenoble, B.P. 87, F-38402 St Martin d'Heres Cedex, France

INTRODUCTION

This paper deals with a qualitative illustration of the NMR approach to the characterization of elementary chain properties in a polymeric gel. Investigations start from partly swollen gels ; the corresponding swelling process of elementary chains is analysed from the relaxation of the transverse magnetization of nuclear spins bound to these chain segments.

Despite the very short range of magnetic interactions (< 5 Å), the scale unit of observation in space may be as long as 50 Å, because it is governed by coupling junctions or cross-links : it is a semi-local space-scale.

In the present paper, it is shown how model end-linked networks can be used to give a molecular understanding of the swelling effect. Without giving many details about NMR, it may be worth describing a specific NMR effect induced by the chain structure of polymer molecules.

A SPECIFIC NMR EFFECT

1 BASIC ASSUMPTIONS

Structural units of polymeric gels are supposed to be elementary chains whatever their exact physical definition. For the sake of simplicity, elementary chains will be described as freely jointed segments made of Ne skeletal bonds of equal length a ; calculations of most NMR properties are conveniently carried out considering ideal chain segments.

Also, it is assumed that all elementary chain end points in a gel can fluctuate in space. Accordingly, it is convenient to picture a polymeric gel as a field of forces applied to all elementary chain end points.

2 ORIENTATIONAL COMPLIANCE of SKELETAL BONDS

The mean fluctuation of the end-to-end distance $|Re|$ of an ideal chain experiencing a force $\vec{f}$, is given by the well-known state equation :

$$\langle\vec{Re}\rangle/aNe \;=\; \mathcal{L}(u)\vec{f}/|\vec{f}| \tag{1}$$

$\mathcal{L}$ is the Langevin function and $u = af/kT$. Any skeletal bond obeys a probability distribution function of orientations expressed as :

$$\mathcal{P}(\vec{f},\vec{a},Ne) \;=\; (I_{1/2}(u))^{-1} \sum_{\ell,m} I_{\ell+1/2}(u)\; Y_\ell^m\;(\Omega(\vec{f}))\; Y_\ell^m\;(\Omega(\vec{a})) \tag{2}$$

where $I_{\ell+1/2}(u)$, $\ell = 0,1,2\ldots$ are Bessel functions of the second kind ; $\Omega(\vec{f})$ and $\Omega(\vec{a})$ denote angular variables of $\vec{f}$ and $\vec{a}$, respectively. The function given by (2) may be considered as describing the orientational compliance of skeletal bonds within a chain segment experiencing a force $\vec{f}$ applied to its end points.

3 NMR VARIABLE

The main structural variable, to which NMR is sensitive, is the angle $\theta(\vec{a})$ which a given bond $\vec{a}$ makes with the steady magnetic field direction. This angle may correspond to a (CH) bond, or to the axis of rotation of a methyl group, or to the vector joining two nuclei. This angle is involved in NMR properties through dipole-dipole interactions or quadrupolar interactions. For a two-spin system, bound to a skeletal bond $\vec{a}$ with a fixed orientation, the angle appears through a broad spectrum structure ; the splitting of the corresponding resonance doublet is simply expressed as :

$$\hbar\mathcal{E}(\vec{a}) \propto (3\cos^2[\theta(\vec{a})] - 1)\hbar/b^3 \tag{3}$$

b is the distance between the two nuclei $(\mathcal{E}(\vec{a})/2\pi \simeq 10^4$ Hz).

4 RESIDUAL ENERGY of SPIN INTERACTION

The spectrum structure effect and the orientational compliance of skeletal bonds are now combined with each other. Suppose that a given bond a belongs to a chain segment in the molten state, with a fixed force applied to its end points ; then, such a bond has no longer a fixed orientation in space ; it rapidly fluctuates in time. It is the average value of $\mathcal{E}(\vec{a})$ which is actually observed from NMR :

$$\mathcal{E}e(\vec{f}) = \int \mathcal{E}(\vec{a})\, \mathcal{P}(\vec{f},\vec{a},Ne)\, d\vec{a} \tag{4}$$

$\mathcal{E}e(\vec{f})$ again appears through a spectrum structure ; however, the splitting $2|\mathcal{E}e(f)|$ of the resonance doublet is much smaller than $2|\mathcal{E}(\vec{a})|$ because of the average procedure (4) ; $\mathcal{E}e(\vec{f})$ reads :

$$\mathcal{E}e(\vec{f}) \propto (3\cos^2[\theta(f)] - 1)\; I_{5/2} / I_{1/2} \tag{5}$$

It is a residual energy of spin-spin interactions induced by the force applied to the observed chain segment ; for $|f| = 0$, $|\mathcal{E}e(f)| = 0$.

5 SPACE SCALE TRANSFER of NMR PROPERTIES

The state of stretching of any elementary chain is observed from the splitting $\mathcal{E}e(\vec{f})$ which reflects the orientational compliance of monomeric units ; from (1) and (5), $\mathcal{E}e(\vec{f})$ can be easily expressed as a function of the mean fluctuation of the end-to-end distance $\langle Re\rangle$. For moderately stretched chains :

$$\mathcal{E}e(\vec{f}) \propto (3\cos^2[\theta(f)] - 1)\; u^2(1 - .1\,u^2)/15 \tag{6}$$

The approximate form :

$$\mathcal{E}e(Re) \propto (3\cos^2[\theta(Re)] - 1)\langle Re\rangle^2/a^2Ne^2 \tag{7}$$

describes a transfer of NMR properties from the local space scale defined by a skeletal bond (≃ 3 Å) to the semi-local scale defined by $\langle Re\rangle$, (≃ 30 Å). Throughout the remaining part of this paper, details about monomeric units will be ignored ; these are considered as undistinguishable from one another ; they all are represented by $\mathcal{E}e(\vec{f})$, within a given chain segment.

POLYMERIC GELS : QUALITATIVE NMR ASPECTS

There is no problem to experimentally prove the presence of a residual energy of spin-spin interactions in a polymeric gel (1-3). Consequently, there are non-zero forces $\vec{f}$ applied to all elementary chain end points. More precisely, the force applied to a given elementary chain either is a constant or fluctuates in time, but its mean value is not equal to zero over the time scale of NMR measurements. Furthermore, the probability distribution function of orientations of forces $\vec{f}$ throughout an isotropic gel must be isotropic, too ; consequently, the average value of the residual energy calculated over a whole gel sample is equal to zero :

$$\overline{\mathcal{E}e(f)} = 0 \ ; \ \overline{[\mathcal{E}e(f)]^2} = \Delta e^2 \neq 0 \tag{8}$$

The nuclear magnetic resonance spectrum actually observed is not a doublet but a single line, resulting from the angular average of all resonance doublets associated with individual properties of elementary chains.

The usual order of magnitude of observed line-widths is about 10^2 Hz ; in most experiments, including gel stretching effects, the parameter u should range within the interval $.3 \lesssim u \lesssim 1.$; note that for an ideal chain segment, the approximate relationship :

$$\langle Re \rangle^2 \simeq \langle Re^2 \rangle$$

holds on for u values larger than .3 ; the resulting expression

$$\mathcal{E}e \propto \langle Re^2 \rangle / a^2 Ne^2 \tag{9}$$

may help picturing NMR properties observed on gels and now discussed from experimental results previously reported in several papers [4-6].

SCALING NMR PROPERTIES : CONCENTRATED SOLUTIONS

As long as the spin-system response is governed by a residual energy $\hbar\,\Delta e$, pure symmetry properties of random segmental motions are involved in the relaxation function $Mx(\Delta et)$ of the transverse magnetization. The deviation of monomeric motions from isotropic rotations is closely related to both the average size of the chain segment and its stretching state. Accordingly, whenever the size of any chain structural unit is only governed by scaling properties, the expression of the relaxation function $Mx(\Delta et)$ must be invariant under variations of the corresponding scaling parameter ; it must only exhibit a shift of its time scale Δe^{-1}, illustrated by a superposition property concerning the whole relaxation function $Mx(\Delta et)$.

It has been recently shown that the proton magnetic relaxation function observed in high molecular weight ($Mw \simeq 10^6$) polyisobutylene concentrated solutions well obeys a superposition property induced by adding small amounts of solvent [7] ; the residual energy was found to vary as $\hbar\,\Delta e \propto c/\rho$ in the concentration range going from $c = \rho$ (pure polymer) to $c \simeq 0.5$ g/cm^3. This dependence was identified with the well-known scaling property of viscoelastic submolecules defined by entanglements and necessarily frozen in the time scale of NMR measurements [8].

SCALING NMR PROPERTIES : POLYMERIC GELS

In this section, it is shown how the residual energy of interactions $\mathcal{E}e(\vec{f})$ may be applied to the investigation of elementary chain properties. End-linked polydimethylsiloxane (PDMS) networks are well appropriate to such an illustration ; scaling NMR properties are observed from methyl groups (three-spin systems) ; these tetrafunctional covalent gels were made from chains characterized by the average molecular weight $Mn \simeq 10^4$; three polymer volume fractions of synthesis Vc were analysed : Vc = 0.46, 0.74 and 0.84 [9].

1 A TWO-STEP SWELLING of ELEMENTARY CHAINS

In a first approach, small controlled amounts of a good solvent are progressively added to a given dry gel. Although the volumes of polymer and solvent obey a simple addition law, the variation of the residual energy $\hbar\,\Delta e$ may be shown to exhibit a typical behaviour observed as a function of the swelling ratio $Q = V/Vo$ (Vo and V are the volumes of the dry gel and the partly swollen gel, respectively). Starting from Q = 1., two well-defined swelling ratio ranges are observed.

1.a $1 \leq Q \leq Qd \simeq 4$.

In the first one, hereafter called I, there is a decrease of $\hbar\,\Delta e$, usually observed whenever a good solvent is added to a polymer melt ; Δe varies from about 10^3 to 10^2 rad sec^{-1}.

1.b $Qd \leq Q \leq Q^* \simeq 12$

In the second range, hereafter called II, there is an unusual increase of the residual energy, observed until the maximum swelling ratio Q^* is reached, when the gel is in equilibrium with the liquid good solvent.

1.c The swelling ratio Qd corresponding to the minimum of the residual energy $\hbar\,\Delta e$ was identified with the maximum swelling ratio Qo induced by a theta-solvent ; at Qd, elementary chains obey both a Gaussian statistics and a packing condition, although the gel is partly swollen by a good solvent.

2 AFFINE DEFORMATION OF END-TO-END DISTANCES

2.a In the concentration range II, the residual energy varies as $Q^{2/3}$, in accordance with (6) and (9), provided that both a packing condition and an affine deformation apply to partly swollen elementary chains :

$$\hbar\,\Delta e \propto \langle Re^2\rangle/Ne^2 \propto Q^{2/3} Ne^{-4/3} \qquad (10)$$

2.b According to the Q^*-theorem proposed by DE GENNES [10], the maximum swelling ratio Q^* is expressed as a simple function of the average number of bonds in an elementary chain : $Q^* \propto Ne^{4/5}$. Correspondingly, the residual energy is predicted from (10) to vary as :

$$\Delta e^* \propto 1/Q^*$$

The product $\Delta e^* Q^*$ is experimentally found to be constant [6].

3 TRAPPED TOPOLOGICAL CONSTRAINTS

Following the hypothesis proposed by CANDAU, BASTIDE et al [11], it is considered that the actual mesh size of a gel is governed by screening effects occurring through overlapping chains in the solution prepared before the cross-linking reaction. Consequently, the average number of bonds Ne in an elementary chain must vary in a known way as a function of the concentration of synthesis Vc : $Ne \propto Vc^{-5/4}$.

In the range II, the residual energy $\hbar\,\Delta e$ must vary according to the formula :

$$\hbar\,\Delta e \propto Q^{2/3}(Vc)^{5/3} \qquad (11)$$

This scaling property has been well observed considering three concentrations of synthesis and two swelling agents [6].

4 DESINTERSPERSION of ELEMENTARY CHAINS

In the concentration range I, the residual energy of spin-spin interactions is experimentally shown to obey the scaling law :

$$\hbar\,\Delta e \propto (Ne)^{-1}(Qd/Q)^{2/3} = (Q^*)^{-5/6}\,Q^{-2/3} \qquad (12)$$

In this concentration range, elementary chains are necessarily swollen by one another. When larger and larger amounts of solvent are added to the dry gel, a desinterspersion of elementary chains occurs until they obey a packing condition, corresponding to the swelling ratio Qd. It is postulated that in the range I, elementary chains obey a Gaussian statistics ; their mean dimension is kept constant. However, there is a screening effect induced by the overlap of these chain segments ; it is associated with submolecules which are supposed to govern the mesh size of the network actually observed from NMR. Any submolecule is supposed to consist of n bonds (n < Ne) ; from (6) and (9) :

$$\Delta e \propto \langle Rn^2 \rangle / n^2 \propto n^{-1} \qquad (13)$$

and :

$$n^{3/2} \propto Ne^{3/2}\, Q(Qd)^{-1} \qquad (14)$$

Formula (14) describes the statistical size of the space domains span by the submolecules in the range I, until all elementary chains are fully separated from one another (Q = Qd).

UNIAXIAL STRETCHING

The uniaxial stretching effect on a polymeric gel can be described within the same framework of interpretation. Under stretching, the distribution of forces applied to elementary chain ends is not isotropic anymore. Consequently, the mean value of the residual energy calculated over the whole gel sample is not equal to zero : $\overline{\mathcal{E}e(f)} = De \neq 0$. Accordingly, a resonance doublet must be observed instead of a single line. NMR measurements were recently performed by observing quadrupolar interactions instead of dipolar ones [12,13] . The splitting De of the doublet was shown to vary as :

$$De \propto (Q^*)^{-1}(\lambda^2 - \lambda^{-1}) \qquad (15)$$

where λ is the stretching ratio.

This formula is given a simple interpretation by only assuming that the distribution of forces is shifted toward larger values of fx (direction of stretching) and toward smaller values of fy and fz (in the plane perpendicular to the stretching direction). The shift-factors are λ and $(\lambda)^{-1/2}$, respectively. Also, considering that (6) or (9) still apply to gels under stretching, the mean value of the residual energy is now expressed as :

$$\hbar \Delta e \propto (2\overline{fz^2} - \overline{fx^2} - \overline{fy^2}) = -A(\lambda^2 - \lambda^{-1})$$

The front factor A is given by (12) for Q = 1 ; it is proportional to $(Q^*)^{-5/6}$ which is nearly equal to the experimentally determined function (15).

CONCLUSION

The present paper illustrates an unusual application of NMR, leading to the investigation of scaling static properties of polymeric gels. This non-conventional approach is based on the characterization of quantum coherence properties of the spin-system instead of the observation of energy exchanges between the spin-system and the neighbouring medium. Pure symmetry properties of random segmental motions are consequently observed ; these are governed by the statistical size of structural chain segments,defined over an appropriate pertinent space scale.

The NMR investigation reveals a threshold of affinity which is identified with the average size of actual elementary chains, taking all screening effects into consideration ; these are involved both in the polymer solution before the cross-linking reaction and in the resulting polymeric gel, observed at a low swelling ratio. NMR can be used to get a deep insight into static elementary chain properties in addition to measurements of the moduli (shear, osmotic pressure, longitudinal modulus, elongation) and to neutron scattering experiments [14,15]. It is not the purpose of the present paper to compare scaling NMR properties with main features predicted from numerous models of polymeric gels [10], [16-18]. However, it may be worth emphasizing that NMR properties were analysed according to model picturing a polymeric gel as a field of forces applied to volumeless effective elementary chain ends. Scaling properties were applied to the forces. Such a model resembles a phantom network ; by contrast, the fluctuations of a junction about its average position are not independent of the strain. The present approach can be easily extended to studies of effects induced by the functionality or dangling chains.

REFERENCES

1 J.P. Cohen-Addad : J. Chem. Phys. 60, 2440, (1974)
2 A.D. English and C.R. Dybowski : Macromolecules 17, 446, (1984)
3 B. Schneider, D. Doskocilova and J. Dybal : Polymer 26, 253, (1984)
4 J.P. Cohen-Addad : J. Physique 43, 1509, (1982)
5 J.P. Cohen-Addad, M. Domard and J. Herz : J. Chem. Phys. 76, 2744, (1982)
6 J.P. Cohen-Addad, M. Domard, G. Lorentz and J. Herz : J. Physique 45, 575, (1984)
7 J.P. Cohen-Addad and A. Guillermo : J. Polym. Sci. 22, 931, (1984)
8 W.W. Graessley : Adv. Polym. Sci. 16, 3, (1974)
9 A. Belkebir-Mrani, G. Beinert, J. Herz and P. Rempp : Eur. Polym. J. 13, 277, (1977)
10 P.G. De Gennes : Scaling Concepts in Polymer Physics (Cornell University, Ithaca 1979)
11 S. Candau, A. Peters and J. Herz : Polymer 22, 1504, (1981)
12 B. Deloche and E.T. Samulski : Macromolecules 14, 575, (1981)
13 A. Dubault, B. Deloche and J. Herz : Polymer, in press
14 S. Candau, J. Bastide and M. Delsanti : Adv. Polym. Sci. 44, 27, (1982)
15 M. Beltzung, J. Herz and C. Picot : Macromolecules 16, 580, (1983)
16 J.E. Mark : Adv. Polym. Sci. 44, 1, (1982)
17 A.J. Staverman : Adv. Polym. Sci. 44, 73, (1982)
18 P.J. Flory : Principles of Polymer Chemistry (Cornell University, Ithaca 1953)

Part VI **Elastic and Dielectric Properties of ill-connected Media**

Scaling and Crossover Considerations in the Mechanical Properties of Amorphous Materials

S. Alexander
The Racah Institute of Physics, The Hebrew University, Jerusalem, Israel

The peculiarities of weakly bound free energy solids such as rubbers, gels and glasses are discussed. The role of the crossover from short range liquid like to molecular, Born von Karman behaviour is discussed. It is shown that this crossover leads to internal stresses on the mechanical net - and therefore to scalar elasticity in the absence of external forces - as for a foam. Using a specific gel model, it is shown that the Born von Karman crossover-length (b) is independent and can be much smaller than the Cauchy length where the system becomes affine. In the intermediate range, the mechanical properties are fractal and the eigenmodes fractons. The role of the breakdown of hyperscaling and the consistency of mean field theory when b is large are analysed in detail for a model.

1. INTRODUCTION

The intuitive distinction between solids and fluids is based on their mechanical properties. Solids have shear rigidity and fluids do not. For crystalline materials this is usually a very clear and unambiguous distinction. For amorphous materials the distinction is more delicate. We believe that such materials should be thought of as weakly cross-linked liquids. This means that there are two basic length scales in the problem. The short length scale is the crossover length from liquid to solid-like behaviour. We shall call it the Born-von Karman length because BVK lattice dynamics only becomes meaningful above this length scale. The second length scale is that at which elastic deformations become affine. We call this the Cauchy length. Continuum elasticity can only be defined on larger length scales. Following recent work[1] we use fractal scaling considerations to describe the intermediate, non-affine, regime. In that context, the Cauchy length shows up as the crossover length between fractons and phonons. The distinction between affine and non-affine mechanical behaviour has been discussed extensively in the context of rubber elasticity but not, we believe, for other systems. Our point of view is, however, somewhat different.

We discuss the emergence of different length scales in a systematic treatment of mechanical properties in section 2.

For tightly bound solids the elastic energy is usually explicitly rotationally invariant. This means that the (Born von Karman) harmonic expansion can be expressed in terms of 2-body central 3-body angular and some four-body force constants, and that elasticity only involves the linear symmetric strains. In amorphous materials scalar elasticity, resulting from internal stresses, is important and often dominant[2]. The physical origin of these terms and the reasons why one expects them to be important are discussed in section 3.

The fact that classical mean field indices do not obey hyperscaling is well known. We discuss the geometrical implications[3] in the context of a simple rubber model in 3-dimensions in section 4. The model is unrealistic for vulcanisation but soluble, and illustrates important aspects of rubber elasticity in a simple way.

Our basic point of view is that fractal scaling concepts can be extremely useful in getting insight into the mechanical properties of disordered materials. As in previous work[1-3] our main purpose is to emphasize the physical features and implications of this approach, which seems to fit many empirical observations. We make no attempt to construct detailed realistic microscopic models, and also disregard complicating aspects, such as viscoelasticity, which are undoubtedly important.

2. LENGTH SCALES IN FREE ENERGY SOLIDS

There is a long tradition in physics for handling the elastic and vibrational properties of solids. One writes down an energy

$$U = U(\vec{r}_i) \tag{1}$$

which depends on the positions $(\vec{r}_i)$ of the particles constituting the solid material and expands it around the equilibrium positions $(\vec{R}_i)$

$$\vec{r}_i = \vec{R}_i + \vec{u}_i \tag{2}$$

We shall call this a Cauchy parametrisation.

For a self-bound system U can only depend on the relative positions (r_{ij}) and from this follow the translational and rotational invariance properties of the expansion coefficients. The technique goes back to Cauchy[4] and is the basis for the theory of elasticity for the Born-von Karman theory of lattice dynamics, and for that matter for the treatment of molecular vibrations. If one wants to generalize, one can replace the energy (U) by a parametrized free energy[5] (F)

$$F = F(r_{ij}) \tag{3}$$

which implies some statistical averaging. It is obvious from the form of equation (3) that the free energy is invariant under rigid translations and rotations. If we expand around some positions $\vec{r}_i$ this must hold to each order in the expansion separately, in $\vec{u}$ when

$$\vec{r}_i \to \vec{r}_i + \vec{\ } \tag{4a}$$

and in $\vec{\omega}$ when

$$\vec{r}_i \to \vec{r}_i^{\,1} = O(\vec{\omega})\vec{r}_i \tag{4b}$$

There is of course a slight difference between 4a and 4b because the rotation matrix $(O(\vec{\omega}))$ and therefore the deviation $(\vec{r}_i^{\,1}-\vec{r}_i)$ is not linear in ω. For mechanical equilibrium positions $(\vec{R}_i)$ one must also have

$$\left(\frac{\partial F}{\partial \vec{r}_i}\right)_{R_i} \equiv 0 \tag{5}$$

The Cauchy parametrisation of the free energy (equation 3) is only meaningful when it is stable under external stresses and thermal exitations. The relative positions (R_{ij}) have to change smoothly and slowly under further statistical averaging. The most common test is the Lindemann Criterion. This can be regarded as the definition of a solid. Obviously one cannot write down a (stable) Cauchy expansion for a liquid, in spite of the fact that it is a dense strongly correlated and self-bound state of matter. As a result, one has to use different techniques (i.e. statistics) in describing the mechanical properties of a liquid.

For crystalline solids and for molecules one can usually assume that a Cauchy parametrisation is possible at the microscopic level so that the indices (i,j) label atomic positions. The effective force constants are then also of microscopic origin. There are, however, many systems which exhibit solid rigidity only on larger length scales. Ordered examples are the Luzzati phases in the binary water-surfactant phase diagrams, the blue phases of Cholesteric liquid crystals and colloidal crystals. In all these cases the microscopic properties at the atomic level are liquid-like, and the Cauchy parametrisation only becomes meaningful at much larger length scales. Typically this happens in the range 10-1000 nm. In such cases the

Cauchy parametrisation of the energy (U) is meaningless and an elastic free energy ($F(r_{ij})$ - eqn. 3) only emerges after extensive statistical averaging over microscopic degrees of freedom. The same holds true for many amorphous solids such as vulcanized rubbers, Gels and at least some glasses. The systems are not rigid at the atomic level and relative positions at this level are not stable. Thus, one typically[6] treats the polymer chain connecting two cross-linking points in a rubber-statistically. The Cauchy expansion is then defined in terms of the positions of the cross-linking points. The distance between such points - directly connected by a chain - defines a length scale which we shall call the Born von Karman length (b). It defines the scale up to which one has to perform "liquid" statistical averaging on the system before one can write down a Cauchy expansion of the free energy. b can be large on an atomic scale.

The Born von Karman length is, however, by no means the only length scale in this problem. One notes that for a typical rubber model b is not even identical to the (much smaller) average separation between crosslinking points (which are not required to be directly linked to each other). Much more important is the length scale relevant to the continuum limit. To define the macroscopic elastic properties we have to replace the lattice dynamics variables u_i by a well behaved field $\vec{u}(\vec{r})$[4,5]. The requirement that $\vec{u}(\vec{r})$ behave smoothly, say under a macroscopic stress, is quite stringent. It certainly does not follow automatically from the fact that we were able to define a Cauchy parametrisation (eqn. 3) at the length scale b. Amorphous materials do obey the equations of elasticity macroscopically, so that one can define strains and elastic constants, and the deformations are affine on sufficiently large length scales. This defines a new length scale. Since the emphasis of Cauchy's work was the derivation of the equations of elasticity, we call this length scale the Cauchy length (c).

To see that c can be much larger than b it is useful to go back to the percolation model of rubbers or gels introduced by de Gennes[7] which we have discussed in detail elsewhere[1]. On scales larger than the correlation length (ξ) or equivalently the Skal Shklovski[8] de Gennes[7] mesh size, one can define elastic moduli and the transformations are affine. Thus the Cauchy length (c) can be identified with ξ. On shorter scales, down to b, one has a random fractal with a scaling form for its rigidity and mass distribution. In particular, the fractal is more rigid on short length scales than on long ones. The relative deformations ($\vec{u}_{ij}$) will therefore depend in a complicated scaling form on $\vec{R}_{ij}$, and in particular one expects to see smaller strains at short distances than at long ones, in response to the same (uniform) macroscopic stress. This seems to be observed[9].

For a single ordered fractal or even for a specific realisation of a random cluster one can, just barely, get away with defining a continuum field ($\vec{u}(r)$) on the fractal. It is, however, evident that this cannot be done, on this length scale ($b < L < c$) for the macroscopic system.

Thus, one has liquid-like behaviour at scales smaller than b. One can define eigenmodes (fractons) and lattice dynamics in the intermediate range($b \ll L < c$). Finally, elastic constants (and phonons) are defined on large scales ($L > c$) where the deformations become affine.

When $b \gg a$ one is, in essence, dealing with some sort of polymeric net in the sense of Cates[10]. The fact that b can be much smaller than the Cauchy length c does not really depend on this "polymeric" nature. As shown by Webman[11] here and in essence also by Webman and Kantor[12] and by Bergman and Kantor[13] one gets similar behaviour (with different indices) also for purely mechanical models for which $b = a$.

3. SCALAR ELASTICITY

The fact that the Born von Karman length(b) can be large on an atomic scale does not change the formal structure of the Cauchy parametrisation. One replaces $U(r_{ij})$ by $F(r_{ij})$ and the identity of the reference points is different, but this obviously has no effect on the form of the Taylor expansion (of F). There is, however, an important effect of these short-range liquid properties on the relative magnitude of the different terms in the expansion.

I have pointed out elsewhere that a Taylor expansion of eqn. 3 (or eqn.1) gives a scalar elasticity term

$$F_s = \frac{1}{2} \sum_{i,j} \frac{\partial F}{\partial R_{ij}} \frac{(\vec{u}_{ij})^2}{R_{ij}} \tag{6}$$

in the harmonic expansion which is of the same order in the $\vec{u}_i$ as the usual harmonic expansion terms

$\left(\frac{\partial^2 F}{\partial r_{ij} \partial r_{\ell k}} \frac{\vec{r}_{ij} \cdot \vec{u}_{ij}}{r_{ij}} \frac{\vec{r}_{\ell k} \cdot \vec{u}_{\ell k}}{r_{\ell k}} \right)$ but has different symmetry properties. Since the expansion is straightforward I do not repeat the full calculation.

Since equation 6 only involves the single pair (i,j) $(\frac{\partial F}{\partial R_{ij}})$ can be interpreted as the force exerted by the two-body central force-spring connecting i and j. It is non-vanishing only when the spring is stretched $(\frac{\partial F}{\partial R} \neq 0)$. In the continuum limit the scalar elasticity term F_s (eqn. 6) becomes

$$F_2 = \sum_{\alpha,\beta,\gamma} S_{\alpha\beta} \partial_\alpha u_\gamma \partial_\beta u_\gamma \tag{7}$$

and thus represents the interaction of internal stresses ($S_{\alpha\beta}$) with the second order term in the local strain[5] ($\partial_\alpha u_\gamma \partial_\beta u_\gamma$). The derivation of eqn. 7 is again straightforward when one substitutes

$$\vec{u}_{ij} \simeq (\vec{\nabla} * \vec{u})_i * \vec{R}_{ij} \tag{8}$$

in eqn. 6. These terms arise naturally. They are also important in some special problems. In considering the vibrations of a stretched string or a stretched membrane, the scalar terms (eqn. 7) are usually the only terms one keeps. F_s is however quite generally neglected in continuum elasticity and in lattice dynamics. This approximation, which is perfectly justified for most solids, also goes back to Cauchy[4].

There are two reasons for this. First, the scalar terms tend to be small for short-range forces. We can compare the scalar coefficient $(\frac{1}{R} \frac{\partial F}{\partial R})$ to the spring constant $(\frac{\partial^2 F}{\partial R^2})$ for the same pair

$$\frac{1}{R} \frac{\partial F}{\partial R} \simeq \frac{\delta}{R} \frac{\partial^2 F}{\partial R^2} \tag{9}$$

where δ is the stretching of the spring(ij) from its equilibrium separation (R_{ij}). Atomic forces are usually short range and therefore almost always $\delta/R \ll 1$. Thus the scalar terms (eqns. 6 and 7) are usually small for strongly bound materials, and can be neglected[14]. It is easy to see that this need not apply to "free energy" solids for which the Born von Karman length is large. There is, for example, no intrinsic reason why a polymer chain of n monomers cannot be stretched to many times its equilibrium end-to-end distance ($\sqrt{n}$). This applies also to other "entropic" free-energy bonds. Thus δ/R can be large and the scalar terms in the Cauchy expansion can be dominant.

In addition, the scalar terms arise from a first derivative of $F(r_{ij})$. For a tightly bound solid the implication is that positive and negative contributions to F_s cancel unless there are external forces[3]. For reasons discussed below, this also does not hold when b is large. For a tightly bound solid one can usually assume that all mechanical effects appear explicitly in the parametrisation. This is no longer necessarily true when the Born von karman length is large. One cannot perform a Cauchy expansion for a liquid, but the pressure (P_L) and compressibility (K_L) are well defined. There can therefore be contributions to the total pressure and compressibility which originate in interactions and degrees of freedom which do not show up explicitly in the Cauchy elastic energy expansion. Consider the large scale behaviour in the affine regime i.e. for length scales large compared to the Cauchy length. Volume changes are described by

$$\Delta = \frac{\Delta V}{V} = \mathrm{div}\, \vec{u} \tag{10}$$

For tightly bound solids one has

$$p = \frac{\partial F}{\partial R_{ij}} \quad \frac{\partial R_{ij}}{d\Delta} = \frac{\partial F}{\partial \Delta} = 0 \tag{11}$$

unless there are some external body or surface forces acting. For the situation we are interested in (large b) there are two contributions to the pressure. The explicit dependence of $F(r_{ij})$on the strains $\partial_\alpha u_\beta$ through the change in the relative positions (r_{ij}) is not the only effect of a change in density. There is also the effect on the microscopic interactions on scales smaller than b which would be present even in a liquid when there is no cross linking. Thus

$$p = (\frac{\partial F}{\partial r_{ij}})\ (\frac{\partial r_{ij}}{\partial \Delta}) + p_L = 0 \tag{12}$$

where p_L is the "liquid" contribution to the pressure which cannot be parametrized properly in terms of the $\bar{r}_i$. In equations 11 and 12 we have indicated the network contribution to the pressure by writing the derivatives in a rather combersome way. It is more convenient to separate the free energy (per unit volume) explicitly into the network part (F_N)and the liquid part (F_L):

$$F = F_N + F_L \quad \text{where} \tag{13}$$

$$F_N = F_N(r_{ij}) \tag{14}$$

as before,but the liquid free energy can be written:

$$F_L = F_L^o + p_L\Delta + \frac{1}{2}K_L\Delta^2 \tag{15}$$

where p_L and K_L are the liquid contributions to the pressure and to the compressibility respectively. One can have these additional contributions to the mechanical properties which are not really sensitive to the positions of the Cauchy reference points (i,j). Defining the network pressure

$$p_N = \frac{\partial F_N}{\partial \Delta} \tag{16}$$

one has for the total pressure (p)

$$p = p_N + p_L \tag{17}$$

and for the compressibility

$$K = K_N + K_L \tag{18}$$

For a Gel, a rubber or a glass there is no reason to assume that p_L and K_L vanish. Where there are no external forces, one must also have

$$\vec{\nabla} p = \vec{\nabla} p_N + \vec{\nabla} p_L = 0 \tag{19}$$

Thus, one can have mechanical forces between the "liquid" and the "network" parts of the mechanical free energy. p_L $\vec{\Delta} p_L$ and K_L need not vanish. Whether they are actually important in a specific system depends on the physical details. The only general restriction is that the liquid stress must be hydrostatic (p_L).

The implication is that the springs of the network can be stretched ($\frac{\partial F_N}{\partial r_{ij}} \neq 0$) so that equation 17 is obeyed. The result is an elastic energy contribution (from eqn. 7)

$$F_S = p_N \cdot \sum_{\alpha,\gamma} (\partial_\alpha u_\gamma)^2 \tag{20}$$

which contributes to the compressibility and to shear rigidity.

We consider some specific examples. Consider first the standard Flory model for a rubber i.e. a crosslinked polymer melt[6]. One parametrizes in terms of the crosslinking points and writes

$$F_N = T \sum r_{ij}^2/N \tag{21}$$

where the summation is over all directly linked pairs (ij). A 3-dimensional mechanical system described by eqn. 21 is manifestly unstable and would collapse[10]. To maintain a finite density, as required physically, it is customary to add a constraint on the density. In our formulation, this is equivalent to adding a liquid free energy (F_L eqn. 15) with a suitable value of p_L. The result is that $p_N(\simeq -p_L)$ is large and all springs in F_L (eqn. 21) are stretched. The scalar elasticity terms (eqns. 6 and 7) are the only terms which ever show up in the theory of rubber elasticity. Standard elastic central force (and angular) terms are not even considered.

A second example is provided by the elasticity of gels. It is well known[15] that crosslinking is equivalent to an attraction, and reduces the (osmotic) pressure. This is equivalent in our formulation to saying that the network exerts a negative pressure ($p_N < 0$) cancelling part of the liquid p_L. Thus, the bonds must be stretched both below and above the percolation threshold. If e.g. one were to cut the infinite cluster into the Skal Shklovski-deGennes[7-8] links, the osmotic pressure would increase. The difference in pressure can be expressed as a net force stretching these links. The reduction in the osmotic pressure (π) is continuous as gelation proceeds, so that this is not sufficient for describing the local stretching of the chains. Since π is not strongly singular at p_c its actual dependence on p only gives corrections to scaling. Thus one can describe the leading scalar contributions to gel elasticity as though the chains were uniformly stretched - neglecting the dependence of $p_N(\pi_N)$ on $p-p_c$. This implies conductivity indices for the shear modulus (but not for the compressibility, which has a direct, non-singular contribution K_L.

A third, somewhat more intuitive, example is a foam. In three dimensions a foam consists of a continuous elastic membrane which partitions space into small cells. In the cells, one has a fluid exerting a pressure on the enclosing membrane. For simplicity, we consider a two-dimensional model. Consider a square lattice with nearest neighbor central force springs. At the equilibrium separation, this model has no mechanical rigidity. We make it into a foam by introducing a (2-dimensional) fluid into the squares (considered as closed cells) at some pressure p_L. All the springs are now stretched, and the foam has both (scalar) shear rigidity and a finite compressibility. One notes that all springs are stretched, and there are stresses exerted on the cubic net (by the fluid). There are of course no external forces acting on the foam as a whole. This is analogous to the way the liquid free energy (F_L) affects the elastic network.

We believe we have shown unambiguously that scalar elasticity contributions have to be considered and can be important when b is large. Just how important they are in a specific situation will depend on the competition with standard, rotationally invariant contributions. The effect of the latter on tenuous nets has been studied extensively recently, in particular by the authors of refs. 11-13 and by Sen[16] and his collaborators. This work is discussed in great detail elsewhere in this volume[11,17,18] where detailed references are also given. The main conclusion, from our point of view, is that tenuous nets tend to be much softer when they are not stretched. This certainly implies that scalar elasticity will tend to dominate the macroscopic behaviour close to criticality i.e. when the Cauchy length is large compared to b. The most obvious signature of this behaviour is that the shear modules should scale with conductivity indices[2,7,9] rather than with the larger indices found recently. The details of the crossover in the non-affine regime ($b < L < c$) and for the elastic constants must of course depend on the physical details.

4. BREAKDOWN OF HYPERSCALING

We have already noted that one has to consider at least two length scales - the Born von Karman length (b) and the Cauchy length in the description of the geometry and mechanical properties. The fact that b can be large on an atomic scale also has other implications. de Gennes[15] noted long ago, for a percolation model of vulcanisation that b plays the role of a bare correlation length (ξ_0) in the critical behaviour. Thus,a large b means that the Ginzburg critical region is very narrow and mean field theory is adequate except very close to the critical point p_c. We note that this applies also to other more detailed and realistic models of gelation,such as those described here by Burchard[19] and Dusek[20]. In such situations the standard procedure is to check for a Ginzburg criterion for the consistency of the mean field treatment. Following ref. 3 I want to discuss a model of gellation from the same point of view. The technique can obviously be generalized. A curious aspect of mean field behaviour is the breakdown of hyperscaling - which has curious implications for the geometry of the crosslinked network. We would like to illustrate the technique and implications on a simple model.

Consider a gel formed by crosslinking a melt of polymer chains of length n. The concentration of crosslinks, per chain, is p(p/n per monomer). One has

$$b = n^{\nu} \simeq n^{1/2} \tag{22}$$

We want to consider the consistency of mean field theory, using a Flory type argument[3]. Consider first a single cross-linked cluster. In mean field one has the Flory Stockmayer value of the fractal dimension D = 4. Thus a cluster of size R has, using eqn. 22:

$$S(R) \propto n(R/b)^4 \propto R^4/n \tag{23}$$

monomers. The internal density of the cluster is

$$C(R) \propto \frac{S(R)}{R^3} \simeq R/n \tag{24}$$

Thus, the internal density of the cluster only becomes large ($\simeq 1$) for very large clusters ($R \simeq n \gg b$). This in spite of the fact that we have a cluster with a fractal dimension (D = 4) larger than a dimension of space (d = 3).

Excluded volume effects show up for much smaller clusters:

$$\frac{S(R)^2}{R^3} \simeq \left(\frac{R^4}{n}\right)^2 \frac{1}{R^3} = \frac{R^5}{n^2} \simeq 1 \tag{25}$$

which becomes important for $R \simeq n^{2/5} (< b)$. However,for a dense system, excluded volume effects are screened and can be discarded. As we have argued elsewhere[3], this still leaves a dimensional effect due to the crowding of branch points - i.e. to the probability of three point contacts in the cluster, one of which is a polyvalent group. The probability for this, in the Flory approximation is

$$\frac{1}{n} \cdot \frac{S(R)^3}{R^6} = \frac{R^6}{n^4} \tag{26}$$

which becomes important for

$$R \simeq R_c \simeq n^{2/3} \gg b \tag{27}$$

Clusters smaller than R_c will be mean field-like.

We now consider the percolation problem explicitly. One has a percolation correlation length

$$\xi \simeq b\left|\frac{p - p_c}{p_c}\right|^{-\nu} \simeq n^{1/2} \left|\frac{(p - p_c)}{p_c}\right|^{-1/2} \tag{28}$$

For the present argument we restrict ourselves to p < 1 i.e. at most one crosslink per chain. We note that this is not a proper vulcanisation model where p, as defined here, can be much larger ($\simeq n$). ξ is, as usual the size of the largest clusters and the mesh size of the infinite cluster. Comparing equations 27 and 28 gives a Ginzburg criterion for crossover from mean field to 3-dimensional behaviour

$$\xi \sim R_c; \quad (\frac{p - p_c}{p_c} \propto n^{-1/3} << 1 \qquad (29)$$

One of the manifestations of the breakdown of hyperscaling in mean field is that there are many ($N(\xi)$ fractal clusters of size ξ in a piece of the infinite cluster of this size. Since $\beta = 1$ one has

$$N(p - p_c) \cdot S(\xi) = (p - p_c) \cdot \xi^3$$

$$N = n^2/\xi^3 \qquad (30)$$

The crossover to 3-d hyperscaling behaiour ($N = 1$) is given by equation 29($\xi \simeq R_c \simeq n^{2/3}$).

The argument is also consistent in the sense that each of these N fractals has a probability of order 1 to have one branch point connecting it to the infinite cluster:

$$\frac{1}{n}S(\xi) \cdot (p - p_c)^2 \simeq (\xi/b)^4 \; (n/\xi^2)^2 \simeq 1 \qquad (31)$$

One notes that $N(p)$ does not go to one for $p = 1$:

$$N(1) \simeq \frac{n^2}{b^3} \simeq n^{1/2}; \qquad \xi(1) \simeq b \qquad (32)$$

as one expects in this sort of model.

Finally one can calculate the conductivity

$$\sum = \frac{1}{n} \cdot N \cdot (B/\xi)^{d-2} \cdot (B/\xi)^z = p - p_c)^3/n^{1/2} \qquad (33)$$

using $d = 3$ and the mean field value $Z = 2$ for the Skal Shklovski[8] de Gennes[7] index. When scalar elasticity is dominant,this also describes the shear elastic modulus[2].

The most attractive feature of this picture is that it exhibits a crossover between affine and non-affine behaviour, because the Cauchy length (ξ) can be much larger than the Born-von Karman length (b). For scales large compared to ξ one has proper elasticity and deformations are affine. Between b and ξ one has non-affine elasticity and the vibrational modes are fractons[1]. The fractal links have a proper, Born von Karman, harmonic expansion but, because of their scaling structure they are more rigid, and dense, on short length scales and therefore deform in a non-affine way. The fact that $N >> 1$ enhances this effect but is not essential.

The main difference between the model we have described and standard vulcanisation models is in the fact that $C >> b$ so that a large intermediate fractal non-affine regime arises naturally. Our model is unrealistic for rubbers because one is usually in a strongly crosslinked situation ($p >> 1$), for which our model would also give $\xi \simeq b^1 \simeq (n/p)^{1/2}$. We feel our model illustrates the role of the two basic length scales and the type of behaviour one expects when they are different. We do not offer it as a realistic model.

ACKNOWLEDGEMENT

Part of this work was done while the author was a visitor at Exxon Research Labs, Annandale, N.J. Very illuminating discussions with M. Cates and M. Daoud on the elasticity of rubbers are gratefully acknowledged.

REFERENCES

1. S. Alexander, R. Orbach: J. de qhysique (Lett.), 43, L 625, 1982.
   S. Alexander in Percolation Structures and Processes Annals IPS vol. 5, G. Deutscher, R. Zallen, and J. Adler, eds., p. 149 (1983)
   S. Alexander, C. Laermans, R. Orbach and H.M. Rosenberg, Phys. Rev. B28, 4615 (1983)
2. S. Alexander, J. de Physique, 45, 1939 (1984)
3. S. Alexander, G.S. Grest, H. Nakanishi and T.A. Witten, J. Phys. A17, L185 (1984).

4. A. Cauchy, Exercices de mathematiques t.2 (1827), p. 42 t. 3 (1828), p. 213
5. L.D. Landau and E.M. Lifshitz, Theory of Elasticity, vol. 7 of Course in Theoretical Physics (second English Edition) (Pergamon Press, Oxford, N.Y. 1970, chap. 1
6. W.W. Graessley, Macromolecules 8 (1975) 186 ibid 8 (1975) 865 and references there
7. P.G. de Gennes, J. de Physique, Lett. 37 (1976) L-1
8. A.S. Skal and B.I. Shblovskii, Sov. Phys. Semicond 8 1029 (1975)
9. J. Bastide - this conference.
10. M.E. Cates, Phys. Rev. Lett. 53 926 (1984)
11. I. Webman, this conference
12. Y. Kantor and I. Webman, Phys. Rev. Lett. 52 (1984), 1891
13. D. Bergman and Y. Kantor, Phys. Rev. Lett. 53 (1984) 511
14. There are however exceptions - e.g. the one component plasma.
15. P.G. de Gennes, Scaling Concepts in Polymer Physics, Cornell University Press, Ithaca and London (1979), Chap. 5.
16. S. Feng and P.N. Sen, Phys. Rev. Lett. 52 (1984) 216.
17. P.N. Sen, This conference.
18. L. Ben Guigui, This conference and
19. W. Burchard, This conference
20. K. Dusek, This conference

Elastic Properties of Depleted Networks and Continua

P.N. Sen
Schlumberger-Doll Research, Ridgefield, CT 06877, USA
S. Feng and B.I. Halperin
Physics Department, Harvard University, Cambridge, MA 02138, USA
M.F. Thorpe
Physics Department, Michigan State University, East Lansing, MI 48824, USA

Numerical simulations, effective medium theories and scaling arguments are used to examine the elastic properties of depleted networks and continua. The simplest model that embodies rotationally invariant forces is the central force only model. The numerical results for the central force model gives a universality class which is different from the conductivity problem on the same networks. The cental force threhold p_{cen} is much greater than the usual connectivty threshold p_c. Rotationally *non invariant* bond bending forces give the conductivity universality class, but a strong cross-over to the cenral force like behavior is observed near p_{cen}. Two dimensional bond percolation networks involving both central and rotationally *invariant* bond-bending forces were studied by numerical simulations and finite size scaling arguments. A critical exponent f (about 3.2), which is much higher than t (about 1.3), the conductivity exponent, is found. The effective exponent was found to depend on sample size L for small L. The scaling arguments based on the nodes-links-blobs picture can explain the sample size dependence of the effective exponent. Experimental data of Benguigui on elastic sheets with holes punched in them gives an elastic exponent which is in good agreement with simulations. The initial slope of the Young's modulus vs. the fraction of holes was found to be in good ageement with the effective medium approximations (EMA). The EMA for elastic continua give classical exponents but predict that at p_c the ratio of the bulk and shear modulii approach a constant value which is independent of the modulii of the starting medium. However the ratio depends on the shape of the holes and the type of EMA. For continuum percolation when the holes are punched on a regular grid the exponent is the same as that in a network. When the holes are punched randomly such that the neck thickness of the transport media can vary over a wide range, new universality classes were found.

1. Introduction

The purpose of this paper is to summarize our previous work [1-4] which was presented in two talks given by one of the authors at Les Houches. Four topics were covered :

1. Percolation with central forces and Born forces [1].
2. Percolation with rotationally invariant bond bending forces [2]
3. Effective Medium Approximations and Comparison with Experiments [3]
4. Differences between lattice and continuum percolation transport exponents [4]

Ideas of *scalar* percolation have been extremely useful in the study of composite and porous media. In scalar percolation one has scalar voltages at the nodes and scalar currents through links or bonds that connect these nodes. In elastic response one has *vector* displacements of particles and vector forces. Consider a 2D- square lattice. The *rotationally invariant* potential energy of the lattice is taken to be,

$$V = \frac{\alpha-\beta}{2}\sum_{ij}\left[(\vec{U}_i - \vec{U}_j)\cdot\hat{r}_{ij}\right]^2 g_{ij} + \frac{\gamma}{2}\sum_{i}(\delta\Theta_{jik})^2 g_{ij}g_{ik} \tag{1}$$

Here $\vec{U}_i$ and $\vec{U}_j$ are displacements of node i and node j; $g_{ij} = 1$ for the bonds that are occupied, with a probability p and, $g_{ij} = 0$ for the bonds that are empty with a probability $1 - p$; and, $\hat{r}_{ij}$ is the unit vector from node i to node j. The bond-bending forces between two connected nearest neighbour occupied bonds ij and ik is given in terms of the change in angle $\delta\Theta_{jik}$ at node i. This model is due to Kirkwood [5,6], and includes force constant for bending of 180^0 bonds. This model is to be contrasted with the *rotationally non invariant* Born model, which is used in describing the angular forces [1,7] by adding a term,

$$V_{Born} = \frac{\beta}{2}\sum_{ij}\left[(\vec{U}_i - \vec{U}_j)\right]^2 g_{ij} \tag{2}$$

Next we proceed to describe elastic properties of depleted networks.

2. Percolation on Elastic Networks : New Exponent and Threshold [1]

The simplest model that embodies vector percolation is the central force only model, $\beta = \gamma = 0$ in Eq.s (1) and (2). The bulk modulus K and shear modulus N of the entire network go to zero as the fraction of the bonds present p falls below a critical value. For the case of purely central forces, $\beta = \gamma = 0$, we find [1], as p approaches p_{cen} from above, that,

$$K, N \sim (p - p_{cen})^{f_{cen}} \quad , \quad \beta = 0 = \gamma . \tag{3}$$

We find

$$p_{cen}=0.58, \qquad f=2.4\pm0.4 \qquad \text{2D Triangular lattice} \tag{4}$$

$$p_{cen}=0.42, \qquad f=4.4\pm0.6 \qquad \text{3D fcc lattice.}$$

This universality class is characterized by exponents f_{cen} which are substantially greater than the corresponding conductivity exponents t. The percolation threshold for central force only p_{cen} is also greater than the connectivity threshold p_c. In this model one can have rigid structures connected by stuctural pieces which have no rigidity towards angular distortions. This structre as a whole does not transmit elastic rigidity, although it is connected. This is clearly different from percolation of connectivity via all of the links.

Our estimates of threshold and exponent differ substantially from those obtained by Tremblay et. al. [8]. This may be due difficulties in obtaining reliable results for the central force only models. The threshold obtained in Ref. [8] is close to $p_c = 2/3$, the value given by the effective medium theories [9]. At this point, we cannot rule out the possibility that the bond-bending model and the central force only model belong to two distinct universality classes. In the central force only model the rigid backbone is rather different from the backbone of the connectivty problem.

It has been argued [10] that a Born like term is appropriate for describing the elastic properties of gels [7]. When angular forces are included, there are no floppy regions, which are present in the purely central force model and the elastic threshold moves down to p_c, the connectivity threshold. In that case one observs an interesting cross over near p_{cen}. We show in the figure below moduli for $\gamma = 0$, $\beta \neq 0$.

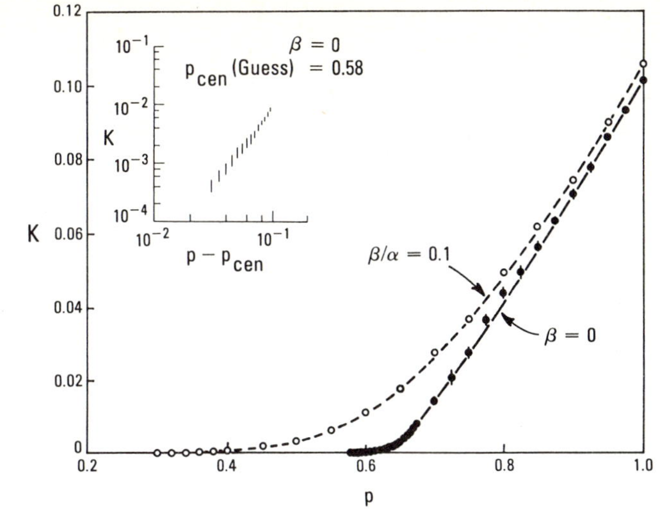

Fig. 1 Bulk modulus vs p for the case when the forces are purely central $\gamma = \beta = 0$ and when the Born force is small $\gamma = 0$, $\beta/\alpha = 0.1$

3. Percolation on 2-D Elastic Networks with Rotationally Invariant Bond Bending Forces ($\beta = 0$) [2]

A more realistic model involves both central and rotationally invariant bond-bending forces. From here on we assume that $\beta = 0$. Two dimensional bond percolation networks were studied by numerical simulations and finite size scaling arguments.

For the non zero angular forces, $\beta = 0$, $\gamma \neq 0$, we now find that

$$K, N \sim (p - p_c)^f \quad , \quad \beta = 0, \gamma \neq 0. \tag{5}$$

A critical exponent f (about 3.2), which is much higher than t (about 1.3), the conductivity exponent, is found. The bond bending forces were also studied independently in a similar context by Bergman [11] by numerical simulation and finite size scaling. Bergman also discusses a cross-over to asymptotic value, as the sample size increases. This model was studied earlier by Kantor and Webman [5] by a scaling analysis. They predicted accurately the values of f, but more importantly, povided a basis of understanding physically why f is greater than t. Below, we use this physical picture of Kantor and Webman [5]. Recently Benguigui [13] reported simultaneous measurements on the conductivity and the elastic exponent of two dimensional metallic sheets with holes punched in them. The critical exponent for the Young's modulus was found to be in good agreement with the exponents obtained by numerical simulations, and from the node link picture of percolation.

As p approaches p_c, the coherence length $\xi \sim (p - p_c)^{-\nu}$ diverges. We employ the finite size scaling calculational scheme following the work of Lobb and Frank [12] on the conductivity problem. In this aproach, we fix p at the exact value of p_c, and vary the sample size L.

When the ratio γ/α is not small, we find an interesting feature that the lines have one slope for smaller values of L, and another slope for the higher values of L. This is explained below.

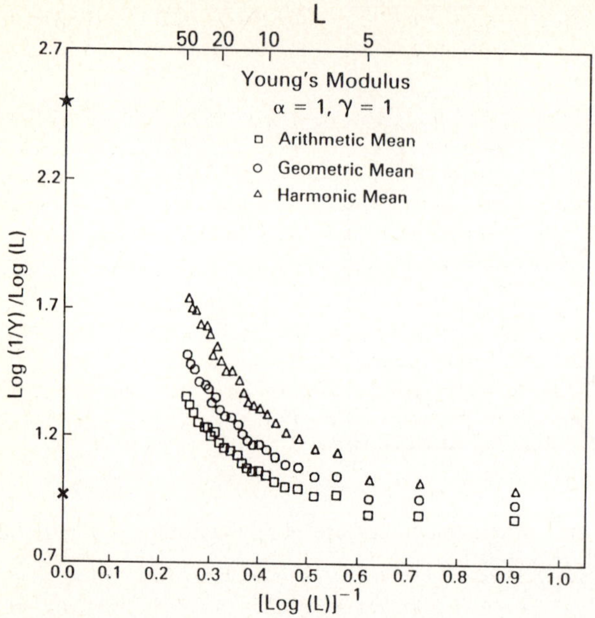

Fig. 2 Variation of log Young's modulus and log conductivity vs log L. $\alpha=\beta$, 400 realisations for each L, $3<L<50$. Three different lines correspond to three different averages, arithmetic mean (squares), geometric mean (circles) and harmonic mean (triangles) For $L< 10$, the central forces provide percolative non rigidity and for $L>10$, the angular foces provide percolative non rigidity, but the sample size is not great enough for the scaling behavior to set in.

Near the percolation threshold, strong regions are connected by tenuous weak one dimensional chain like regions. The strong regions can be regarded as perfectly rigid, so the elastic properties near p_c are dominated by that of the weak regions. According to Ref. 5, the effective force constant is dominated by the bond bending force, which falls off as $\gamma/\xi^{2+1/\nu}$. As p approaches p_c, the coherence length $\xi\sim(p-p_c)^{-\nu}$ diverges, and the elastic constant of the chains, as well as that of the entire system go to zero as $\gamma/\xi^{2+1/\nu}\sim\xi^{-f/\nu}$, giving $f = 2\nu+1$, which is Eq. (16) of Ref. 5. The central force term is unimportant, because it falls off more slowly, as α/ξ.

On the other hand, when α/ξ is smaller than $\gamma/\xi^{2+1/\nu}$ the central force contribution will determine the elastic behavior. For finite size sample, $\xi\sim L$, the α term is the weaker one, if $L<c\ (\gamma/\alpha)^{\nu/1+\nu}$. Here c is a numerical constant. In Fig.2 , with $\alpha=\gamma$, we see one slope for $L < 10$, and another for $L > 10$. This we interpret as a cross over from soft α , i. e. , central force dominated, to soft γ, i.e. , bond bending force dominated behavior. This implies that $c\sim10$. This value of c is consistent with the simulation with $\beta/\alpha=0.1$, as shown in Fig. 2 of Ref. 2.

In other words, the results of the previous section imply that the percolative behavior is expected only when $L>>c\ (\gamma/\alpha)\nu/(1+\nu)$

4. Elastic Moduli of Two Dimensional Composite Continua with Elliptical Inclusions: Effective Medium Approximation[3]

Consider the "swiss- cheese" model where elliptical holes are cut randomly in an isotropic two dimensional elastic continuum. The major and minor semi-axes of the ellipses are a and b. The centers and axes of the ellipses are randomly positioned in the plane and of course the ellipses can overlap one another. The reason for choosing ellipses (rather than say rectangles) is that the problem of a *single* elliptical inclusion can be solved exactly. The effective medium theory then uses this exact result for one inclusion to derive approximate results for many inclusions.

Two distinct self-consistent mean field approximations have been discussed in the literature. In the first, the host and the inclusions are treated symmetrically. We refer to this as SCA-S. In the second (referred to as SCA-A) the host and the inclusions are treated asymmetrically. In the case of elliptical holes in a continuum, the host and the inclusions are manifestly different and the SCA-A is generally to be preferred to the SCA-S.

Two dimensional isotropic elastic systems are characterised by two elastic constants, the bulk modulus K and the shear modulus μ. When holes are introduced, these two constants become K^* and μ^*. Both K^* and μ^* are functions of the volume fraction of holes (1-p) and the aspect ratio (b/a). Effective medium theories do not treat overlap effects between different ellipses. The effective medium results can be found in Refs. 3 and 14. Here we emphaise two aspects only : (1) the flow of the ratio K^*/μ^* to its critical value and (2) comparison with experiments of Benguigui [13].

First consider the flow to the fixed point : Figure 3 illustrates this.

For circles both effective medium theories are identical but they differ from each other for general ellipses. *Both* effective medium theories (SCA-S and SCA-A) lead to the result:

$$p_c + \sigma_c = 1 \tag{6}$$

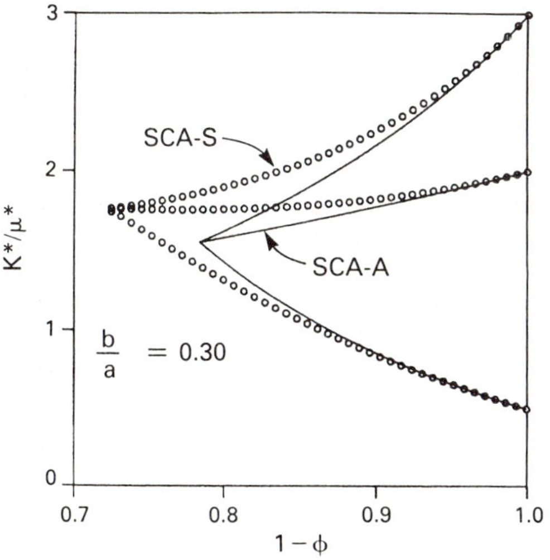

Fig. 3. Flow of K^*/μ^* as the concentration of holes are increased for three different starting values of the ratio. Two different effective media approximations are shown.

where σ_c is the value of the Poisson ratio at percolation

$$\sigma_c = \frac{K_c - \mu_c}{K_c + \mu_c} \tag{7}$$

Here K_c, μ_c are the values of the bulk, shear modulus at the critical point. The value of p_c and hence those of σ_c and K_c/μ_c depend on the geometries but not on the initial value of K/μ. The percolation threshold in SCA-A is given by

$$p_c = [1+ab/(a^2+b^2)]^{-1} \tag{8}$$

The SCA-S gives

$$p_c = 2\left[1+\sqrt{2(a+b)^2/(a^2+b^2)}\right]^{-1} \tag{9}$$

Thus the continuum model has a fixed point within effective medium theory and one expects that exact results will eventually yield universal behavior. It is curious that both approaches lead to (6) even though the p_c (and hence σ_c) are rather different.

Next consider the comparison with experiments [13]. We find that the intial slope of the data for the Young's modulus E* and conductivity versus void fraction ϕ, agree well with the effective medium approximations. The *initial* slope at $\phi=0$ is given *exactly* by the effective medium theory [3] The effective medium result for circular holes is

$$E^* = E\,(1 - 3\,\phi), \tag{10}$$

$$(\sigma^* - 1/3) = (\sigma - 1/3)\,(1 - 3\,\phi). \tag{11}$$

For the Young's modulus the line, which joins the initial data points near $\phi \sim 0$, will intercept the abcissa at $\phi = 1/3$, which agrees well with the data [12]. The effective medium electrical conductivity Σ^* is

$$\Sigma^* = \Sigma(1-2\phi), \tag{12}$$

has an intercept at $\phi=1/2$, for the initial slope with the abscissa, which is also in close agreement with the data. Experiments verifying the flow of Fig. 3 will be interesting.

5. Differences between lattice and continuum percolation transport exponents [4]

Next we consider transport properties near the percolation threshold of the "swiss-cheese" continuum models of the previous section. We find [4] that the exponents governing the behavior of electrical conductivity and elastic constants in such media can be quite different from the corresponding ones in the conventional discrete lattice percolation models. The results of our analysis are summarized in Table below, for the swiss-cheese models in two and three dimensions. The exponents $\bar{t}$ and $\bar{f}$ are defined by the assumption that the macroscopic electrical conductivity Σ and the shear modulus N vanish as the volume fraction ϕ of holes approaches a critical value ϕ_c, according to the power laws $\Sigma \sim (\phi_c - \phi)^{\bar{t}}$, and $N \sim (\phi_c - \phi)^{\bar{f}}$.

The fluid permeability is described in Ref [4] and will not be considered here. While the permeability and conductivity exponents are identical to each other in the standard lattice percolation model, we find that the permeability exponent is dramatically larger than $\bar{t}$ in our continuum model, in both two and three dimensions.

TABLE

Estimates of the differences between the transport percolation exponents in the swiss-cheese continuum model and the corresponding exponents on a discrete lattice.

	Conductivity $(\bar{t}-t)$	**Elasticity** $(\bar{f}-f)$
d=2	0	3/2
d=3	1/2	5/2

The analysis consists of three steps. First we map the contnuum problem on to a network problem. Next we find the strength of bonds and distribution thereof. Finally, we use a scaling analysis to compute the macroscopic conductivity of a network with a bond strength distribution.

The mapping of the swiss-cheese model onto a discrete random network was described by Elam, Kerstein and Rehr [15]. In two-dimensions, a bond is present if the two neighboring holes do not overlap, but the "strength" of the bond i depends crucially on the channel width δ_i . It is important to note that δ_i has a continuous probability distribution $p(\delta)$ which approaches a *finite* limit $p(0)$, for $\delta \rightarrow 0^+$. In higher dimensions, the construction corresponds to a "Voronoi tesselation", and the bonds are the edges of the Voronoi polyhedra.

Next, we estimate the strength of a bond in the two-dimensional example. For the elasticity problem, the bond-bending force constant γ_i, associated with a narrow neck of width δ_i, is defined such that $\frac{1}{2}\gamma_i\theta^2$ is the energy necessary to bend the neck by a small angle θ. For small δ_i it is given, up to a constant of order unity, which we ignore, by

$$\gamma_i \sim Y_0\delta_i^{5/2}/a^{1/2} \tag{13}$$

where Y_0 is the two-dimensional Young's modulus of the constituent material and $\underline{a}$ is the hole radius. This result may be understood if we approximate the neck by a thin rectangle of width δ_i and length $l_i \approx (\delta_i a)^{1/2}$, and use the classical result $\gamma_i \sim Y_0\delta_i^3/(12 l_i)$ for a two-dimensional bent-beam problem. The corresponding result for the 2D electrical conductivity is given by $g_i = \sigma_0\delta_i^{1/2}/a^{1/2}$, which has a much weaker dependence on δ_i. Results for the three-dimensional Swiss-cheese model, is described in Ref. [4] and will not be discussed here.

Next we consider how these bonds are connected in the macroscopic system. In the nodes-links-blobs picture of percolation backbones, the conducting "backbone" of the infinite cluster is imagined to consist of a network of quasi-one-dimensional string segments ("links"), tying together a set of "nodes" whose typical separation is the percolation correlation length $\xi \sim (\phi_c - \phi)^{-\nu}$. Each string is supposed to consist of several sequences of singly-connected bonds, in series with thicker regions, or "blobs", where there are two or more conducting bonds in parallel [16].

For the elastic problem, we define a force constant K for a string such that $\frac{1}{2}Ku^2$ is the energy cost to displace one end of the string by a small distance u, when the other end of the string is clamped in position and orientation. If one only considers the compliance of the singly-connected bonds in the string, one finds

$$\frac{1}{K} = \sum_{i=1}^{L_1} \zeta_i^2/\gamma_i \tag{14}$$

where γ_i is the bending force constant of bond i, and ζ_i, the moment-arm of the i th bond, is a length of order ξ. The sum is restricted to the L_1 singly connected bonds on the string. Typical value of L_1 is proportional to $(\phi_c - \phi)^{-1}$ [16].

If all bonds have the same bending constant, as is the case in the conventional lattice percolation model, we find $K \sim \gamma/L_1\xi^2$. Since the macroscopic elastic constants are proportional to $\xi^{2-d}K$, this implies the relation $f \approx 1+d\nu \equiv f_1$, a result first obtained by Kantor and Webman [5].

Now we must estimate a *typical* value of the string force constant K for our case, where there is a distribution of bond strengths. If a string contains many singly connected bonds, we should be able to replace the sum in Eq (19) by an integral over the probability distribution $p(\delta)$, provided that we properly control the contribution of the weakest bonds. In particular, using Eq (13) we replace Eq (14) by

$$\frac{1}{K} \approx \frac{a^{1/2}}{Y_0}\xi^2 L_1 \int_{\delta_{min}}^{\infty} p(\delta)\,d\delta/\delta^x \tag{15}$$

where δ_{min} is the minimum value of δ for the singly connected bonds on the string, and $x = 5/2$ for $d=2$ ($x = 7/2$ for $d=3$, and Y_0 replaced by 3d Young's modulus). It may be seen that for large values of L_1, the typical value of δ_{min} is equal to δ_0/L_1, where $\delta_0 \equiv 1/p(0)$. We see then that most of the time K is determined by the weakest singly connected bond on the string; i.e., that $K \sim \xi^{-2}L_1^{-x}$. If we use this estimate to determine the shear modulus, via $N \sim K\xi^{2-d}$, we find $\bar{f} \approx d\nu + 5/2 = f_1 + 3/2$, for $d=2$, and $\bar{f} \approx d\nu + 7/2 = f_1 + 5/2$ for $d=3$.

The elastic system studied by Benguigui [12] with circular holes at randomly selected sites of a regular lattice, has no narrow necks, and is expected to have the same exponents as a lattice elastic model. The differences among various continuum models arise [4] from the differences in the probability distribution and geometry of the narrowest channels, and all can be analysed by the methods of Ref. [4]

Acknowledgements: We are grateful for useful discussions with E. Guyon, C. Lobb and L. Schwartz. The work at Harvard was supported in part by the NSF, through the Harvard Materials Research Laboratory and Grant DMR 82-07431. The work at the Michigan State University was supported in part by the ONR and the NSF.

References

1. S. Feng and P. N. Sen, Phys. Rev. Lett. *52*,216 (1984)
2. S. Feng, P. N. Sen, B. I. Halperin and C. J. Lobb, Phys. Rev., *B30* , 5386 (1984)
3. M. F. Thorpe and P. N. Sen, J. Acost. Soc Am. (in press)
4. B. I. Halperin, S. Feng and P. N. Sen (preprint)

5. Y. Kantor and I. Webman, Phys. Rev. Lett. *52*,1891 (1984)

6. J. G. Kirkwood, J. Chem. Phys., *7*, 506 (1939), see also, P. N. Keating, Phys. Rev. *152,* 774 (1966).

7. P. G. deGennes, J. de Physique, *37*, L-1, (1976).

8. M. A. Lemieux, P. Breton, and A. M. S. Tremblay, J. de Physique, *46*, L-1, (1985).

9. S. Feng, M. F. Thorpe, and E. Garboczi Phys. Rev. *B, 31*, 276 (1985)

10. S. Alexander, J. Physique, *45*, 1939 (1984)

11. D. J. Bergman, Phys. Rev., *B31* , 1696, (1985)

12. C. J. Lobb and D. J. Frank, J. Phys C *12*, L827(1979); Phys. Rev.*30*, 4090 (1984).

13. L. Benguigui, Phys. Rev. Lett., *53* , 2028 (1984); P. N. Sen and M. F. Thorpe, Phys. Rev. Lett., *54* , 1463 (1984); L. Benguigui, Phys. Rev. Lett., *54* , 1464 (1984).

14. S. Feng, L. Schwartz, P. N. Sen and M. F. Thorpe (preprint)

15. W. T. Elam, A. R. Kerstein and J. J. Rehr, Phys. Rev. Lett., *52* , 1516 (1984)

16. A. Coniglio, Phys. Rev. Lett., *46* , 250 (1981); R. Pike and H. E. Stanley, J. Phys., *A14* , L169 (1981)

The Elastic Properties of Fractal Structures

Itzhak Webman
Department of Physics and Astronomy, Rutgers University,
Piscataway, NJ 08854, USA

1. Introduction

Small particles may be aggregated to form low-density macroscopic materials, characterized by a fractal geometry over a range of length scales [1-4] between the particle size a, and an upper scale ξ. Examples of such materials are gold and silica colloidal aggregates [2], such as those described in the paper by Schaefer [4], and the highly porous metallic composites formed by sintering a powder of submicron silver particles [3]. The macroscopic elastic properties of such materials are determined by the details of the structure over the fractal regime, which is typically rather tenuous and consists of many thin tortuous chains and lamellae.

In this paper I discuss some of the quite unusual elastic properties of such materials. Section II describes a theory for the linear elastic behavior of the macroscopic elastic moduli near the limit where ξ becomes very large. In a composite made up of fused hard grains as well as very soft grains or voids, this limit occurs near the percolation threshold of the system. The dynamical properties, such as the spectrum of vibrational modes, and the nature of mechanical stability are discussed in section III. The linear elasticity in the presence of external stress, and the nonlinear elastic response are addressed in section IV. The work described in sections II and III was carried out with the collaboration of Y. Kantor and G. Grest.

II. Linear Elasticity

Lattice models for the physical properties of disordered fractal materials can be constructed by generating lattice clusters of sites or bonds which possess an appropriate stochastic geometry, together with a definition of a lattice Hamiltonian relevant to the property studied. For concreteness I concentrate mostly on a bond percolation elastic model, which represents a random composite made up of hard particles and very soft particles or voids. An elastic lattice Hamiltonian for such a system has to obey several criteria:

a) Rigid connectedness: Lattice clusters should have finite rigidity. For a percolation model the lattice should have finite macroscopic moduli above the percolation threshold, which vanish as $p \to p_c$.

b) The tensorial aspects of elasticity of thin elements should be properly reproduced.

c) In the absence of significant external interactions, the elastic Hamiltonian describing a system which retains an intrinsic mechanical stability, has to be rotational-invariant.

The simplest Hamiltonian which obeys these criteria has the following form for a two-dimensional lattice [5]:

$$H = G \sum^{c}_{\substack{i,j,k \\ (j,k \text{ nn of } i)}} \delta\phi^2_{jik} + \frac{Q}{a^2} \sum^{c}_{\substack{i,j \\ nn}} (\vec{u}_i - \vec{u}_j)^2_{\parallel} \qquad (1)$$

where $(\vec{u}_i-\vec{u}_j)_{\parallel}$ is the difference of displacements of site i and the site j in the direction parallel to the bond (i,j), and $\delta\phi_{jik}$ is the change in the angle between the bonds (i,j) and (i,k) connected to site i. The summation $\sum^c$ is over lattice sites which belong to the cluster. G and Q are local elastic constants and a is the lattice constant.

In contrast to Eq. (1), a Hamiltonian which contains only the nearest neighbor central force term does not obey the elastic connectedness criterion, and has a rigidity threshold p_r which is higher than p_c [6]. For cubic lattices $p_r=1$.

The scalar Born Hamiltonian which has been used to represent the elasticity of gels [7], and which leads to an analogy between the elastic moduli and the conductivity, does not possess rotational invariance. It may be an appropriate model for systems, which are mechanically stabilized due to interactions not included in the Hamiltonian of the elastic frame, such as osmotic pressure and excluded volume interactions in gels [8,9].

A basic element which appears in tenuous fractal structures is a thin tortuous chain. The overall elastic response of such a chain can be calculated exactly for a continuum chain as well as for the analogous lattice cluster. The elastic behavior of the chain can be described by a tensorial force constant K which depends on both the chain length N and on its geometrical configuration: For long chains, the change in the elastic energy associated with a change $\delta\vec{R}$ of the end-to-end vector is given by:

$$E = \frac{1}{2}\,\delta\vec{R}\;\hat{K}\;\delta\vec{R} \qquad (2)$$

$$\hat{K} = G\,\frac{\hat{Z}\hat{S}^{-2}\hat{Z}}{N}$$

Here $\hat{S}^2$ is the tensor of gyration of the chain defined as: $\hat{S}^2 = \frac{1}{N}\begin{pmatrix}\int x^2 ds & 0 \\ 0 & \int y^2 ds\end{pmatrix}$, where the integration is along the chain length ($\hat{S}^2$ is of the order of the squared chain size), and $\hat{Z}$ is a 90° rotation operator. The constant G depends on the width of the chain as: $G\sim a^d$. A generalization of these two-dimensional results to higher dimensions is straightforward.

I have omitted a term in K due to the central force part of Eq. (1), since it becomes negligible for large N. This result can be generalized for the case of an inhomogeneous chain, in which the local bending strength G(s) varies along the chain. The constant G in Eq. (2) is now replaced by $1/\int^N G^{-1}(s)ds$.

The most striking feature of this result is the dependence of the small strain elastic behavior on the chain configuration. In contrast, for the scalar Born model $K \sim 1/N$ and the elastic behavior depends on the chain length only. A simple example for this difference is the elasticity of a long thin rod of length ℓ. The tensor K consists of a logitudinal component which scales as ℓ^{-1} and a transverse component which scales as ℓ^{-3}. In the scalar Born model, both components scale as ℓ^{-1}, as does the conductance.

The backbone of a percolating cluster at and above six dimensions contains almost no multiply connected regions up to size ξ. Thus the results for a convoluted chain can be applied to obtain the stiffness K(L) of a region of the backbone of size L. Since the Hausdorff dimensionality of the backbone in d=6 is $D_B=2$, using Eq. 2 with $\hat{S}^{-2} \sim L^{-2}$, and $N(L) \sim L^{D_B}$:

$$K(L) \sim L^{-\zeta_E} \qquad (3)$$

$$\zeta_E = D_B + 2 = 4 \qquad (4)$$

The size scaling can be applied to obtain the exponent τ which describes the behavior of the macroscopic elastic moduli near the percolation threshold:

$$K(p-p_c) = K_o(p-p_c)^\tau \quad (5)$$

$$\tau = [(d-2) + \zeta_E]\nu = 4$$

Note the difference between this mean field value of $\tau=4$ and the value of the corresponding conductivity exponent: $t=3$.

Generally, the elastic constants of a region of size L of the fractal depend on the structure of the backbone in that region, which consists of both singly connected and multiply connected parts. Assuming that the softness of a region is determined by the singly connected bonds and that the multiply-connected parts are completely rigid, one can use the above generalized result for an inhomogeneous chain [5], and obtain the following upper bound for the rigidity of a region of size L of the backbone:

$$K(L) < \frac{1}{N_s(L)L^2} \quad (6)$$

$N_s(L)$ is the number of singly connected bonds in a region of size L, $N_s(L) \sim L^{\frac{1}{\nu}}$ [10,11]. These lead to $K(L) \sim L^{-\zeta_E}$ where $\zeta_E = 2 + \frac{1}{\nu}$. The following lower bound for the exponent τ results:

$$\tau = d\nu + 1 \quad (7)$$

The values of $\tau = 3.6$ in d=2 and $\tau = 3.55$ in d=3. These values are very different from the corresponding values for the conductivity exponent: $t = 1.28$ in d=2 and $t = 2.05$ in d=3.

Since Ref. 5 was published, several numerical calculations of τ were carried out. The values obtained for two-dimensional lattices are $\tau = 3.5 \pm 0.2$ (Ref. 12), $\tau = 3.3 \pm 0.5$ (Ref. 13,14) and $\tau = 3.5 \pm 0.4$ (Ref. 15).

Two recent experiments also obtain values of τ which agree quite well with the theory. Benguigi [16] has studied the elasticity and conductivity of metallic sheets with holes, and obtained for aluminum and copper sheets $\tau = 3.3 \pm 0.5$, and $\tau = 3.5 \pm 0.4$ respectively. More recently, Deptuck et al. [3] have studied the Young modulus of porous beams made by sintering submicron silver particles, near the percolation threshold. They obtain $\tau = 3.8 \pm 0.5$ in good agreement with the theoretical value for three dimensions.

In both experiments the electrical conductivity was also measured, resulting in values of the exponent t in agreement with numerical values, and demonstrating the difference between the two types of critical behavior. It is interesting to speculate why the lower bound for τ based on the assumption that the multiply-connected parts are completely rigid is also a very good approximation, while a similar approximation does not work as well for the conductivity exponent (e.g. a lower bound of 1 vs. a numerical value of 1.28 in d=2). The contrast of the relevant local property between the blobs and the singly connected regions is larger in the elastic case: Consider an inhomogeneous chain with two different local widths A and a. The contrast in the local conductance scales as $(\frac{A}{a})^{d-1}$, while the contrast in the local elastic stiffness scales as $(\frac{A}{a})^{d}$. The relative difference is most pronounced at d=2.

III. Dynamical Properties

The low-frequency vibrational modes of a fractal object are determined by the structure of both the backbone and the branches. The backbone determines the elastic constants while the mass contribution is determined by both. Consider a region of size L of fractal structure of Hausdorff dimension D. A dilation of the length by a factor λ gives $K(\lambda L) = \lambda^{-\zeta_E} K(L)$, where ζ_E depends on certain geometrical features of the backbone of the specific fractal structure. The mass in a region of size L scales with L as $M(\lambda L) = \lambda^D M(L)$. From these relations we obtain the scaling property for the vibrational frequencies:

$$\omega(\lambda L) = \lambda^{-(\zeta_E+D)/2} \omega(L) \tag{8}$$

The density of vibrational states is approximately given by $\rho(\omega,L) \sim 1/L^D \Delta\omega$, where $\Delta\omega$ is the spacing between the frequencies of the low vibrational eigenmodes of a structure of size L. The spacing $\Delta\omega$ is of the order of the frequency of the lowest eigenstate, and it scales in the same manner as $\omega(L)$ in Eq. (8). These scaling relations can be combined to yield the exponent $\tilde{d}_E$ which describes the low-frequency behavior of the density of vibrational modes $N(\omega)$ at low frequencies:

$$N(\omega) \sim \omega^{\tilde{d}_E - 1}$$

$$\tilde{d}_E = \frac{2D}{\zeta_E + D} \tag{9}$$

The mean field (d = 6) value of $\tilde{d}_E$ is 1. For d=2 and d=3, $\tilde{d}_E$ = 0.8 and 0.9, respectively. Thus $N(\omega)$ for a percolating cluster at $p=p_c$ is slightly divergent as $\omega \to 0$. In contrast, $N(\omega) \sim \omega^{\tilde{d}-1}$ with $\tilde{d}$ = 4/3 for the scalar Born model [17].

The backbone of branched fractals such as DLA [18] and cluster-cluster aggregates [19] contain no loops. The elastic properties can be thus obtained from Eq. (2), $K(L)\ 1/(L^2 L^{D_B})$. The fractal dimensionality of the backbone D_B is larger than unity and smaller than the fractal dimensionality of the object. These bounds together with the relation $\zeta_E = 2+D_B$ gives

$$\frac{D}{D+1} < \tilde{d}_E < \frac{2D}{3+D} \tag{10}$$

Using the value of D = 1.7 for DLA in d=2 [18] we find $0.62<\tilde{d}_E<0.71$ in d=2. Analogous arguments for the Born model give $1 < \tilde{d} < 2D/(D+1)$.

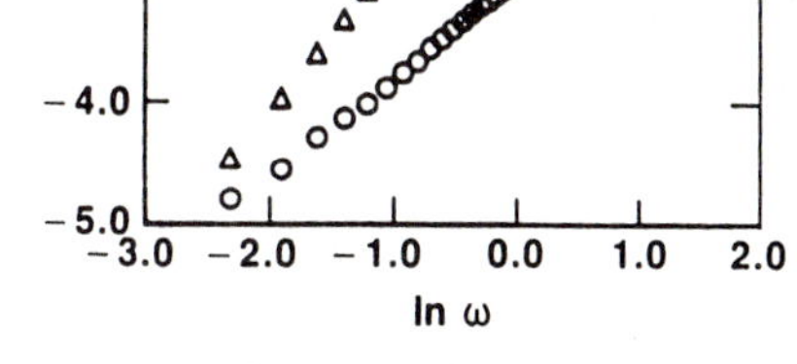

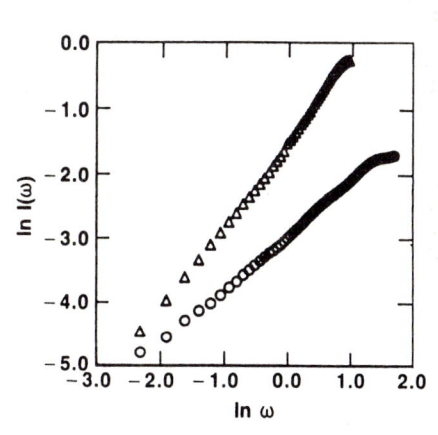

Fig. 1. Integrated density of states for a percolating cluster at $p=p_c$: Circles: Tensorial Elasticity. Triangles: Born Model.

The density of vibrational states has been numerically calculated by Webman and Grest [15] for percolating clusters and for diffusion-limited aggregates. The integrated density of states ($I(\omega) = \int N(\omega)d\omega \sim \omega^{\tilde{d}_E}$) calculated for percolating clusters of ~ 1000 sites on a 64x64 lattice is shown in Fig. 1.
For comparison, the density of states of the scalar Born model is also plotted. The density of states for DLA clusters is qualitatively similar. The numerical results for the elastic spectral dimensionality are $\tilde{d}_E$ = 0.82 ± 0.05 for percolation clusters, and $\tilde{d}_E$ = 0.6 ± 0.05 for DLA. Both values agree well with the prediction of the scaling arguments.

The most interesting aspect of these results is that an elastic spectral dimensionality that is smaller than unity dominates the low-frequency dynamical behavior of fractals analyzed here, though quite different from each other, are qualitatively similar, suggesting that this phenomena is rather general for a tenuous fractal system.

The divergence of the density of states leads to questions about the mechanical stability of fractal objects. Indeed, our results imply that if the size of the object is sufficiently large, so that the frequency of the lowest mode falls below some critical value, the object will not retain its original shape and will become unstable with respect to thermal fluctuations. The argument leads to a criterion for the stability of aggregates which is analogous to the one suggested by Kantor and Witten [20]. It sets an upper limit on the size of nonequilibrium fractal aggregates. On scales larger than this critical size, the aggregate configuration will be determined by relaxation to thermal equilibrium, like in a large branched polymer, resulting in a crossover to a different fractal dimension. As a result, the divergence of the density of states will be cut off below a corresponding critical frequency.

IV. Elasticity at Large Strains

A large fractal object is very susceptible to external perturbations, which might cause pronounced deformation of the original geometry. Examples of such effects are the coupling to thermal fluctuations discussed in section III, or the effect of gravity. Both interactions would undermine the stability of a large object. The change in structure over large length scales implies that the linear elastic behavior of the system will change under fixed external stress. Another, closely related, consequence of the small elastic moduli of fractal networks is the onset of non-linear elastic response at relatively small stress, although the general condition of a strain of order unity for this transition is still valid. Also, in the case of tenuous materials, one can associate the non-linear behavior with the deformation of the system over large length scales, rather than with changes over atomic length scales, as in the case of dense solids. The discussion in this section attempts to make a connection between the tensorial elasticity and the scalar elasticity discussed by Alexander [9].

The present approach is based on the following intuitive picture: As stress is applied to a fractal object, deformation occurs mostly on length scales beyond a certain length which decreases with increasing the stress. Very little deformation takes place on smaller scales. A similar idea was invoked by Pincus [21] and de Gennes [22] to study the elasticity of macromolecules. Consider a tortuous thin chain of size L and length N, subject to a stretching force T. The chain will react like a large random spring. The spring constant for linear response depends on the configuration according to Eq. (2): $K(L) \sim 1/NS_{\perp}^2$, where $S_{\perp}^2$ is the squared radius of gyration perpendicular to the end-to-end vector. Let the stress be sufficiently large so that the overall strain is much larger than unity. The chain may then be viewed as a quasi-linear sequence of blobs defined by a subset of points $\{R_i\}$ on the chain. One may construct the following "blob Hamiltonian" [23]:

$$H = \sum_i \left(\frac{T}{\chi_{\parallel}} (\delta \vec{R}_i^{+})^2 + \frac{G}{g\chi_{\perp}^2} (\delta \vec{R}_i^{\perp})^2 \right) \qquad (11)$$

Here $\chi_{\parallel}$ and $\chi_{\perp}$ are the dimensions of the blob parallel and perpendicular to the direction of the stretch T. $\delta\vec{R}_i$ are the displacements of the points R_i, and g is the mean chemical length of each blob. One can simplify and assume $\chi_{\parallel} = \chi_{\perp} = \chi$.

The size χ is determined by the condition that the local strain of a blob is unity:

$$\chi \sim \left(\frac{T}{G}\right)^{-\frac{1}{D+1}} \tag{12}$$

Here D is the Hausdorff dimensionality of the chain (which remains unchanged up to scales of order χ). The corresponding force constant of a blob is:

$$K_B(T) \sim G^{\frac{1}{D+1}} T^{\frac{D+2}{D+1}} \tag{13}$$

One may now rewrite Eq. (11) in the form

$$H = \frac{1}{2} \sum_i K_B(T) \left(\delta \vec{R}_i^{\parallel 2} + \delta \vec{R}_i^{\perp 2}\right) \tag{14}$$

The effect of the external stress is thus to transform the elastic energy from the form given by Eq. (2) into a renormalized scalar type energy. The crossover to the scalar behavior occurs when the whole chain is a single blob. This condition leads to a crossover stress

$$\bar{T} \sim GL^{-(D+1)} \tag{15}$$

For external stress smaller than $\bar{T}$, the elasticity remains linear. For fixed $T > \bar{T}$, the behavior for small superimposed strains follows the Hamiltonian given by Eq. (14). Concentrating on the response to a large stress at large extensions, one finds that the length of the chain depends on T in a non-linear manner:

$$R \sim \frac{N}{g} \chi \sim \left(\frac{T}{G}\right)^{\frac{D-1}{D+1}} \tag{16}$$

This picture can be applied to a fractal network, such as a percolating cluster above p_c. Given a fixed external stress T, a crossover from non-scalar to scalar elasticity will take place when $\xi = \xi_c$ is of the order of the blob size. The corresponding relation between ξ_c and $\bar{T}$ is:

$$T = G\, \xi_c^{-(D+1)} \tag{17}$$

where D is the Hausdorff dimensionality of the backbone on scales smaller than ξ. The ξ dependence of a macroscopic linear modulus K is given by:

$$K(\xi) \sim \begin{cases} \xi^{2-d} \xi^{\zeta_E} & \xi < \xi_c(T) \\ \left(\frac{T}{G}\right)^{(\zeta_E - \zeta_s)/(1+D)} \xi^{2-d} \xi^{-\zeta_s} & \xi \gg \xi_c(T) \end{cases} \tag{18}$$

where ζ_s describes the size scaling of a scalar elasticity (and is equal to the corresponding exponent for the conductance). For a percolating cluster, at d<6 the relevant Hausdorff dimensionality, is that of the set of the singly connected bonds, so that $D = \frac{1}{\nu}$. In d=6 this value is identical to the Hausdorff dimensionality of the backbone.

Finally, we apply these results to obtain the $p-p_c$ dependence of the linear macroscopic modulus of a percolating system above p_c subject to a fixed stress T. Sufficiently close to p_c, at $p=p_*$ a crossover to scalar elasticity occurs. p_* is given by:

$$p_* - p_c = A \left(\frac{T}{G}\right)^{\frac{1}{1+\nu}} \tag{19}$$

The behavior of $K(p-p_c)$ in the two regimes is:

$$K(p-p_c) \sim \begin{cases} (p-p_c)^{\tau} & p > p_* \\ (\frac{T}{G})^{(\tau-t)/(1+\nu)} (p-p_c)^{t} & p_c < p < p_* \end{cases} \qquad (20)$$

Thus, for a network under fixed external stress, the macroscopic elastic behavior is scalar very close to p_c and crosses over to a tensorial behavior as p increases away from p_c.

Several comments are in order:

This picture of elasticity at large strains is very simplified. At large distortions pieces of the network may collide and entangle with each other, tending to make the system more rigid. In this sense, the above arguments can be expected to hold better for a network which consists mostly of a sparse backbone with few or no dead ends. Additional study is required in order to assess how broad the non-linear regime in various systems is, before the rupture limit of the network is approached. For this, more detailed information on fluctuations in the structure of backbone may be needed.

The term external stress used in this section is meant in the broad sense. It includes any interactions which dilate the original stable configuration of the network [9]. It is clear, however, that in order to crossover to a scalar elastic behavior, this dilation should be quite large on the macroscopic scale. This suggests the possibility that in polymer networks made out of chains of relatively high local rigidity to bending and a large persistence length, a crossover from scalar elasticity to tensorial elasticity may occur at some concentration above the gel point. In this case both exponents, τ and t, could appear in the corresponding elasticity regimes.

Acknowledgement: I would like to thank S. Alexander and D. J. Bergman for valuable discussions.

References

1. Kinetics of Agreggation and Gelation, edited by F. Family and D.P. Landau (North Holland, Amsterdam, 1984).
2. D.A. Weitz and M. Oliveria, Phys. Rev. Lett. 52, 1433 (1984); D.W. Schaefer, J.E. Martin, P. Wiltzius and D.S. Cannell, Phys. Rev. Lett. 52, 2371 (1984).
3. D. Deptuck, J.P. Harrison, and P. Zawadski, Phys. Rev. Lett. 54, 913 (1985).
4. D.W. Schaefer, this conference.
5. Y. Kantor and I. Webman, Phys. Rev. Lett. 52, 1891 (1984); I. Webman and Y. Kantor in: Ref. 1.
6. S. Feng and P.N. Sen, Phys. Rev. Lett. 52, 216 (1984).
7. P.G. de Gennes, J. Phys. (Paris) Lett. 37, L1 (1976).
8. S. Alexander, J. Phys. (Paris) 45, 1939 (1984).
9. S. Alexander, this conference.
10. A. Coniglio, Phys. Rev. Lett. 46, 250 (1981), and this conference.
11. R. Pike and E.H. Stanley, J. Phys. A 14, L169 (1981).
12. D.J. Bergman, Phys. Rev. B 31, 1696 (1985).
13. S. Feng, P.N. Sen, B.I. Halperin and C.J. Lobb, Phys. Rev. B 30, 5386 (1984).
14. P.N. Sen, this conference.
15. I. Webman and G.S. Grest, Phys. Rev. B 31, 1689 (1985).
16. L. Benguigi, Phys. Rev. Lett. 53, 2028 (1984), and this conference.
17. T.A. Witten and L.M. Sander, Phys. Rev. Lett. 47, 1400 (1982).
18. M. Kolb, R. Botet and R. Jullien, Phys. Rev. Lett. 51, 1123 (1983); P. Meakin, ibid. 51, 1119 (1983).
19. S. Alexander and R. Orbach, J. Phys. (Paris) Lett. 43, L625 (1982).

20. Y. Kantor and T.A. Witten, J. Phys. (Paris) Lett. 45, L675 (1984).
21. P. Pincus, Macromolecules 9, 386 (1976).
22. P.G. de Gennes, Scaling Concepts in Polymer Physics, Cornell University Press, 1979.
23. I. Webman (to be published).

Elasticity of Percolative Systems

L. Benguigui
Solid State Institute and Department of Physics, Technion, Israel Institute of Technology, 32000 Haifa, Israel

We determined the elastic behavior of a 2D-system made of metal and voids. Near the percolation threshold, we found that C_{11} and C_{44} go to zero with the same exponent, $T = 3.5 \pm 0.4$, much larger than the conductivity exponent. Near $p \sim 0$, our results are in agreement with the effective medium theory (EMT).

The elasticity of percolative systems (for example a solid with voids) is actually a very active field of research. The purpose of this paper is to report experiments on a 2D system and to compare them with the recent theoretical results. But before recalling them, we want to give the various elasticity equations of a two-dimensionnal isotropic solid.

ELASTICITY EQUATIONS.

It is well known that in an isotropic solid, two parameters are sufficient to characterize its elastic behavior. However, there are different ways to express them. If one uses the crystallographic notation, one writes

$$\begin{aligned} X_1 &= C_{11}x_1 + C_{12}x_2 \\ X_2 &= C_{12}x_1 + C_{11}x_2 \\ X_4 &= C_{44}x_4 \end{aligned} \qquad (1)$$

with the condition $C_{11} - C_{12} = C_{44}$ valid for isotropic solids. $[x_i]$ and $[X_i]$ are respectively the strain and stress tensors. It is also possible to use the Young modulus E and the Poisson ratio σ,

$$\begin{aligned} x_1 &= \frac{1}{E}(X_1 - \sigma X_2) \\ x_2 &= \frac{1}{E}(-\sigma X_1 + X_2) \\ x_4 &= \frac{1+\sigma}{E} X_4 \end{aligned} \qquad (2)$$

with the correspondence

$$E = C_{11} - \frac{C_{12}^2}{C_{11}} \qquad C_{11} = \frac{E}{1-\sigma^2}$$

$$= \frac{C_{11} - C_{14}}{C_{11}} = \frac{C_{12}}{C_{11}} \qquad C_{44} = \frac{E}{1+\sigma} \qquad (3)$$

Finally the bulk modulus K and the shear modulus are related to the above constants by

$$K = \frac{E}{2(1-\sigma)} = \frac{C_{11} + C_{12}}{2}, \qquad C_{11} = K + \mu$$

$$\mu = \frac{E}{2(1+\sigma)} = \frac{C_{44}}{2}, \qquad E = 4\,\frac{K+\mu}{K-\mu} \qquad (4)$$

HISTORY AND THEORY

The first mention of the elastic behavior of a percolation system near the percolation threshold was probably made by de Gennes[1]. He proposed that the elastic constants will go to zero with the difference $|p - p_c|$, with the same exponent as the conductivity exponent t (p is proportion of voids or of missing bonds in a lattice system). This result was confirmed by measurements[2] of the shear modulus of gels.

However, recent studies [3-7] on various lattice models showed that if one considers non-isotropic force constants, one finds $T > t$. For systems including bond-streching and bond-bending forces[4-7], it is found that in 2D $T \sim 3.5$, whereas it is known that $t \sim 1\text{-}1.3$. The two elastic constants go zero with the same exponent $\sim |p-p_c|^T$. Bergman[7] predicted that the ratio C_{11}/μ goes to a universal value around 3.5.

Thorpe and Sen[8] studied a 2D solid with circular holes, using the EMT theory, which is correct when the number of holes is small. They showed that E and σ are decoupled and behave like

$$E = E_1 \frac{(p - p_c^*)}{1 - p_c^*}$$

$$(\sigma - \sigma_c) = (\sigma_1 - \sigma_c) \frac{(p - p_c^*)}{1 - p_c^*} \qquad (5)$$

where the index 1 refers to the solid. p_c^* is the threshold of the EMT theory, and $\sigma_c = 1 - p_c^*$. In the case of circular holes $p_c = 2/3$ if p is the fraction of the solid part. (In our experiments, we shall take ϕ, as the fraction of the voids and in order to compare with the results of Thorpe and Sen, one has to make the correspondence $\phi \rightarrow 1-p$, $\phi_c^* \rightarrow 1 - p_c^*$).

EXPERIMENTS

We performed experiments on metallic sheets of Copper and Aluminium (thickness 0.2 to 0.3 mm) in measuring the Young modulus and the shear modulus. We punched holes on a square lattice(21 x 21 cm) with the hole size slightly larger than the distance between two nearest sites. In fig. 1, we see the positions of the holes punched randomly from a random number generator.

One sees in fig. 1a and fig. 1b that the holes are isolated and this corresponds to the regime for which the EMT is valid. In fig. 1c and 1d, in which the holes are about 20% and 40% respectively of the total surface, they are in clusters or isolated but close to each other. It is the beginning of the "critical region" and it is no longer possible to neglect the interactions between holes.

The experimental details are given in ref.9. Here, we show only the apparatus for the measurement of C_{44} (fig. 2). We used two identical sheets (identical number of holes at the same places) connected together as shown in the figure.

In a first experiment, we measured the extension of the sheet when one applied a force at its ends, i.e. , $C = E = C_{11} - C_{12}^2/C_{11}$. We also measured at the same time the conductivity of the sheets. The results are given in Fig. 3.

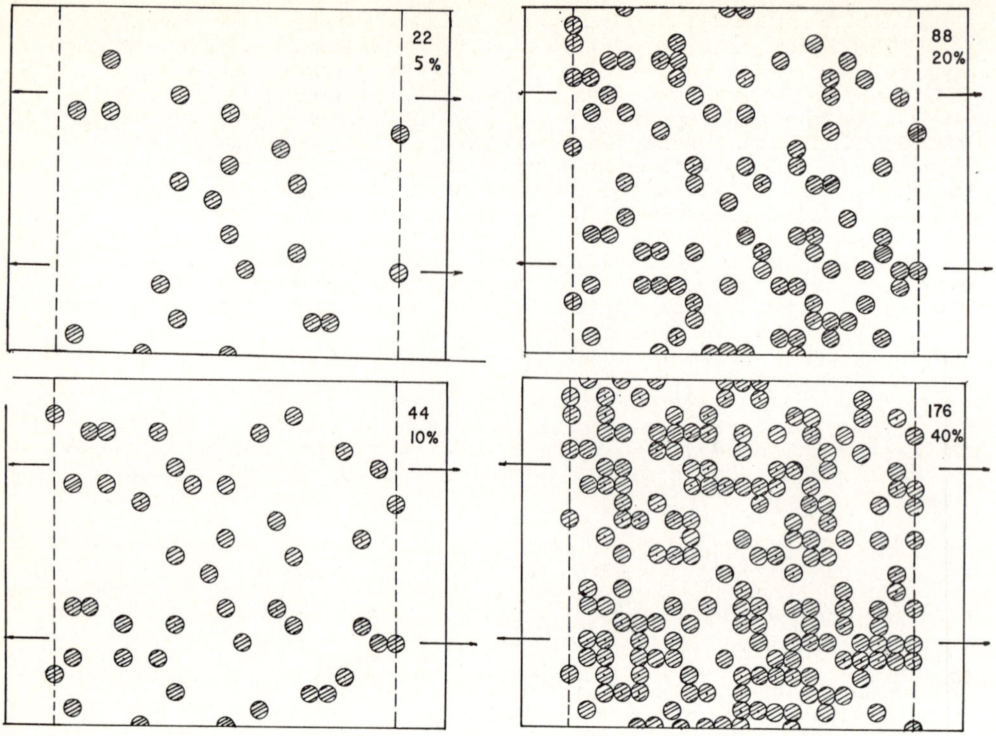

Fig. 1. Holes punched randomly on a metallic sheet. a) 22 holes which represent 5% of the surface b) c) d) 44 holes, 88 holes and 176 holes representing approximately 10%, 20% and 40% of the surface. The arrows indicate the applied force in the extension experiment.

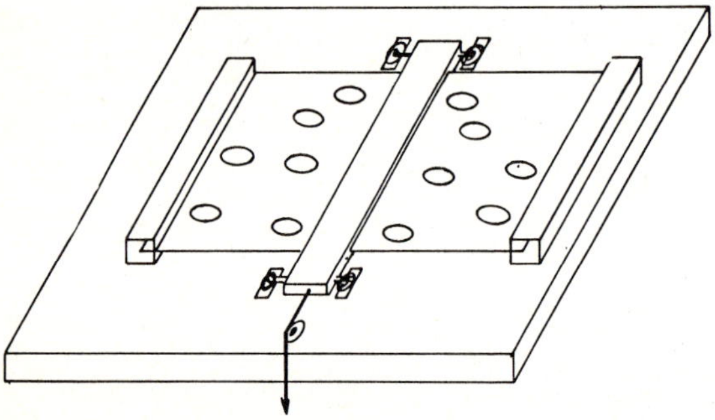

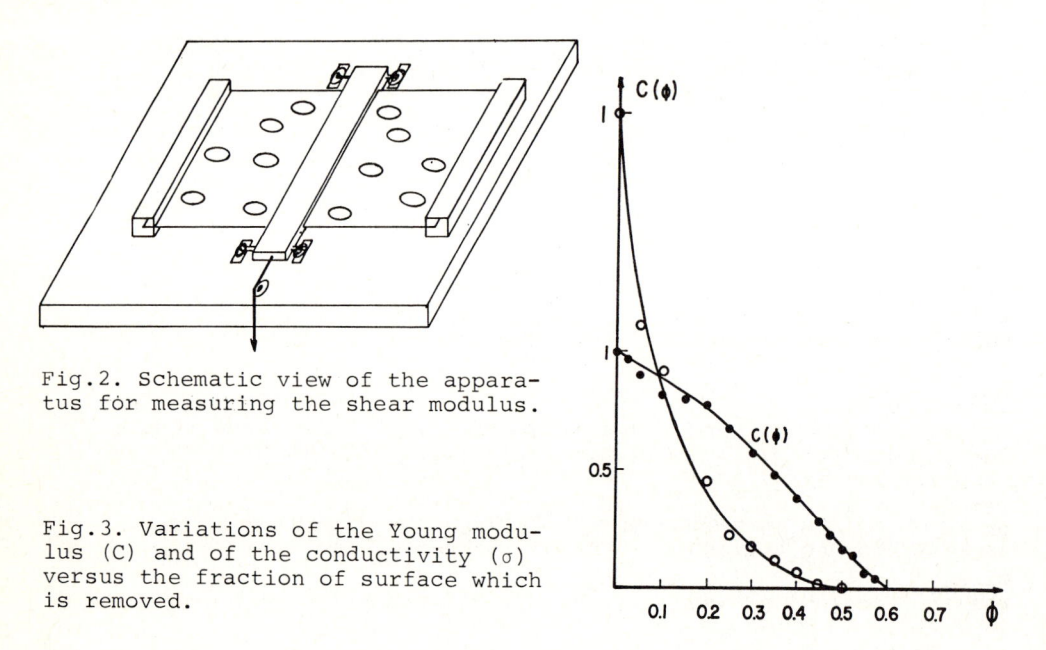

Fig.2. Schematic view of the apparatus for measuring the shear modulus.

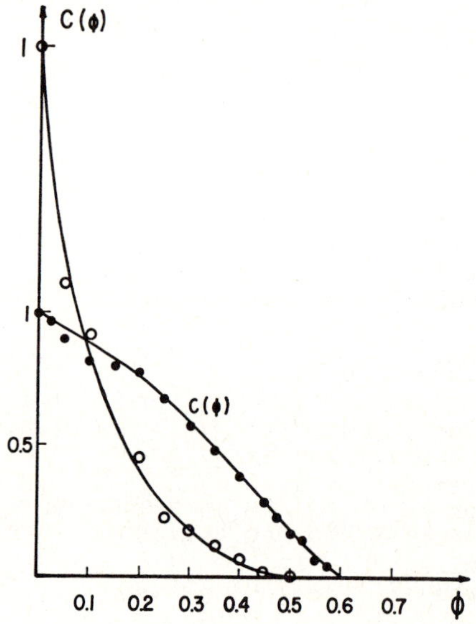

Fig.3. Variations of the Young modulus (C) and of the conductivity (σ) versus the fraction of surface which is removed.

The two quantities (C and σ) behave very differently near $\phi \sim 0$ as well as near $\phi \sim \phi_c$. From a log-log plot we found T = 3.5 ± 0.4 and t ≃ 1 - 1.2. The value of ϕ_c is taken 0.6. very near the value of the threshold of a square lattice. Near $\phi \sim 0$, the variation of C is in agreement with $C \sim (\phi_c^* - \phi)$ ($\phi^* = 1/3$) if one excludes the point $\phi = 0$. We believe that the measurement of the sheet without holes is not "correct", in this sense that the sheet had an original bending. This point was confirmed by measurements on an Al sample[10].

In fig. 4a. we show C_{44} as a function of the removed surface and from a log-log plot (fig. 4b), we get $C_{44} \sim (\phi_c - \phi)^T$, $T \sim 3.6$.

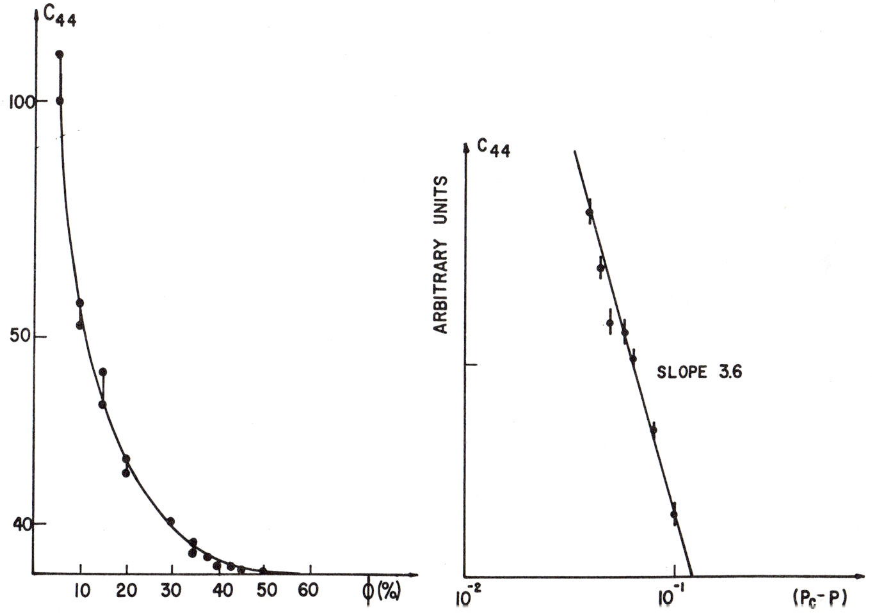

Fig. 4. C_{44} versus ϕ a) in a regular plot. b) in a log-log plot

Near $\phi = 0$, it is very difficult to measure C_{44} since the system is rigid, but it is clear that the original slope is different from the preceeding case, as predicted by Thorpe and Sen.

It is also difficult to calculate precisely the ratio C_{11}/μ, but what we can say is that $C_{11}/\mu = 3 \pm 1$. Clearly, more precise measurements are needed to verify the theoretical value of this ratio and if it is a universal value.

CONCLUSION.

These very simple experiments permitted us to verify some important points concerning the elasticity of percolative systems. First, it is well established than $T > t$, by direct measurements of an elastic quantity and of the electrical conductivity on the same sample. Secondly, the exponent T was found to be equal to 3.5, in very good agreement with the theoretical determination made independently by different researchers. Thirdly, whatever the two parameters, we choose (C_{11}, C_{44} or K, μ), they go to zero with the same exponent. However, the question of their ratio is not experimentally resolved.

As a final remark, we want to mention that solid systems like those described in this paper differ from gels in two ways. First, in gels only the shear modulus vanishes at p_c but in solid system both moduli go to zero. Secondly, as mentioned by de Gennes, solid systems are purely mechanical and their temperature is equal to zero, but a gel is at a finite temperature.

REFERENCES.

1 - P.G. de Gennes, J. de Phys. Lett. 37, 1, (1976).
2 - B. Gauthier-Manuel and E. Guyon, J. de Phys. Lett. 41, 503, (1980).
B. Gauthier-Manuel, this conference.
M. Tokita, R. Niki and K. Hikichi, J. Phys. Soc. Jap. 53, 480 (1984).
3 - S. Feng and P.N. Sen, Phys. Rev. Lett. 52, 216, (1984).
4 - D.J. Bergman and Y. Kantor, Phys. Rev. Lett. 53, 511, (1984)
5 - Y. Kantor and I. Webman, Phys. Rev. Lett. 52, 1891, (1984).
6 - S. Feng, P.N. Sen, B.I. Halperin and C.J. Lobb, Phys. Rev. B 30, 5386, (1984).
7 - D.J. Bergman, Phys. Rev. B 31, 1696 (1985).
8 - M.F. Thorpe and P.N. Sen, Preprint (1984).
9 - L. Benguigui, Phys. Rev. Lett. 53, 2028, (1984).
10 - L. Benguigui, Phys. Rev. Lett. To be published.

Electrical Properties of Percolation Clusters: Exact Results on a Deterministic Fractal

J.P. Clerc, G. Giraud, and J.M. Laugier
D.P.S.D. Université de Provence, F-13397 Marseille Cedex, France
J.M. Luck
SPHT, CEN Saclay, B.P. 2, F-91191 Gif-sur-Yvette Cedex, France

Abstract : We study general electrical properties of percolation clusters through a Deterministic Fractal Lattice, generalizing a model proposed by Kirkpatrick. The model admits an exact renormalization group transformation, and hence most physical quantities are exactly computable. We focus our attention on the frequency-dependence of the impedance and the loss angle, on the transient response to an arbitrary input signal, and to the amplification of resistor noise (Flicker or 1/f noise).

1- INTRODUCTION

We present a model of percolation clusters in which a large variety of electrical properties are exactly analytically tractable. The plan of this report will follow closely that of our original publication [1], but we shall concentrate on the results and their physical interpretation, without repeating the detailed derivations and mathematical digressions.

Our model of a Deterministic Fractal Lattice (DFL) is a generalization of Kirkpatrick's model for static conduction properties of percolation clusters [2] and also described by Mandelbrot [3]. It is built on purpose to admit an exact renormalization group transformation, and hence allows for analytical results. We shall focus successively on the frequency-dependence on the impedance and the loss angle, on the transient response to an arbitrary input signal, and on the critical amplification of resistor (Flicker) noise. The quantitative comparison to actual percolating systems is made through finite-size scaling arguments.

2-THE DETERMINISTIC FRACTAL LATTICE

The original Kirkpatrick's construction is obtained by cutting off a quarter of a square conducting sheet, then a randomly chosen quarter of the 3 remaining squares, and so on (see figure 1).

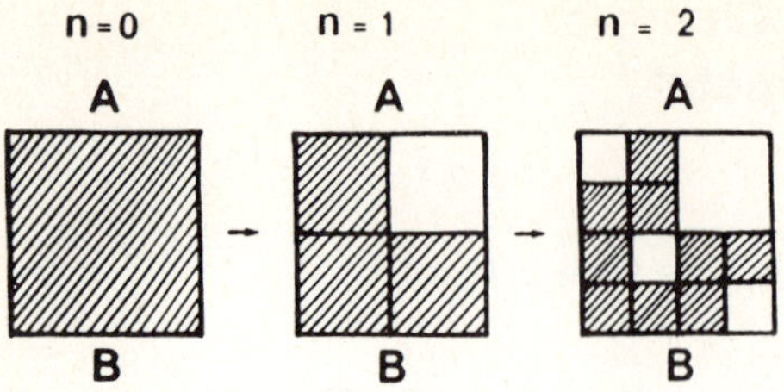

Figure 1

In order to modelize the AC properties of percolation clusters, let us replace the holes of the DFL by perfect capacitors. The equivalent circuit at generation n = 0, 1 and 2 is shown on Figure 2.

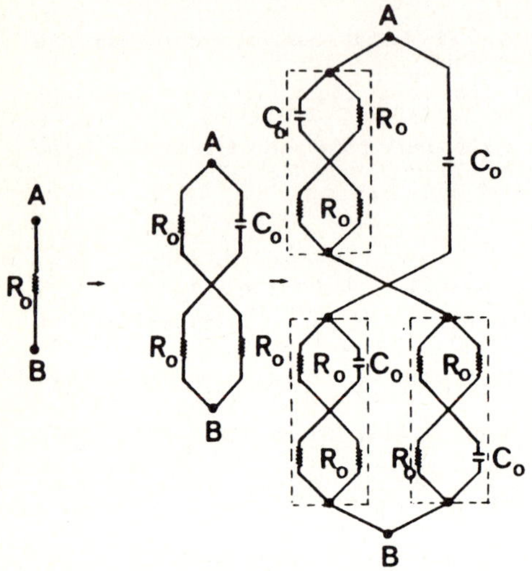

Figure 2

Our DFL is unambiguously defined through Figure 2. Let us generalize this construction, and allow for a conducting fraction f different from Kirkpatrick's original choice f = 3/4 : this is readily done by replacing each bond at each step by a cell which contains 4(1-f) capacitors instead of one. We assume that the upper half of each cell is made of (4f-2) resistors and 4(1-f) capacitors in parallel, and the lower half always contains two resistors in parallel. The conducting fraction f will be judiciously chosen in the following.

Our construction can be easily extended to dimensions higher than two. We shall discuss mainly the two-dimensional DFL hereafter, and refer the reader to our original publication [1] for more details on the 3D model.

3-FREQUENCY-DEPENDENCE OF THE IMPEDANCE

3-1 Renormalization group transformation

Let R_o, C_o denote the resistance of a conducting bond, and the capacitance of an insulating one. The impedance $Z_n(\omega)$ of the DFL between A and B then satisfies :

$$Z_n(\omega) = \frac{1}{2} Z_{n-1}(\omega) + \left[\frac{4(1-f)}{\zeta(\omega)} + \frac{4f-2}{Z_{n-1}(\omega)} \right]^{-1} \qquad (1)$$

where $\zeta(\omega) = (iC_o\omega)^{-1}$ is the impedance of an insulating bond. Equation (1) can be iterated down to $n = 0$ ($Z_o=R_o$) and yields easily :

$$Z_n(\omega) = (iC_o\omega)^{-1}\, T^n\,(i\omega / \omega_o) \qquad (2)$$

where $\omega_o = (R_oC_o)^{-1}$ is our microscopic frequency scale, and where T^n is the n-th iterate ($T^n= ToT.o..oT$, n times) of the following rational transformation :

$$T(x) = x\,\frac{(1-f)x + f}{2(1-f)x + 2f-1} \qquad (3)$$

The study of the impedance $Z_n(\omega)$ and more generally of every physical property of the DFL (as we shall see after), is therefore closely related to that of the rational function T, which acts as a renormalization transformation on the impedance ratio $x = R_o/\zeta(\omega)= i\omega/\omega_o$.

We shall not present here the mathematical properties of iterations of T but rather discuss their physical consequences, with a minimal appeal to the theory of rational transformations. The physical values of the variable x are purely imaginary ($x = i\omega/\omega_o$). It can be shown that $T^n(x)$ goes to 1 as n gets large for such values of x. The point 1 is the only stable fixed point of T, corresponding to a pure (capacitive) medium. The points $x = 0$ and $x = \infty$ are unstable fixed points of T, which are responsible for the existence of critical behaviour of the DFL as $\omega \to 0$ and $\omega \to \infty$

3.2 DC response

The DC (static) impedance of our DFL is clearly given by equation (2) in the $\omega \to 0$ limit where the transformation T can be replaced by the linear relation

$$x \to \mu_o x \text{ with } \mu_o= \frac{dT}{dx}(0) = \frac{f}{2f-1} \qquad (4)$$

We then obtain, as expected, a pure resistance :

$$Z_n(\omega=0) = \mu_o{}^n R_o \qquad (5)$$

In order to interpret the prefactor $\mu_o{}^n$, let us introduce the total length of the DFL between probes A and B :

$$L = 2^n \qquad (6)$$

In terms of L, the resistance of the DFL reads :

$$Z(\omega=0) = R_o L^{t/\nu} \qquad (7)$$

with $t/\nu = \ln\mu_o/\ln 2$ (8)

We have identified the exponent in equation (7) with t/ν through a standard finite-size scaling argument. Equation (8) therefore relates our parameter f to the quantity t/ν which is known with a good accuracy from several numerical works : $t/\nu = 0.977$. If we insert this value in equations (4)-(8), we obtain $f = 0.670$ which is in some sense the optimal value of our parameter. A more natural criterion to choose a particular value of f is to require that the DFL is selfdual, just as 2D regular lattices (implying in particular the equality $s = t$). The DFL shares this remarkable property for $f = 2/3$, where the transformation T satisfies $T(1/x) = 1/T(x)$ for all x. We shall keep f as a free parameter in the following, but use the selfdual value $f = 2/3$ for numerical estimations. This value yields : $s/\nu=t/\nu=1$, which is only 2 or 3 percent overestimated.

3-3 Scaling behaviour at low and high frequency

Let us now consider the behaviour of $Z_n(\omega)$ in the critical regime where $\omega \to 0$ and the size $L = 2^n \to \infty$ simultaneously. This region exhibits a crossover between the limits $Z_n(\omega) \approx \mu_o{}^n R_o$ for $\omega \to \infty$ first, and $Z_n(\omega) \approx (iC_o\omega)^{-1}$ for $n \to \infty$ first. The complete study of this crossover is done in our paper [1], where we prove the existence of a scaling function G_o such that :

$$Z_n(\omega)_{\omega \to 0} \approx R_o L^{t/\nu} G_o\left[\frac{\omega}{\omega_o} L^{t/\nu}\right] \qquad (9)$$

The essential feature of G_o is that it is regular at the origin but singular at infinity :

$$G_o(x)_{x \to \infty} \approx \frac{1}{ix}(1 + K_o x^{-\Delta_o} + \dots) \qquad (10)$$

with $\Delta_o = -\ln f/\ln \mu_o$ (11)

A very similar argument shows that $Z_n(\omega)$ also admits a scaling limit for $\omega \to \infty$ namely :

$$Z_n(\omega) \underset{\omega \to \infty}{\approx} R_o L^{-1} G_\infty \left[\frac{\omega}{\omega_o} L^{-1} \right] \quad (12)$$

where the function G_∞ has a singular component at the origin :

$$G_\infty(x) \underset{x \to 0}{\approx} \frac{1}{ix} (1 + K_\infty x^{\Delta_\infty} + \ldots) \quad (13)$$

with $\Delta_\infty = -\ln f/\ln 2$ (14)

The exponent (-1) of the size L in equation (12) can be identified with the exponent ratio s/ν, since the $\omega \to \infty$ limit corresponds to a regime where the capacitors become very good conductors. The DFL has therefore the particularity that t/ν depends continuously on the parameter f [see equations (4)-(8)], while $s/\nu=1$ independently of f. The selfdual value f=2/3 gives consistently $s/\nu=t/\nu=1$.

Another particularity of the DFL is the fact that the exponents of L in the arguments of the functions G_o, G_∞ in equations (9)-(12) are respectively t/ν and $s/\nu=1$, while the value $(s+t)/\nu$ is expected in both cases for actual percolation clusters. This pathology of DFL can be rephrased in the following more physical way: the static dielectric constant ϵ remains finite as the size $L \to \infty$, while it diverges as $p \to p_c$ in real systems. One possible explanation of this difference is that our DFL modelizes a bare backbone, rather than a whole percolating lattice with its collection of finite clusters in interaction with it.

3-4 The loss angle

We use the standard definition $\tan \delta = -\operatorname{Re} Z/\operatorname{Im} Z$. The preceding results give 4 different power laws when ω is varied, namely :

$$\begin{aligned}
\tan \delta &\approx \omega^{-1} && \text{for } \omega \ll \omega_o L^{-t/\nu} \\
\tan \delta &\approx \omega^{-\Delta_o} && \text{for } \omega_o L^{-t/\nu} \ll \omega \ll \omega_o \\
\tan \delta &\approx \omega^{\Delta_\infty} && \text{for } \omega_o \ll \omega \ll \omega_o L \\
\tan \delta &\approx \omega && \text{for } \omega_o L \ll \omega
\end{aligned} \quad (15)$$

If f assumes its selfdual value f = 2/3, then both intermediate slopes are equal : $\Delta_o = \Delta_\infty = 0.585$.

4-RESPONSE TO AN ARBITRARY INPUT SIGNAL

Let us now determine the electrical response of the DFL to an arbitrary potential difference V(t) between its endpoints A and B. We define the response as the voltage $F_n(t)$ across one of the capacitors which appear at generation $n = 1$. Figure 3 illustrates these definitions.

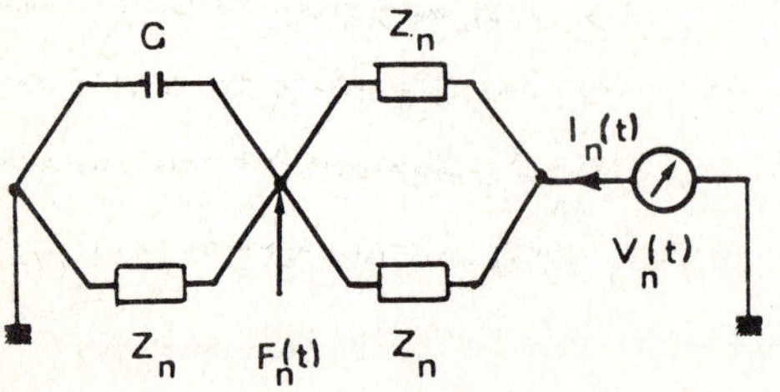

Figure 3

The voltage $F_n(t)$ can be computed explicitly in terms of the renormalization mapping T, just as the impedance $Z_n(\omega)$. The reader is again referred to [1] for a complete derivation. In the case of a step input signal $V(t) = V_o\, \Theta(t)$, the response reads :

$$F_n(t) = \int_{-\infty}^{+\infty} \frac{d\omega}{2\pi i} \frac{e^{i\omega t}}{\omega - i0} \frac{V_o}{2\left[(1-f)T^n\left[\frac{i\omega}{\omega_o}\right] + f\right]} \qquad (16)$$

This integral is readily evaluated by closing the contour in the lower half plane for $t < 0$: $F_n(t) = 0$ as expected. For $t > 0$, the residues of the poles in the upper half plane yield :

$$F_n(t) = \frac{V_o}{2f} + \frac{V_o}{2(1-f)} \sum_{1 \leq a \leq 2^n} \frac{e^{x_a \omega_o t}}{x_a \cdot \frac{dT^n}{dx}(x_a)} \qquad (17)$$

where, apart from the constant part, the signal is a sum of 2^n decreasing exponentials. The decay rates x_a are the 2^n preimages of the point $y=-f/(1-f)$ by T^n, i.e; the 2^n solutions of the equation $T^n(x)=y$. Without any pretence at mathematical rigour, we mention that these 2^n points form, as $n \to \infty$, a Cantor set of negative real numbers, stretching from 0 to $-\infty$. This set of points, termed the Julia set associated to the transformation T, plays a central role in the mathematical theory of iterations. In our model, it has the physical interpretation of being the collection of intrinsic relaxation times in the transient response to all input signals.

5-AMPLIFICATION OF RESISTOR NOISE

The observable noise spectrum of the whole DFL is also exactly computable in terms of the transformation T, and of the microscopic resistor noise. Since the thermal noise is always simply related to the real part of the impedance through Nyquist theorem, and hence exactly known in our model, we shall only consider the Flicker noise due to resistance fluctuations. Assume each resistor has a small time-dependent random component $\Delta_o(t)$

$$R(t) = R_o[1 + \Delta_o(t)] \qquad (18)$$

characterized by a common spectral density

$$S_o(\omega) = \int dt\, e^{i\omega t} \langle \Delta_o(t)\, \Delta_o(0) \rangle \qquad (19)$$

The noise spectrum of the DFL at generation n = 1 is obtained as follows : starting from equation (18) for each of the 4f resistors, one looks for a renormalized noise $\Delta_1(\omega)$ such that the impedance Z_R of the DFL reads :

$$Z_R(\omega) = Z_1(1 + \Delta_1(\omega)) \qquad (20)$$

where Z_1 denotes the impedance in the absence of noise, according to equation (2). The resulting spectral density, defined in analogy with (19), then reads :

$$S_1(\omega) = g\left[\frac{i\omega}{\omega_o}\right] S_o(\omega) \qquad (21)$$

where the amplification or gain g(x) is the following function :

$$g(x) = \frac{2f-1 + |2(1-f)x + 2f-1|^4}{8\,|(1-f)x + f|^2 \cdot |2(1-f)x + 2f-1|^2} \qquad (22)$$

This evaluation is easily pursued up to an arbitrary generation n, where the noise spectrum of the DFL reads :

$$S_n(\omega) = S_o(\omega)\, g\left[\frac{i\omega}{\omega_o}\right] g\left[T\left[\frac{i\omega}{\omega_o}\right]\right] \cdots g\left[T^{n-1}\left[\frac{i\omega}{\omega_o}\right]\right] \qquad (23)$$

Since we are mostly interested in the frequency-dependence of the noise amplification, let us define a <u>relative</u> amplification by normalizing properly

$S_n(\omega)$:

$$\Gamma_n\left[\frac{\omega}{\omega_o}\right] = \left[\frac{4}{f}\right]^n \frac{S_n(\omega)}{S_o(\omega)} \qquad (24)$$

which has a smooth limit $\Gamma_\infty\,(\omega/\omega_o)$ as n gets large. In particular, we find after some algebra that Γ_∞ obeys the following scaling behaviours at low and high frequency :

$$\Gamma_\infty \underset{\omega \to 0}{\approx} \left[\frac{\omega}{\omega_o}\right]^{-Y_o} \quad \text{with } Y_o = \frac{\ln 4g(0)/f}{\ln \mu_o} \qquad (25)$$

$$\Gamma_\infty \underset{\omega \to \infty}{\approx} \left[\frac{\omega}{\omega_o}\right]^{Y_\infty} \quad \text{with } Y_\infty = \frac{\ln 2/f}{\ln 2} \qquad (26)$$

Y_o and Y_∞ are two new critical exponents.

If we take for f the selfdual value 2/3, then their values read :

$$Y_o = 2.392 \quad ; \quad Y_\infty = 1.585 \qquad (27)$$

The fact that Γ_∞ diverges at $\omega \to \infty$ with the positive exponent Y_∞ is not surprising if we remember that we deal with the amplification of the dimensionless quantity $\Delta(t)$, normalized with respect to the total impedance of the DFL, which falls off as $1/\omega$.

6-CONCLUSION

We have presented in this short report the most important results concerning our DFL model of percolation clusters in two dimensions. The paper [1] contains more detailed derivations, as well as the three-dimensional results. The main advantage of the DFL is the existence of rather simple analytic expressions for all the quantities of physical interest, in terms of one single renormalization group transformation T. The basic difference between the DFL and actual percolating systems is the non-divergence of the static dielectric constant, which seems to be due to the absence of finite clusters. This difference also affects some of the physical quantities we have discussed here, such as the value of the loss angle in the critical regime. Let us finally mention that one of us has examined the very same kind of properties of percolation clusters using the Migdal-Kadanoff renormalization group method [4].

REFERENCES

[1] J.P. CLERC, G. GIRAUD, J.M. LAUGIER, and J.M. LUCK, J.Phys. A (1985, to appear).

[2] S. KIRKPATRICK, Les Houches Summer School " Ill Condensed Matter" eds R. Balian, R. Maynard and G. Toulouse (North-Holland, Amsterdam, 1979).

[3] B.B MANDELBROT, "Fractals : Form, Chance and Dimension" (Freeman, San Francisco, 1977).

[4] J.M. LUCK, J. Phys. A (1985, to appear)

N.B. A more complete bibliography can be found in ref. [1]

Introduction to the Ionic Transport in Semi-Crystalline Complexes

A. Le Mehauté
Laboratoires de Marcoussis, Energetics Department, Route de Nozay,
F-91460 Marcoussis, France

What about ionic-polymers for solid state electrolyte?

The advantages of associating a plastic electrolyte and a solid (organic or inorganic) electrode were first mentioned in earlier studies (1).

Later, PEO and PPO salts complexes were proposed as solid electrolytes [SE] (2). The practical interest of a lithium (Li^+) conductor was, of course, emphasized.

As pointed out by Armand (3) and Berthier (4), the conductivity is achieved via the amorphous phase and the major limitation of the lithium conductivity is related to the crystallization of the polymer salt complex : $[Li^+A^-]_nP(EO) \rightarrow {}_n\,[Li^+,P(EO)],A^-$ that is why such complexes were first used at a temperature close to 80° C.

In order to improve the ambient temperature conductivity, the use of polyether-polyurethane crosslinked networks was proposed by Cheradame (5) and, since 1980, alloyed complexes have been used by the author to make lithium solid state power sources (6).

As far as the ambient temperature conductivity is concerned, the influence of the crystallisation upon the conductivity may enlight the ionic dynamics in heterogeous media.

Results

The device used to determine the ionic conductivity is a symmetric Li/SE/Li cell, which is tested by mean of sinusoïdal signal of current down to the low frequencies (≈ 0,1 milli Hertz). The electrical impedance is then computed and plotted in a Nyquist diagram (fig.1).

On storing, increase of the impedance is displayed. It is related to the growth of a non-conductive crystallized complex,meanwhile the diagram keeps all over the crystallization and down to low frequencies,its major property : the "Cole and Cole" characteristics of the spectra (7).

In fact, the electrical impedance $Z(\omega)$ displays two distinct "Cole and Cole" relaxation modes,each of them being related to a specific carrier.

$R\alpha$ is the bias irreversible term calls "transfer resistance". The local time-constant is related to the ionic mobility (8). According to the TEISI-model

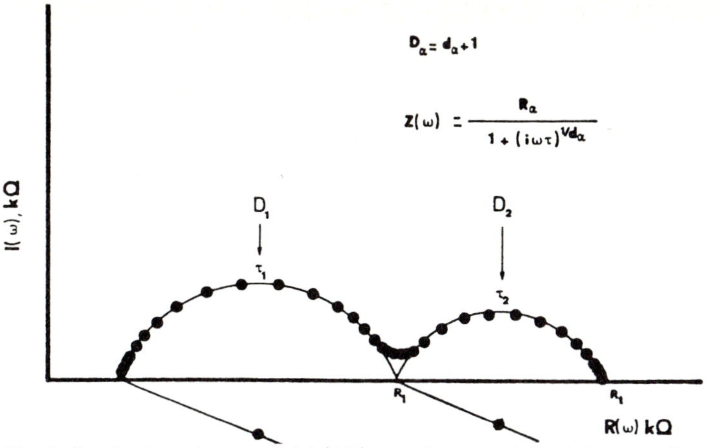

Fig.1:Ionic impedance of Li/SE/Li cell at ambient temperature

given elsewhere (9) ,the "Cole and Cole relaxation may be understood in terms of a δ-ionic first order transfer distributed by a ($d\alpha$+1) fractal metric. $d\alpha$+1 is a fractal dimension related to the amorphous phase sowed with the crystalline stoichiometric complex (Fig.1).

From this model the imaginary term discloses the presence of dynamic ionic-space-charges which involves a fractance behavior (9). This capacitive behavior in fractal media originates in the existence of two carriers characterized by opposite charges and having a quite different mobility. The free energy of the screening is given by $\Delta F = 2\chi e\Delta\phi$ where $\Delta\phi$ is the electrical potential and e the charge of the two carriers. χ is the transport number of the lowest carrier. The non-integer scaling metric is given by the parameter $d\alpha$. The constant in equation (1) is justified by the presence of an electrical driving force able to cross the charges across the screening interface.

On storing, the electrical tests display the following properties:

a) the ageing leads to an increase of the transfer resistance (Fig.2). This observation stresses the decrease of the carriers concentration in the amorphous phase during the crystallisation.

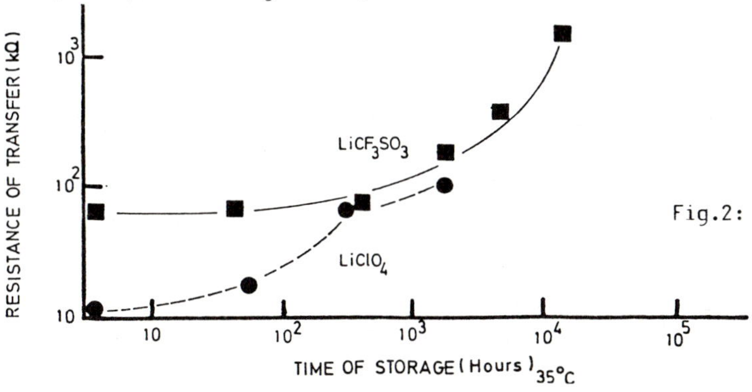

Fig.2: Evolution of the bias resistance on storing

b) by contrast, during the same period the relaxation time stays almost constant. The temperature of activation displays in general an Arrhenius VTF law (3). That result confirms the conservation of the carriers mobility already pointed out by Berthier and al.(4) (fig.3).

c) The most original evolution on storing concerns the fractal dimension. Depending on the nature of the anion, the ageing involves either a drastic drop of the dimension ($LiClO_4$) or a pinning of the initial metric ($LiCF_3SO_3$).

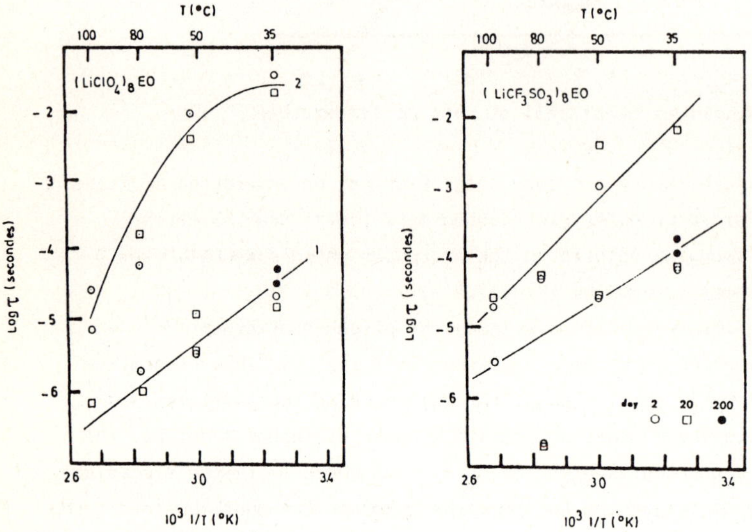

Fig.3:Activation energy on mobility

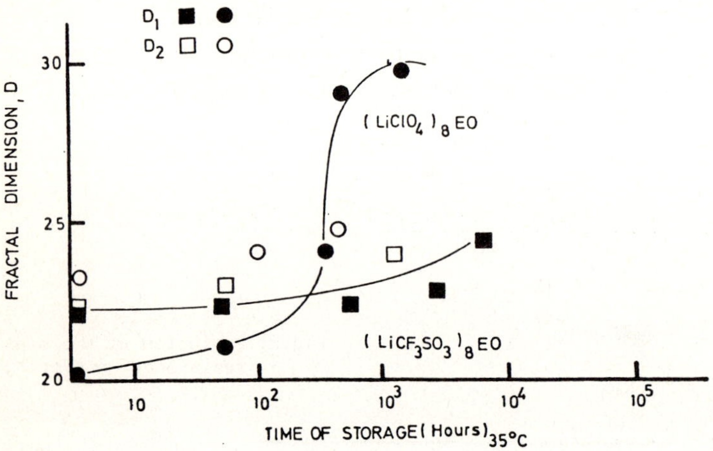

Fig.4: Evolution of the fractal dimension on storing

CONCLUSION

The analysis of the ionic transfer function of the lithium conductive polymers points out the "Cole and Cole" form of the impedance. That form is especially interesting because it stresses with a high accuracy the preservation of the scaling properties down to the bias behavior.

The TEISI thermodynamics fractal model (9) seems adequate to handle the physical meaning of the parameters disclosed. It seems especially efficient to give access to the local activated process which drives the ionic transport across the polymers. Such mode does not require any reference to an "effective" medium but refers to a local behavior. The results obtained point out the stability of the mobility on crystallizing.

The values of the fractal dimensions are given. It is clear that such dimensions are related to the metrics of the amorphous medium sowed by the crystalline phase: meanwhile it is today impossible to precise the mathematical content of these dimensions. According to the initial schematic view of the TEISI model (9) they can be either the capacitive, the Hausdorff or the Kolmogorov's dimension, but they can also be related to the Tricot's packing dimension of the conductive amorphous phase (11).

BIBLIOGRAPHY

(1) ARMAND M.B. in "Fast Ion Transport in Solid",(W.van GOOL Ed.,665 (1973), North Holland,N.Y.)

(2) French Patent, 2 442 513 (1980)

(3) ARMAND M.B., Solid State Ionics, 9/10,745 (1983).

(4) BERTHIER C., GORECKY W., MINIER M., ARMAND M., CHABAGNO M. and RIGAUD P., Solid State Ionics 11 91 (1983) : J. de Physique 45,739 (1984) : GORECKY W., PhD Grenoble 1984.

(5) French Patent 800 71 35 (1980).

(6) European Patent 00 78 505 (1981).

(7) K.S. COLE and R.H. COLE J. Chem.Phys. 9, 341 (1941).

(8) G. FEUILLADE, Electrochim. Acta 14, 127 (1969).

(9) LE MEHAUTE A., CREPY G., Solid State Ionics 9/10 17 (1983):
LE MEHAUTE A., J. of Stat. Phys. 36 (5-6) 665 (1984).

(10) B. MANDELBROT , "The Fractal Geometry of nature" (Freeman, San Francisco 1983).

(11) C. TRICOT, Thesis Geneva 1979 : PhD Orsay 1984.

Williams-Watts or Cole-Cole: The Universal Response of Polymers

A. Le Mehauté and L. Fruchter
Laboratoires de Marcoussis, Route de Nozay, F-91460 Marcoussis, France

Earlier studies [1,2,3] have pointed out the "anomalous" behaviour of some kinetic laws that describe the evolution of various physical or chemical parameters $\Phi(t)$ related to irreversible processes taking place in heterogeneous media. More recent analyses have established the widespread occurrence of such "anomalies" in chemistry [4,5], electrochemistry [6,7,8], viscoelasticity [9], electronics [10], etc.
The introduction of percolation [11], renormalisation group theory [12], fractal [13] and of scaling concepts in general has stressed kinetic laws which require time exponents to account for the dissipative behaviour in heterogeneous media.
In the present note we emphasize the consequences of two different points of view which may be adopted in regard to these "anomalies". In the first one, time is assumed to be a fractal variable [14] and the scaling properties are expressed in terms of this parameter. In the second one, the scaling properties are expressed in the Fourier space, i.e. in terms of frequency.

According to the first point of view, chemical diffusion on fractal media has led to short time-decay laws of the Williams-Watts form [15] for reaction of species with traps on fractal network [16,17,18]. The relaxation function is then:

$$\Phi(t) \propto \exp[-\mathrm{const}(t/\tau)^{\tilde{d}/2}] \qquad (t \to 0)$$

where $\tilde{d}$ is the spectral dimension of Alexander and Orbach [19].
An analogous expression has been obtained at long times [16,20,21] with:

$$\Phi(t) \propto \exp[-\mathrm{const}\ t^{\tilde{d}/\tilde{d}+2}] \qquad (t \to \infty)$$

Shlesinger, in an extension of the CTRW model [22] to a fractal time distribution [14], gives a similar expression:

$$\Phi(t) \propto \exp[-\mathrm{const}\ t^{\alpha}] \qquad (t \to \infty)$$

where α is related to the time distribution.

Scaling properties are therefore directly assigned to time, and the major problem involved is then the physical interpretation of α [17]. Montroll and Bendler [23] have shown that such a type of relaxation may always be written as a sum of exponentials:

$$\Phi(t)=\int_0^\infty \rho(\tau)\ \exp(-t/\tau)\ d\tau$$

For α less than 2, the $\rho(\tau)$ function is related to Levy (stable) distributions of time-constant. This point of view is extensively used, for instance, to describe dielectric relaxation [24] or dynamic elastical properties of polymers [23]. Obviously, when $\alpha=1$, $\rho(\tau)$ is a constant: the logarithmic rate of the irreversible process is then proportional to the time, and the transfer function (TF), defined as the Fourier transform of $\Phi(t)$, follows the Debye-Kelvin model for any kind of first order process. This result links up with the traditional point of view.

One can also apply scaling to the frequency. The analog of the previous decay laws at short times and high frequencies are then of the form:

$$\mathrm{TF}\left[\Phi(t)\right]_{t\to 0} \propto 1\ /\ (i\omega)^{\alpha} \qquad (\alpha < 1)$$

Some other forms were introduced long ago, such as the Cole-Cole one:

$$\mathrm{TF}\left[\Phi(t)\right] \propto 1\ /\ \left[1 + (i\omega\tau)^{\alpha}\right] \qquad (\alpha < 1) \quad [25]$$

where α is a non-integral parameter.

This last form has been shown to be an accurate description of many impedance spectra for heterogeneous media. In the case of electrochemistry, Le Mehauté relates α to non integer metrics of electrodes [6,7].

It may be noticed that the two points of view lead obviously to the same asymptotic form at high frequencies. As a result, in order to distinguish between them, it becomes necessary to analyse the long time $\Phi(t)$ behaviour, or the low-frequency behaviour of the related fourier transform (FT). This can be done most easily using the FT of the Williams-Watts function. Figure 1 provides a Nyquist's representation of the transfer function associated to this type of relaxation [23].

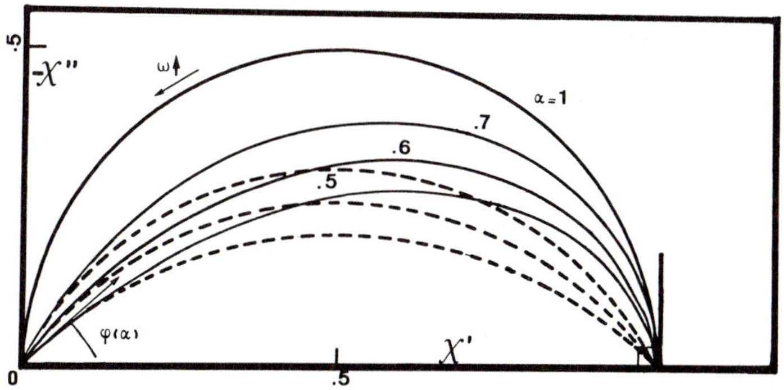

Figure 1 Transfer function associated with Williams-Watts relaxation
Broken line: Cole-Cole for (from top to bottom) $\alpha=0.7$; 0.6; 0.5

It appears to be divided into two domains. For high frequencies, according to the preceding comments, it asymptotically identifies with a Cole-Cole function. At low frequencies, it looks like a Cauchy spectrum with a null influence of the non-integral parmeter α. Such a shape is close to the Cole-Davidson form [26]:

$$1 / (1 + i\omega\tau)^{\alpha}$$

sometimes used to describe dieletric relaxation.

Attempts to fit relaxation curves with $\exp(-t^{\alpha})$ functions at long times should therefore be considered with caution, as it may lead to controversible conclusions. An example is shown in Figure 2, where viscoelastic data, often claimed as a typical Williams-Watts example [23], clearly show a deviation from this behaviour at low frequencies. Such a deviation [27,28,29] should not be disregarded with respect to the current models.

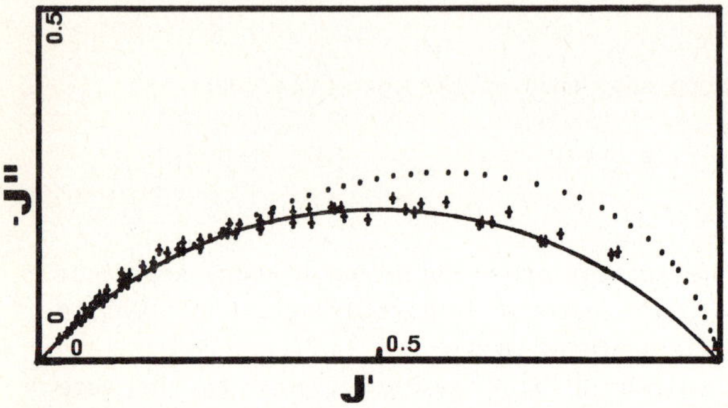

Figure 2 Experimental complex compliance for a high vinylic compound (30)
Black line: Cole-Cole function with $\alpha=0.52$
Dots: transfer function associated to Williams-Watts relaxation with $\alpha=0.52$

In conclusion, the present note stressed that the application of scaling properties to time or frequency may lead to results that cannot be differentiated by short time asymptotic considerations.
In particular, $\exp(-t^{\alpha})$ relaxations involve a loss of scaling properties at long times that might be significant only if it is probed by low-frequency experiments.
For instance, it is claimed that experimental data for polymers at the transition zone are better fit by a Cole-Cole function than by the associated Williams-Watts transfer function.

1 C. Zener, Trans ASM 41 (1949) 1057
2 W.A. Johnson and R.F. Mehc, Trans AIME 135 (1939) 416
3 J.B. Austin and R.L. Rickett, Trans AIME 135 (1939) 396
4 R. Kopelman, P.W. Klymko, J.S. Newhouse and L.W. Anaker, Physical Review 29 (6) (1984) 3747-3748
5 J.S. Newhouse, P. Argyrakis and R. Kopelman, Chem. Phys. Let. 107 (1) (1984) 48-52

6 A. Le Mehauté, G. Crepy, J. of Solid State Ionics 9/10 (1983) 17
7 A. Le Mehauté, J. of Stat. Phys. 36, 5/6 (1984) 665
8 R.M. Brady and R.C. Ball, Nature 309 (1984) 225-229
9 G. Marin, J.P. Montfort and P. Monge, Rheol. Acta 21 (1982) 449
10 H. Sher, to be published (1984)
11 G. Deutscher, R. Zallen and J. Adler,
"Percolation structures and processes" Adam Hilger ltd Haifa (1982)
D. Stauffer "Scaling theory of percolation clusters"
Physics reports 54, 1 (1979) 1-74
12 S.K. Ma, "Modern Theory of Critical Phenomena"
BENJAMIN Ed., New York (1976)
13 B. Mandelbrot "The fractal geometry of nature"
FREEMAN Ed., San Francisco (1983)
14 M. Shlesinger J. of Stat. Phys. 36, 5/6 (1984) 639
15 G. Williams, D.C. Watts, Trans. Faraday Soc. 66 (1970)80
16 R. Rammal, J. of Stat. Phys. 36, 5/6 (1984) 547
17 P.G. De Gennes C.R. Hebd. Sce. Acad. Sci. 296, II (1983) 881
18 J. Klafter, R.Silbey, J. of Chem. Phys. 74, (1981) 3510
19 S. Alexander, R.Orbach, J. Physique Lettres 43, L625 (1982)
20 J. Klafter, A.Blumen, G.Zumofen J.of Stat. Phys. 36(5/6) (1984)561
21 I. Webman, J. of Stat. Phys. 36, 5/6 (1984) 603
22 E. Montroll, G. Weiss, J. of Stat. Phys. 6 (1965) 167
23 J.T. Bendler, J. of Stat. Phys. 36, 5/6,(1984) 625
24 R.M. Fuoss, J.G. Kirkwood J. Am. Chem. Soc. 63 (1941) 385
25 K.S. Cole, R.H. Cole J. Chem. Phys. 9 (1941) 341
26 C. Lindsey, G. Patterson J. Chem. Phys. 73 (1980) 3348
27 G. Marin, Thesis, Université de Pau (1977)
27TH international symposium on macromolecules II (1981) 832
28 G. Marin, W.W. Graessley, Rheol. acta 16 (1977) 527
29 J.P. Monfort, G. Marin, J. Arman, PH. Monge, Rheol. Acta 18 (1979)623
30 L. Fruchter Rapport de DEA (1984)

Part VII **Aggregation and Instability**

Random Kinetic Aggregation

T.A. Witten
Exxon Corporate Research Laboratory, Annandale, NJ 08801, USA

Colloidal aggregation and other stochastic aggregation and growth phenomena produce structures that behave qualitatively differently from ordinary bulk matter in many respects. This pedagogical review is intended to account for this new behavior and to survey the ways of making these new structures. A "fractal" spatial scale invariance property allows much of this behavior to be predicted. Special scaling properties of phenomena confined to such structures, such as electrical conductivity, are noted. Complementary phenomena, such as diffusing, hydrodynamic or electric fields in the space around the structure, are treated on a common basis using the geometric notion of the intersection of two fractals. The notion is extended to treat the thermodynamic interactions of such structures. When a certain "opacity" condition is met, these tenuous structures interact with their environment as though they were solid objects. A method of determining the fractal dimension D using displacement of a polymer solution by the fractal is described. Then two basic mechanisms producing these structures are treated. First, the reaction rates governing colloidal aggregation are discussed. This leads to a solvable schematic model. Corrections to the model due to excluded-volume and polydispersity effects are estimated. The accretion of individual diffusing particles to make a large aggregate is treated separately. Its connection to viscous fingering and to solidification are noted. Certain morphological scaling properties give evidence that these structures are locally homogeneous and isotropic. Diffusing particle aggregation is contrasted with the cluster aggregation responsible for colloidal aggregates.

I. Introduction

Figure 1 shows an electron micrograph of a typical colloidal aggregate, made by David WEITZ [1] of our laboratory. The aggregate is made in aqueous solution starting from a dilute mist of individual gold particles of 100-Angstrom size. The chemical reaction which forms these particles also leaves them with a net electric charge, so that they repel one another and make a stable dispersion.

The space around one of these aggregates is clearly a "finely divided medium" in the sense of this workshop. The aggregate contains large open spaces, but also has smaller and smaller spaces, down to the size of the constituent particles. We expect various properties of this aggregate to reflect this fine structure. Various co-operative phenomena should be modified by this structure in interesting ways, as it is in gels and in percolating clusters.

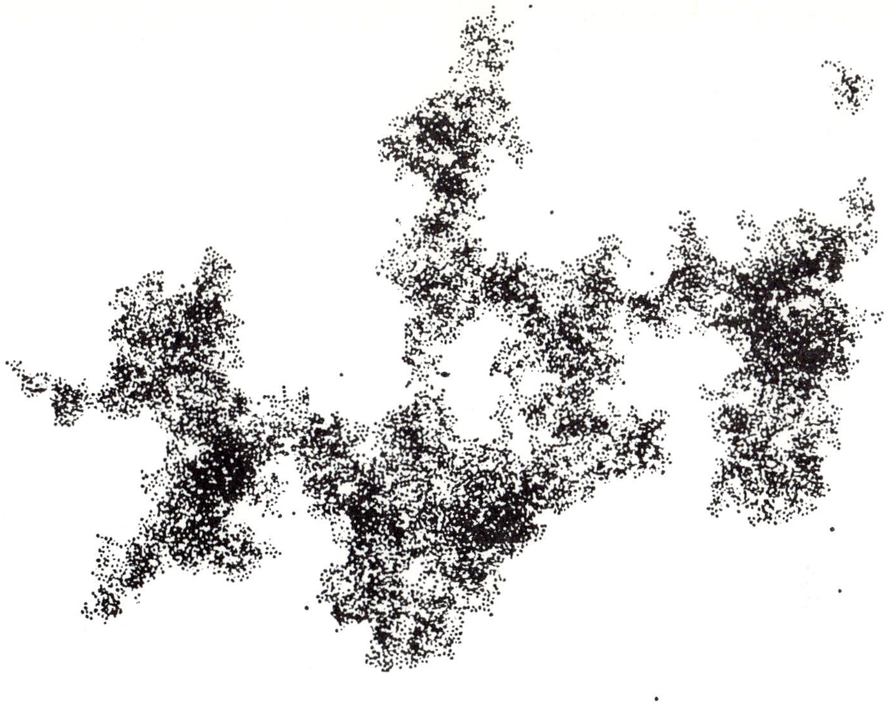

Figure 1

When the reagent pyridine is added, it removes this charge and the particles then attract each other via van-der-Waals forces. Since the particles are many atoms in size, this attraction energy is much larger than the average kinetic energy kT; thus the particles stick almost irreversibly on contact. After a few minutes or hours, one finds that the particles have flocculated to form large, wispy-looking aggregates like the one shown here. (The micrograph was made by putting a drop of the solution on a microscope slide and letting it dry.)

In this talk I want to give an overview of recent progress in understanding aggregates like these. First, I will give a phenomenological description of these tenuous structures and observe that their wispy appearance reflects a pervasive scale-invariance property: they appear to be fractals characterized by a fractal dimension D. Using this D and one other scaling exponent, one may predict how various properties of these structures contrast with those of ordinary matter. Next I will describe the special features that distinguish kinetically formed aggregates from "equilibrium" structures like the percolation clusters discussed by earlier speakers. A rich variety of aggregation phenomena with different scaling properties has been discovered to date. I want to describe these briefly in order to show the relations between them and their connections to other growth phenomena. Other speakers are going to describe some of these in more detail.

II. Properties of Scale-Invariant Structures

The aggregate of Figure 1 has a fascinating self-referent property; each small section of the object appears to embody all the complexity of the whole. Mounting evidence suggests that this appearance reflects the statistical scale-invariance properties of a fractal object [2]. To define scale invariance, we may describe our object by its density profile $\rho(r)$. The statistical properties of an enormous aggregate can then be specified by the correlation functions of this profile:

$$<\rho(0)\rho(r_1)\ldots\rho(r_n)>,$$

where r_1, r_2, etc. indicate specific points in space, and the $<\ldots>$ indicates an average over all positions and orientations of the structure. We take the separations between the r_i to be much smaller than the size of the aggregate. Statistical scale invariance, or dilation symmetry, means that these correlation functions are unaffected by a dilation of space: $r_i \to \lambda r_i$. Thus for a dilation-symmetric object

$$<\rho(0)\rho(\lambda r_1)\ldots\rho(\lambda r_n)> = \lambda^{A(n)}<\rho(0)\rho(r_1)\ldots\rho(r_n)> . \quad (1)$$

The only effect of the dilation is a possible multiplicative factor, controlled by the powers A(n). Certain well-studied random structures--random walks, self-avoiding walks [3], random animals, and percolating clusters [4] are known to be dilation symmetric in this sense. Further, the powers A(n) are related:

$$A(n)=n\,A(1)\equiv n\,A. \quad (2)$$

The simplest test for a dilation-symmetric or scale-invariant object is to examine the simplest of these correlation functions, $<\rho(0)\rho(r)>$, or equivalently, the pair distribution function C(r)--the average density of points on the object at a distance r from an arbitrary point. Evidently in a dilation-symmetric object the correlation function falls off as r^{-A}. This power is essentially MANDELBROT's [2] fractal dimension D. This D may be defined by considering the average mass within a distance "a" of an arbitrary point on the object. One may obtain this mass by integrating the pair distribution function C(r) within a sphere of radius a. Evidently, the mass scales as a^{d-A}, in d-dimensional space. This power also defines the fractal dimension D. Measurements of C(r) and variants of it [5] suggest that colloidal aggregation and other random growth processes are fractals in this sense. More detailed studies [6, 7] give indirect evidence that the higher correlation functions of some of these objects are also consistent with the dilation symmetry defined above.

The fractal dimension of these structures is useful for describing their properties in large part because they are "homogeneous" in addition to being fractal. What I mean by homogeneous is that the environments in the vicinity of various points of the structure are similar. Thus, the mass within distance "a" of a typical point of the fractal is of the same order of magnitude as the average mass, whose scaling is given by D. Otherwise put, the distribution of masses around various points of the structure is not too broad. A simple random walk (D=2) on a three-dimensional lattice is homogeneous in this sense. The distribution of masses is peaked; it rises faster than any power for small masses, and falls off faster than any power for large masses [8]. Thus all positive and negative moments of the mass distribution scale with the average mass. The distribution of masses is narrow enough that D gives the correct scaling behavior for almost any purpose.

A contrasting case is the Levy flight [2], a random flight with a broad, power-law distribution of jumps. Levy flights with D<2 (in two or more dimensions) are inhomogeneous. The average mass is of order D, but the probability of small masses is not exponentially small; it is of the same order as the probability of the average mass. Many of the scaling arguments mentioned below assume implicitly that aggregates are homogeneous. This assumption is widely believed, though to my knowledge it has not been tested explicitly.

Aggregates are different from ordinary bulk matter in many qualitative respects. A major difference concerns phenomena confined to the structure, such as electric conduction, diffusion along the structure, or mechanical vibration. ALEXANDER and ORBACH [9] have suggested that all these properties may be described in terms of a single new exponent called the spectral or fracton dimension. Their arguments appear to apply as well to aggregates as to other scale-invariant structures.

Such properties may be discussed even more simply for aggregates than for other fractals, because they are geometrically simpler: they appear to have no loops except at the smallest length scales. We may illustrate how this simplifies these properties by considering the conductance (inverse resistance) G between an arbitrary point on a loopless fractal and a surrounding conducting sphere of some large radius R. (This is the conductance property relevant for the Alexander-Orbach reasoning.) Figure 2 shows a schematic view of such a structure, with all the nonconducting dead ends removed.

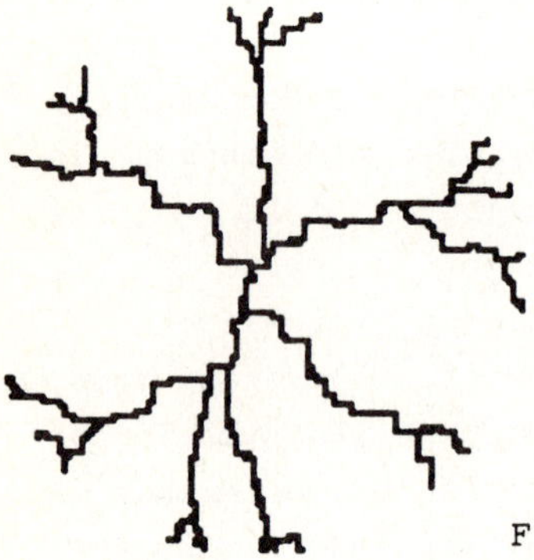

Figure 2

The remaining structure has relatively long branches at the center connecting to progressively shorter branches near the bounding sphere. The conductance G from center to sphere is evidently larger than that of any single path from center to sphere. For the structure illustrated, this conductance is dominated by the few long branches near the center. For branched structures generally, one may show, with mild additional assumptions, that the conductance of the whole structure is of the same order as that of one path to the boundary. Thus, the conductance scales with the average length of such a path [10-12], which we denote as L. Equivalently, L is the average number of particles in the path connecting any two points at distance R. Thus the scaling of conductivity in branched structures can be expressed in terms of a purely geometric scaling property. In the numerical models we consider later [13], the path length L scales as the distance R to a power δ between 1 and 2.

A very important internal response of fractal aggregates is their elastic response. Itzhak Webman described these in detail in his talk. Aggregates become less and less rigid as they grow, and ultimately attain a "floppy" state in which the branches are completely flexible. The scaling properties of these floppy aggregates have only begun to be explored [14]

Complementary to these "internal" phenomena, confined to the structure, are "external" phenomena in the space around the structure. In this category are such questions as how a conducting fractal modifies the electric field around it, or how a fractal modifies fluid flow or the local concentration of diffusing chemical reactants. The basic scaling properties governing these phenomena, unlike the internal phenomena, can be expressed in terms of the fractal dimension D. As a simple example of this, we consider the interaction of a fractal with a ray of light. We imagine a tiny light source somewhere in the middle of the fractal, and ask what is the probability that it will be visible from a given point outside. This is simply the probability that the (arbitrary) light ray from the source to our eye intersects the fractal. A parallel problem is the escape of a random walker from an absorbing fractal. The random walker is released from somewhere in the interior. Its probability of escape is one minus the probability of intersection between the fractal and an arbitrary infinitely long random walk.

These intersection questions are readily answered in terms of the fractal dimension. Indeed, the mean number of intersections [2] $M_{12}(a)$ between any two fractals 1 and 2 of radius R scales as $R^{D_1+D_2-d}$. This law applies to fractals placed independently of each other in the same volume of space, like the light ray or the random walk above. Evidently, if the power is negative, the probability of intersection decreases indefinitely as the size of the structure increases. Thus in three dimensions the ray ($D_2 = 1$) almost always emerges from a sufficiently large fractal provided $D_1 < 2$. The random walker almost always escapes from a sufficiently large fractal provided $D<1$. We may say that the two fractals are transparent to each other.

In the complementary case, where the number of intersections grows with R, we expect the two fractals to avoid each other with a probability approaching zero. The two are opaque to each other. If the absorption probability p per contact is small, the walker may travel some large distance λ through the fractal before $M_{12}(\lambda) = p^{-1}$ and there is a substantial net probability of absorption. Still, if the fractal is much larger than λ we expect that the probability of escape goes to zero. This means e. g. that random walkers entering a fractal with $D>1$ are almost always absorbed; an ideal gas of such walkers would be strongly depleted in the interior of the fractal.

Restating this result in more common language leads us to several conclusions about the interaction of fractals with $D>d-2$. The local concentration $u(r)$ of our gas of walkers obeys a steady-state diffusion equation $\nabla^2 u=0$, i. e. Laplace's equation. Since the walkers are absorbed, $u=0$ on the fractal. The distant walkers are unperturbed by the fractal; u is some constant at infinity. Since u is strongly depleted in the whole interior, it must be strongly reduced just outside, as well. The exterior field is thus that of a perfect absorber with radius of order R.

The fractal behaves similarly with respect to hydrodynamic interactions. The interaction with the velocity field v(r) is described by the Navier Stokes equations. In a quiescent fluid these reduce to a diffusion equation for each component of v, since in this regime the momentum density moves by diffusion, the diffusion constant being the kinematic viscosity ν. The fractal interacts with this field as with the diffusing field above. It absorbs the momentum of the fluid, and strongly screens the flow throughout the interior. The exterior flow is like that around a hard Stokes sphere with radius of order R. Thus the fractal diffuses and sediments in a fluid like such a sphere. This behavior is well known for the special case of polymers [15], but it should occur similarly for any fractal of sufficient size with $D>1$. The fractal screens electric fields in the same way if it is a conductor. Its D. C. capacitance to ground is like that of a sphere of radius R.

The notion of opacity may be used also to treat the thermodynamics of mutually interacting fractals. These properties are well known in the context of polymers, an important special case [15-17]. We consider two fractals whose constituent particles cannot interpenetrate. In calculating the partition function of the two, confined to some volume V, we must eliminate all configurations in which the two intersect. If the two fractals have size R and are mutually opaque, then almost all the configurations in which the two occupy the same volume must be eliminated. Each will be excluded from a volume of order R^d on account of the other.

The opacity of a fractal is easily compromised; it must not behave like a solid object in all respects. Thus the mutual excluded volume of a fractal and one of its constituent particles is much smaller than R^d; instead, it clearly scales as the volume actually occupied by the constituent particles, i. e. it scales as the mass M of the fractal. The exclusion of a small fractal by a large one gives a means of measuring the latter's fractal dimension. This may be derived formally by a scaling hypothesis; it may also be seen pictorially by imagining enclosing the large fractal with balls the size of the small fractal. We suppose that the large fractal has dimension D_1 and radius R_1. Evidently the small fractal is excluded from the interior of each ball, but not from the spaces between them; the excluded volume is the sum of the volumes of the balls. Noting that the mass of the large fractal is the sum of the masses in the balls, we deduce the excluded volume V_e.

$$V_e = R_2^{\,d}\ [R_1/R_2]^{D_1}\ .$$

The case where the small fractal is a short random walk was treated by DE GENNES [18]. He pointed out that if a fractal is made to absorb diffusing particles such, as an excited molecular species, for a limited time t, then the amount absorbed should scale as V_e with $R_2 \sim (\zeta t)^{1/2}$, where ζ is the diffusion constant. The counterpart for flow is the time-dependent friction factor [19] for a fractal in a fluid of kinematic viscosity ν. If the fractal is given an impulse, uniformly on all of its particles, at time zero, the applied momentum occupies the volume V_e after a time $t=R_2^{\,2}/\nu$.

A semidilute polymer solution with correlation length ξ would be excluded from a fractal in the same way; the region where monomers are depleted is essentially the region excluded to a small polymer of size ξ. In her contribution to this meeting, Francoise Brochard exploits this to measure fractal surfaces. For fractal objects like colloidal aggregates, one may infer [19] the change of volume fraction ϕ of a polymer solution upon adding a volume fraction ϕ_f of fractal. The solution must be semidilute, so that the polymer chains overlap strongly, but not so strongly that the correlation length ξ becomes smaller than the constituent particles of the fractals. When the fractals are added, each monomer is excluded from a volume $V_e \cong V\phi_f\ (\xi/b)^{d-D_1}$. Here b is the size of the constituent particles. In such a semidilute solution the length ξ scales with volume fraction by the usual fractal law for the polymers: $(\xi/a)^{D_2-d} = \phi$. Combining these, we find a relation for the change of ϕ :

$$\Delta\phi = \phi_f\ \phi^{(D_2-D_1)/(D_2-d)}\ .$$

If D_2 becomes equal to D_1, then the change of ϕ becomes equal to ϕ_f. Adding fractals to the solution becomes equivalent to adding the same volume fraction of polymer.

The above discussion shows that many aspects of a fractal's interaction with its environment can be inferred from its fractal dimension. Now I want to note another aspect, for which new and little-studied scaling properties of the fractal are needed. We have seen that an opaque fractal screens out most of an external field from its interior. The question naturally arises, how much of the fractal is then exposed to this field. For example, where does the charge reside on a conducting fractal at nonzero electric potential? This question and related ones have been considered by several authors [20-22]. Such questions are clearly important for understanding electrical breakdown around a fractal, the distribution of forces on a fractal in flow, or the amount of some chemical reaction occurring at different places on an absorbing fractal.

To make the problem explicit, we imagine a fractal which absorbs diffusing particles from infinity. Each site i of the fractal has a probability p_i of absorbing a given particle. The number of sites for which this probability is greater than the average gives a measure of the absorbing mass. For a solid sphere, only the particles on the surface can absorb, and the absorbing mass scales as the total mass M to the $(d-1)/d$ power. A simple way to estimate the absorbing mass in a fractal is to approximate it as a cloud of density $\rho=M/R^d$ spread uniformly within the fractal's radius R. Then the absorption rate at a point r is proportional to the concentration u of diffusers there, and to the density $\rho(r)$. Thus the diffusing field satisfies

$$\nabla^2 u = (\text{constant})\; u\rho(r) \;.$$

One easily checks that u falls off exponentially in a small distance λ from the outer boundary:

$$u(r) \sim e^{(r-R)/\lambda};\; \lambda/R \sim R^{(D+2-d)/2} \;.$$

The absorbing mass scales as $M\,\lambda/R \sim M^x$, where $x= (D-2+d)/(2D)$.

This naive estimate of the absorbing mass (not the length λ) seems to be correct for a variety of fractal aggregates [22]. But it clearly gives a mistaken picture of the <u>distribution</u> of absorption probabilities p. In a real fractal, there are protruding branches that absorb intensely, and sites well within the outer radius which nevertheless absorb appreciably. GRASSBERGER [21] has argued that the absorption or charge distribution may be a "fractal measure", whose scaling properties can only be specified using an infinite series of exponents. In this way it may resemble the current distribution on model fractals as discussed by REDNER and CONIGLIO [23, 24].

III. Aggregation Cluster by Cluster

The last section dealt in a general way with various properties expected for fractal structures. In this section we consider how such structures may be produced by kinetic aggregation processes. These kinetic processes produce objects that are qualitatively different from well-known "equilibrium" structures such as self-avoiding walks and random animals. I want to illustrate this difference, and then describe the simplest class of aggregation phenomena: aggregation cluster by cluster. Most colloidal aggregation seems to be of this type, as can be seen by recalling classical results [25] on the aggregation kinetics. I want to sketch some varieties of cluster-by-cluster aggregation and note the main effects that influence the fractal dimension D.

In a naive sense the ensemble of aggregates is identical to the well-known ensemble of lattice animals. These are defined as all connected clusters that can be formed on, e. g. a square lattice. A typical large "animal" has a fractal dimension of about 1.56 in two dimensions [26] and 2 in three dimensions [27]. Even though aggregates are the same objects as random animals, their statistical properties are much different. We may illustrate this point using the simple Eden growth model [28]. In this model a cluster is produced kinetically by randomly adding sites next to [29.] the cluster. The probability of a given cluster is markedly different in the Eden model than in the random animal ensemble. To illustrate this, we consider the relative probabilities of two clusters grown from the origin: a line of four particles extending to the right, and a square with the origin in the upper left corner. In the lattice animal ensemble, each distinct structure is represented exactly once, so that these two have equal probabilities. In the Eden model, each structure has a probability proportional to the number of ways it could have grown. For the line of sites, the particles must have been added in sequence: 1, 2, 3, 4 reading left to right. But for the square, there are four possible growth histories: 1, 2, 3, 4; 1, 2, 4, 3; 1, 3, 2, 4; and 1, 4, 2, 3. The compact square configuration is evidently strongly favored. This effect increases progressively with the cluster size: so much so that a typical large cluster is not fractal at all, but instead fills a region of space nearly completely [30]. In other aggregation models, fractal scaling properties emerge, but even these models give a qualitatively different ensemble from that of random animals.

Although a kinetic process may produce an ensemble with a biased weighting as seen above, it can also produce an equilibrium ensemble with equal weighting. Monte Carlo algorithms [31] are kinetic processes designed to sample an equilibrium ensemble without bias. To decide whether a given aggregation process gives a bias, one considers an arbitrary cluster, and asks how many ways the process could have constructed it. If the number of ways is the same for any object, the sampling is unbiased. Below we discuss one aggregation process that proves to be unbiased.

To construct a kinetic growth model appropriate for colloidal aggregates like those of the Introduction, we first consider the microscopic phenomena controlling an individual aggregation event: a cluster A fusing with a cluster B. The rate $S_A(B)$ at which two such clusters combine depends on how the cluster B moves to the vicinity of A from its initial (typically distant) location. It also depends on the rate at which particles fuse together once they are nearby. The simplest problem of this form [8] is the rate of absorption of spherical particles at concentration c_B onto a partially absorbing sphere of radius R_A. The relative separation of any two spheres diffuses with a diffusion constant equal to the sum of those of the spheres, $\zeta_A + \zeta_B$. Since each sphere is opaque to the fluid flow around it, each diffuses according to the Stokes formula: $\zeta \sim R^{2-d}$. If the absorption begins at time t=0, the absorption rate attains a steady state in a time of order R^2/ζ. After that, if the sphere is much larger than the absorption length λ defined above, the sphere absorbs nearly as a perfectly absorbing sphere. The absorption occurs when the distance between the centers of the spheres A and B is less than R_A+R_B. A volume of order R^d is depleted of diffusers in a time R^2/ζ. Thus the absorption rate [8] is proportional to the relative diffusion constant times this radius to the d-2:

$$S_A(B) \sim (R_A^{2-d} + R_B^{2-d})\,(R_A + R_B)^{d-2}\, c_B.$$

Since the same power, d-2, occurs in both factors, the overall R dependence of the reaction rate cancels out, and only a weak dependence on R_A/R_B remains. The cancellation is no accident; the two factors reflect the rate of absorption of diffusing momentum and of diffusing particles. This law should describe the reaction rate not only of spheres, but of fractals as well, provided they are opaque to flow and diffusing fields: $D>d-2$. Our picture of the relative reaction rates is almost unmodified by the fact that they are fractals. Thus we may adopt the classical discussion of SMOLUCHOWSKI [25] to see how the cluster size distribution develops. Fereydoon Family's talk treats this problem in detail. What emerges from this treatment is that throughout the reaction, clusters fuse with other clusters of comparable size. This observation motivates the "hierarchical" models which we discuss below.

The reaction rates $S_A(B)$ can be different from the form treated above under different conditions. One such case of interest is when the clusters move in straight line trajectories. This is what happens in the "Knudsen regime" of long mean free paths. The reaction rate is proportional to the velocity of approach times the cross section:

$$S_A(B) \sim |v_A - v_B| \, (R_A + R_B)^{d-1} \quad c_B \; .$$

Sedimenting clusters, such as composite snowflakes, should follow this law, with $v \sim M \, R^{2-d}$. In thermal equilibrium, the velocities would be mean thermal velocities, proportional to $M^{-1/2}$.

Another situation of importance is the reaction-limited regime, in which the absorption length λ is much larger than the aggregates. In this regime the aggregates touch many times before they react and the probability of two particles fusing is merely the probability that these would touch in the absence of reactions. The reaction rate then takes the form

$$S_A(B) = w \, Q_{AB} \, c_B \; .$$

Here w is the reaction rate for two touching particles, Q_{AB} is the number of ways in which the two clusters can touch (e. g. on a lattice), and the concentration c_B may be thought of as the probability that a given "origin" site of the B cluster occupies any given (lattice) point. For the counting of Q_{AB} we imagine that A and B are in fixed orientations, and treat other orientations as being different clusters. Unlike the previous cases, the scaling of Q_{AB} is strongly influenced by the fractal nature of A and B. Evidently if B is a single particle, then Q_{AB} is the number of perimeter sites of A, which scales as M_A. When A and B are of comparable mass, Q_{AB} appears also to scale [32] roughly as M, but this question has only begun to be explored. It may be [33] that the scaling of Q_{AB} influences the dimension of the aggregates, which in turn influences Q_{AB}. This relationship between structure and aggregation rates in the reaction-limited regime is of great interest and importance.

For most colloidal aggregation processes it is natural to suppose that typical aggregates are formed by the combination of clusters of comparable size. From this observation, one may gain a qualitative understanding for why these aggregates are fractal. A simple, solvable model embodying this idea is the "Sutherland's Ghost" model of Robin BALL [34]. In this model, pairs of single particles are linked to form dimers oriented in all directions (e. g. on a lattice). Then a pair of these dimers is selected at random and a particle of each is linked at random. In this way an ensemble of tetramers is made. Then arbitrary pairs of these are linked to make an ensemble of octomers, and so forth. Each linking is done by selecting a monomer of each cluster at random and linking these together (e. g. on adjacent sites of the lattice) in an arbitrary direction. The particles are allowed to interpenetrate freely.

To calculate the scaling of the mass of such an aggregate with its radius is a straightforward exercise [34]. The average number of bonds q_{2N} separating one particle from another on a 2N-site cluster is expressible in terms of the same average for the constituent N-site clusters, A and B. (q denotes "chemical" distance, along the links explicitly included; it ignores incidental intersections.) If two particles are chosen at random on a 2N cluster, there is a probability 1/4 that they both belong to the A cluster and 1/4 that they both belong to the B cluster. In either case, the average distance is q_N. There is a probability 1/2 that one of the chosen particles belongs to A and the other to B. In this case the distance between them is the distance from the chosen particle on A to the particle on A linked to B, plus the distance from this to the chosen particle on B. This distance averages $2q_N$. Taking the four cases together, one finds $q_{2N} = 3/2\ q_N$. Thus $q_N \sim N^{\log(3/2)/\log(2)}$. Since the paths connecting the points are random walks, the mean square distance between them is proportional to the q-distance. Combining, one finds $N \sim R^{D_s}$, where $D_s = 2\log(2)/\log(3/2) \cong 3.4$. Like ideal branched polymers [35], these structures have a diverging density in three dimensions, but above four dimensions they are well-behaved fractals with power-law density correlations [36].

The model is unrealistic for physical aggregation in two respects. First, it assumes that the reacting masses are exactly equal. Actually, the reacting masses are taken from a certain statistical distribution, controlled by the reaction rates $S_A(B)$ above. This effect by itself seems to be a minor one [34]. THOMPSON and BALL [36] have shown that it increases the fractal dimension D only up to about 3.7.

The main flaw in this model is that it permits the particles to interpenetrate freely. The particles do not avoid each other. This self-avoidance constraint has the same qualitative effect on the aggregates as on familiar equilibrium structures like self-avoiding walks. To account for the self-avoidance in a simple way, we may modify the Sutherland's Ghost model. The two sub-clusters are linked as before, but then the combined cluster is examined for self-intersections. (For concreteness, this can be implemented on a lattice.) If there are any self-intersections, the cluster is discarded. This hierarchical model [32] describes reaction-limited aggregation of equal-mass clusters. The importance of the discarding feature can be understood using the "opacity principle" described in Section II. The principle says that two such clusters will not intersect appreciably if the sum of their dimensions is smaller than that of space. Even when the a particle of one cluster is adjacent to the other as in the present model, the two mutually transparent clusters intersect with only finite probability. Thus the ensemble with intersections discarded has only a finite fraction removed, and the clusters which remain must have the same scaling properties as in the original Sutherland's ghost. Thus the self-avoidance constraint is not crucial in spatial dimensions above $2\ D_s \cong 6.8$; that is, this aggregation model has an upper critical dimension of $2\ D_s$, in the same sense as the standard equilibrium models [3, 4].

The hierarchical model resembles these standard models in a further respect: it is itself an equilibrium model. The various clusters in the ensemble appear with equal probability. This is because there is only one way each cluster can be constructed. A given cluster can only be made from one specific A and B subcluster. There is only one way in which the connecting particle of each can be chosen and one direction in which the connecting link can be made. (The structure must be specified by its bonds, not just its sites.) The ensemble thus produced is not the ensemble of animals, but only that subset which is "binary decomposable:" each cluster can be divided exactly in half by cutting one bond. The effect of the self-avoidance constraint is to straighten out the branches and to favor attachment near the ends. We expect the heuristic Flory model [37 38] to give a reasonable estimate for the fractal dimension below the upper critical dimension. This gives $D=(d+2)/[2(1+1/D_s)]$. The Flory estimate for random animals (or randomly branched polymers [35]) is given by setting $D_s=4$. Thus, reaction-limited equal-mass aggregation is expected to have a dimension slightly lower than that of random animals. Since it is an equilibrium ensemble, its fractal dimension is not affected if the branches are allowed to be flexible; in any case the clusters form an unbiased sample of the allowed configurations.

Other variants of the hierarchical model describe growth by diffusing [39] or ballistically [40] moving clusters. In the diffusive version, the two clusters are chosen, placed at a large distance, and allowed to diffuse until they touch [39]. Equivalently, one may choose a pair of particles on each subcluster and form the aggregate as in the Sutherland's ghost model. But then, one of the subclusters is made to execute a random walk to infinity. If the two subclusters intersect during this walk, the combined cluster is discarded. Now the opacity principle applies to the one cluster and the path swept out by the other: this path is an object constructed by replacing each site of the cluster by the random walk of sites. Its fractal dimension is the dimension of the cluster plus 2. For this process then, the upper critical dimension is $2D_s + 2 \cong 8.8$. For dimensions between 6.8 and 8.8, these clusters must have a lower dimension than their reaction-limited counterparts. We expect this to be true as well in two and three dimensions. Since each combined cluster could have been made with various choices of the random walk, this process does not make an unbiased equilibrium ensemble like the reaction-limited case.

The combined effects of self-avoidance and the distribution of reacting masses can evidently be studied directly by simulation. The fractal nature of cluster-by-cluster aggregation was in fact discovered by such simulations [41]. Botet's talk describes his comprehensive studies, which show that these simulations give a realistic picture of colloidal aggregates in the diffusion-limited regime. The detailed simulations confirm the picture suggested by the simple hierarchical models discussed above. In the reaction-limited regime, there may be some interesting features not treatable by the hierarchical model, as I have suggested.

IV. Aggregation Particle by Particle

This section deals with a type of aggregation complementary to the cluster-by-cluster processes treated above. Here we consider clusters that are constructed one particle at a time. It is convenient to classify these processes, as above, according to how the particle moves to join the cluster. The case where it does not move, but simply appears at a site adjacent to the existing cluster is the Eden model treated in the last section. In ballistic particle aggregation [42], the particles follow straight-line paths to join the cluster. Neither of these models produces a fractal structure [30,43], though both produce surfaces with interesting scaling properties [29, 42].

If each particle executes a random walk to join the aggregate, an apparently fractal structure [44, 45] results. This "diffusion-limited aggregation" (DLA) model is the more interesting because of its relationship to several continuum growth phenomena in nature. All these are varieties of growth controlled by Laplacian fields, as I describe below.

In diffusion-limited aggregation on a lattice, the probability of growth at a perimeter site next to the cluster is the probability that the random walker visits this site before visiting any other perimeter site. The probability u obeys the steady-state diffusion equation $\nabla^2 u=0$, where u is some constant at infinity, and zero on the perimeter sites. The probability that a perimeter site is visited is proportional to the flux of probability onto that site--i. e. the sum of u on neighboring sites. This quantity for a smooth surface of perimeter sites would be the normal gradient of u. The relative rates of growth are thus controlled by a Laplacian field. The same thing occurs [46] in biphasic fluid flow in a Darcy's Law medium, when a low viscosity fluid pushes a high viscosity fluid. The Laplacian field is the pressure; its gradient (times a mobility coefficient) is the local flow velocity. The fluid-fluid interface displays the famous viscous fingering instability [47]. In their talks David Wilkenson and Johann Nittman show two ways in which such flow can lead to an apparent fractal interface with DLA-like scaling properties.

Diffusion-limited solidification is another phenomenon of this type [48]. Here the diffusing field may be heat or a chemical species; the local growth rate is proportional to the flux of this field. BALL and BRADY [49] have shown that such growth can lead to a tenuous structure with fractal scaling behavior consistent with three-dimensional DLA. The electrical breakdown of a insulating medium is controlled by the (Laplacian) electrostatic potential [30, 50]. The connection with the above models is less direct, since the growth rate here is probably not linear in the gradients. Still, the patterns formed in breakdown bear a striking qualitative resemblance to DLA.

DLA has some important morphological features that are not shared by cluster aggregates. Its connecting path L seems to be proportional to geometric distance R in all dimensions [13]. For other types of structure, this path scales as a R to a power between 1 and 2. When DLA is grown from lines or surfaces [51] the density falls with distance h as h^{D-d}. This behavior is familiar in the context of adsorbed polymers [52]. It can be explained [53] by postulating that the structure around a site at height h is independent of h out to distances of order h. A similar line of reasoning [54] predicts the mass distribution of the various branches. Here one postulates that the branches which reach height h or greater must be spaced at distances of order h. This leads to a relative number of branches of mass s varying as $s^{-\tau}$, where $\tau = (d+D-1)/D$. This dependence has been confirmed by simulations [55]. Both τ and the h dependence of the density give indirect evidence that DLA is locally homogeneous and isotropic. The local neighborhood around one point is like the local neighborhood around another, and these neighborhoods have no preferred orientation.

Despite this apparent local homogeneity, DLA clusters grown from a point have a global structure with a distinct center. The average density profile around the center is systematically higher [7] than that around an arbitrary point. In this respect, a DLA cluster resembles a star polymer [56], a well-understood scale-invariant structure. DLA grown on a lattice takes on a global outline that reflects the lattice [57]. Thus DLA on a square lattice takes on a diamond shape. This suggests that the average deposition rate on the outermost sites is the same for different sectors around the perimeter.

Thus far, theoretical understanding of diffusive growth is meager. There is as yet no qualitative picture which shows why the structure should be fractal, even in some imaginary limit. What is clear is that this type of growth is qualitatively different from all the cluster aggregation models discussed in Section III. In those models D always attains a finite limiting value in high dimensional space: there is an upper critical dimension. By contrast, DLA must always have a dimension D that makes it opaque to the diffusing field [30]. Otherwise, the growth would occur indiscriminately throughout the cluster, and it would become like the Eden model. More detailed considerations [43] lead to the stronger bound that D must be greater than d-1.

Thus we are faced with several fundamental unresolved questions about the nature of DLA. The first is to find a formulation valid at least in some limit which is explicitly scale-invariant. The second is to understand the role of randomness in determining the structure. I believe these questions will be resolved in the near future.

REFERENCES

1. D. A. Weitz and M. Oliveria, Phys. Rev. Lett. 52 1433 (1984)
2. B. B. Mandelbrot, The Fractal Geometry of Nature, Freeman, San Fransisco (1982)
3. S. G. Whittington, Advances in Chem. Phys. 51 1 (1982)
4. J. W. Essam, Repts. Progress. Phys. 43 833 (1980)
5. F. Family and D. P. Landau, eds. Kinetics of Aggregation and Gelation North--Holland (1984)
6. M. Kolb, "Renormalization Group for Aggregation", Orsay preprint
7. T. Halsey, private communication
8. S. Chandrasekhar, Rev. Mod. Phys. 15 1 (1943)
9. S. Alexander and R. Orbach J. Phys. (Paris) Lett. 43 L625 (1982)
10. T. A. Witten and Y. Kantor, Phys. Rev. B. 30 4093 (1984)
11. S. Havlin, Z. Djordvic, I. Majid, H. E. Stanley, and G. H. Weiss, Phys. Rev. Lett. 53 178 (1984)
12. F. Family and A. Coniglio, J. Phys. A. 17 L285 (1984)
13. P. Meakin, I. Majid, S. Havlin, and H. E. Stanley, J. Phys. A, 17 L975 (1984)
14. M. E. Cates, Phys. Rev. Lett. 53 926 (1984)
15. P. G. De Gennes, Scaling Concepts in Polymer Physics, Cornell (1980)
16. T. A. Witten, and J. J. Prentis, J. Chem. Phys. 77 4247 (1982)
17. J.-F. Joanny, L. Leibler, and R. Ball, J. Chem. Phys. 81 4640 (1984)
18. P. G. De Gennes, private communication
19. T. Witten, to be published
20. P. Meakin, and T. A. Witten, Phys. Rev. A. 28 2985 (1983)
21. P. Grassberger, Physics Letters A 97 227 (1983)
22. P. Meakin, H. E. Stanley, T. A. Witten, and A. Coniglio, to be published
23. L. de Arcangelis, S. Redner and A. Coniglio, to be published
24. R. Rammal, elsewhere in these Proceedings
25. M. V. von Smoluchowski, Phys. Z. 17 557 (1916)
26. B. Derrida and L. de Seze, J. de Physique 43 475 (1982)
27. G. Parisi and N. Sourlas, Phys. Rev. Lett. 46 871 (1981)
28. M. Eden, Proceedings of the Fourth Berkeley Symposium on Mathematics, Statistics and Probability 4 223 (1961)
29. R. Jullien and R. Botet, "Scaling Properties of the Surface of the Eden Model," to be published
30. T. A. Witten and L. M. Sander, Phys. Rev. B. 27 5686 (1983)
31. K. Binder, ed. Monte Carlo Methods in Statistical Physics, Springer, Heidelberg, (1979)
32. R. Jullien and M. Kolb, J. Phys. A 17 L639 (1984)
33. D. A. Weitz, T. A. Witten, J. S. Huang, R. C. Ball, and F. Leyvraz, Bull. Am. Phys. Soc 30 268 (1985), and to be published
34. R. C. Ball and T. A. Witten, J. Stat. Phys. 36 873 (1984)
35. M. Daoud and J-F. Joanny, J. de Physique 42 1359 (1981)
36. I. Thompson and R. C. Ball, to be published
37. P. J. Flory, Principles of Polymer Chemistry, Cornell, Ithaca NY (1971)
38. T. C. Lubensky and J. Isaacson, J. de Physique, 42 175 (1981)
39. D. N. Sutherland and I. Goodarz-Nia, Chem. Eng. Sci. 26 2071 (1971); R. Botet, R. Jullien and M. Kolb, J. Phys. A. 17 L75 (1984)
40. R. Jullien, J. Phys. A. 17 L77 (1984)
41. P. Meakin, Phys. Rev. Letts. 51 1119 (1983); M. Kolb, R. Botet, and R. Jullien, Phys. Rev. Lett. 51 1123 (1983)
42. D. Bensimon, B. Shraiman and S. Liang, Physics Letters, 102A 238 (1984)
43. R. C. Ball and T. A. Witten, Phys. Rev. A. rapid commun., 29 2966 (1984)
44. T. A. Witten and L. M. Sander, Phys. Rev. Lett. 47, 1400 (1981)
45. P. Meakin, Phys. Rev. A. 27 1495 (1983)
46. L. Patterson, Phys. Rev. Lett. 52 1625 (1984)
47. P. G. Saffman and G. I. Taylor, Proc. Roy. Soc. London A245 312 (1958)
48. J. S. Langer and H. Mueller-Krumbhaar, Acta Mettall. 2 1081, 1689 (1978); H. Mueller-Krumbhaar and J. S. Langer, Acta Mettal. 2 1697 (1978)
49. R. Brady and R. C. Ball, Nature 309 225 (1984)
50. L. Niemeyer, L. Pietronero and H. J. Wiesmann, Phys. Rev. Lett. 52 1033 (1984)

51. P. Meakin, Phys. Rev. A. 27 2616 (1983)
52. P. G. de Gennes, Macromolecules 14, 1637 (1981)
53. T. A. Witten, Proc. Workshop on Dynamics of Macromolecules, P. Pincus and S. Edwards eds., J. Polymer Sci., Polymer Symposia, to be published, (1985)
54. Z. Racz and T. Vicsek, Phys. Rev. Lett. 51 2382 (1983)
55. P. Meakin, Phys. Rev. B 30 4207 (1984)
56. M. Daoud and J.-P. Cotton, J. de Physique, 43 531 (1982)
57. R. C. Ball and R. Brady, to be published

Scaling of Cluster Aggregation

R. Botet, M. Kolb, and R. Jullien
Laboratoire de Physique des Solides, Bât. 510, Université Paris-Sud,
Centre d'Orsay, F-91405 Orsay, France

The model of kinetic clustering of clusters (short CA for cluster aggregation) is introduced, its properties are briefly discussed and related to the experiments. Recent developments then are presented : simplifying features in high dimensions, gelation as described by CA, the rôle of reaction kinetics (chemically limited aggregation), effects due to the readjustment of the clusters, and anisotropy and corrections to scaling related to the growth mechanism.

The Witten-Sander model [1] gives a simple description of the structural properties of kinetic particle growth. This process is relevant to fracture, dielectric breakdown and electro - deposition [2]. Another process was proposed [3] to describe aggregation of clusters in order to describe irreversible growth of aerosols and colloids [4]. In both cases - and many others since - one is interested in the fractal or scaling properties, which only depend on very few parameters, such as the spatial dimension and the kinetics (the type of trajectories of the particles) but not on the details of the interaction. The model of CA discussed here is defined as follows : N_0 identical particles are placed in a volume V_0. The concentration $\rho_0 = N_0/V_0$ is chosen such that the average distance between the particles is much larger than their radius ($\rho_0 << 1$). The particles move independently of each other in a completely random fashion (brownian motion). Occasionally, two particles are sufficiently close that the strong short-range attraction causes them to stick to each other, permanently. Such clusters of two particles continue, of course, to move with the same diffusive motion as single particles (possibly with a different diffusion constant). The growth process defined this way continues indefinitely, as we suppose that the clusters never break up again. Thus, small clusters (and particles) collide, etc... In a finite system the growth stops when all the particles belong to one large cluster.

In a first analysis, the fractal properties of large clusters (with several thousand particles) are determined. To this end, numerical simulations were performed on a simplified version of the process just described : the particles are placed on a d-dimensional hypercubic lattice, move by jumping to neighboring sites and stick when they sit on neighboring sites (they never sit on the same site). The system is periodically bounded in all directions and the clusters do not rotate. These simplifying features are not expected to change the fractal properties of

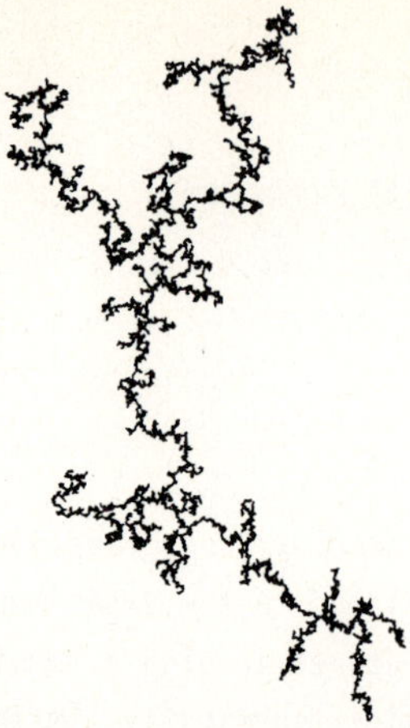

Figure 1 : A cluster of 16384 particles obtained by cluster aggregation in two dimensions.

large clusters.

The fractal analysis is suggested by the stringy appearance of the clusters, as shown in Fig. 1. They become less and less dense with increasing size. A quantitative measure is the radius of gyration

$R^2(m) = \frac{1}{2m^2} \sum_{i,j=1}^{m} (\vec{R}_i - \vec{R}_j)^2$ where the summation is over the m particles of the cluster with coordinates $\vec{R}_i$. One can also study the density correlation function c(r) inside the cluster. The clusters are said to be fractal (without mathematical rigor) if

$$(1) \quad R(m) \sim m^{1/D}, \quad m \to \infty$$

or equivalently

$$(2) \quad c(r) \sim r^{-(d-D)}, \; 1 \ll r \ll R$$

with $D < d$. The equivalence of (1) and (2) can be verified by integrating c(r) over r. D is the fractal dimension. In order to calculate D numerically, one takes an average over a large number of independently grown clusters. Fig. 2 shows a log-log plot of R versus m, from which the exponent D is estimated to be $D = 1.43 \pm 0.03$ in $d = 2$ dimensions. Similar calculations for higher dimensions are quoted in Table 1. The dimension of a brownian trajectory is two ($d_w = 2$). Experimentally, this growth process is realised in aerosols and colloids [4] where a fractal dimension $D = 1.75$ is measured in $d = 3$ dimensions, in good agreement with the theoretical value quoted in table 1 [5, 6]. So far, the dependence of the scaling properties on

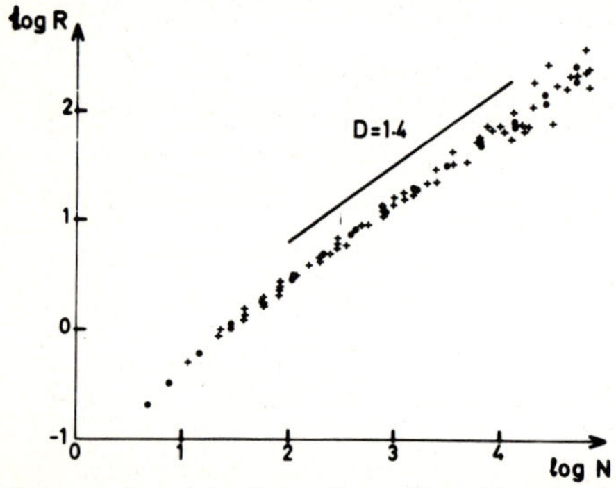

Figure 2 : log-log plot of Radius R versus mass of 2-dimensional cluster-aggregates. The slope determines the fractal dimension D

the diffusion constant has not been considered. In order to take it into account, we suppose that the velocity of a cluster depends on its mass as

$$(3) \quad V(m) = m^{\alpha}$$

where α is a parameter that can be chosen to suit the experiments. In the simulations, this corresponds to moving each cluster by one lattice spacing with probability $V(m)$, per unit time. To within numerical accuracy, one finds that D is independent of α, as long as $\alpha \lesssim 0$. In contrast to particle aggregation [1], time has a direct physical meaning in the cluster aggregation process. Quantities such as the average radius $\bar{R}$ or mass $\bar{m}$ are time-dependent and can be included in the scaling analysis,

	D		
d	$d_w = 0$	$d_w = 1$	$d_w = 2$
2	1.55 ± 0.03	1.51 ± 0.03	1.42 ± 0.03
3	2.00 ± 0.06	1.91 ± 0.03	1.78 ± 0.05
4	2.32 ± 0.10	2.22 ± 0.04	2.04 ± 0.08
5	-	2.47 ± 0.05	2.30 ± 0.20
6	-	2.7 ± 0.3	2.60 ± 0.30

Table 1 : Fractal dimensions D of CA as a function of the spatial dimension d and the dimension of the trajectories d_w ($d_w = 0$ independent of trajectory, $d_w = 1$ linear trajectory, $d_w = 2$ brownian trajectory) [5-7].

(4) $\bar{R} \sim t^{1/z}, \quad \bar{m} \sim t^{\theta}, \quad \theta = D/z$

where z is the dynamic critical exponent. In contrast to D, z depends on the exponent α, eq. (3). In fact for $d > 2$, mean field arguments can be used to relate z to α and D, namely [8]

(5) $z = D(1 - \alpha) - (d - 2)$

This relation is quite well satisfied for $\alpha \lesssim 0$ and $2 < d$. Another quantity which can be expressed in scaling form, is the cluster size distribution. The number of clusters of mass m at time t, $N(m_1 t)$ can be written as $N(m,t) = [N(t)/\bar{m}(t)]P(m/m(t))$ with $N(t) = \sum_m N(m,t)$. The scaling function $p(x)$ then is time-independent for $1 \ll N(t) \ll N_0$. Again, the simulations confirm this prediction and determine $p(x)$ as a function of α.

Is there a relation between the cluster size distribution of CA and the mean field description of cluster growth using the Smoluchowski equation?[2]. The scaling solutions of the Smoluchowski equation depend on the degree of homogeneity of the reaction kernel,just as $p(x)$ depends on α in the simulations. If one matches these parameters such that the dynamics (say $\bar{m}(t)$) corresponds, then also the scaling function has the same form [8, 9]. This shows that while the static aspects, i. e. the fractal dimension, has to be calculated numerically, the dynamics is meanfield.

Another aspect is suggested by the comparison with the Smoluchowski equation : two qualitatively different regimes can be distinguished, dependent on whether gelation (formation of an infinite cluster) occurs in a finite time or not. Indeed, this distinction is also valid for CA [10]. For $\alpha > \alpha_c = 1 - (d-2)/D$, the growth process becomes that of particle aggregation (as seen most clearly for $\alpha \to \infty$, when only the largest cluster moves).

A simplified version of the CA model is the hierarchical model.where at level ℓ clusters of mass $2^{\ell-1}$ are aggregated pairwise to form clusters of mass 2^{ℓ}. This corresponds most closely to the limit $\alpha \to -\infty$ of CA. The fractal exponents agree with the ones from simulations in a box, and can be extracted with a good precision by extrapolating an effective exponent of two subsequent generations,

$D_{eff} = \ln 2/\ln\frac{R(2m)}{R(m)}$, to $m \to \infty$. There are many generalizations possible for the CA model. The particles (and cluster) trajectories play an important rôle in the growth process. Instead of brownian trajectories, one can consider linear trajectories ($d_w=1$) or completely eliminate the trajectories ($d_w = 0$). The fractal dimensions clearly depend on the choice of d_w, as illustrated in Table 1.

Let us now present some more recent and more detailed results of cluster aggregation and related growth processes : CA in high dimensions can be dealt with analytically, as the clusters become transparent ; CA describes a kinetic gelation transition with new fractal properties ; changing the reaction kinetics also leads to a new growth model ; readjusting the clusters has an important effect on the shortrange porperties ; clusters in CA are isotropic as opposed to particle aggregation where anisotropy requires introducing a new length scale.

For CA, there exists an upper critical dimension, above which the clusters can freely interpenetrate and their fractal dimension becomes independent of the spatial dimension d [11]. If one wishes to include the reaction kinetics, it is appropriate to use the Smoluchowski equation, which describes the cluster-size distribution correctly down to dimensions d = 2,3. The degree of homogeneity ω of the reaction kernel K(i, j) that governs the collision of an i-mer with a j-mer, $K(\lambda i,\lambda j) = \lambda^{2\omega}K(i, j)$ is the only relevant parameter for a large class of kernels. Using the scaling function that results for large clusters at long times (depending on ω) a generalised aggregation process can be described, that includes effects such as a change over to a particle aggregation process, as also observed in the simulations [12] The results have been obtained analytically [13]. The fractal dimension which is equal to $\ln 4/\ln(3/2) \simeq 3.42$ for $\omega = -\infty$ increases only very slightly with increasing ω. Its behavior between $\omega = 0$ and $\omega = 1/2$ depends on the scaling function extracted from the Smoluchowski's equation. For $\omega > 1/2$, the fractal dimension is infinite, anticipated because particle aggregation is believed to have $d_c = \infty$. The existence of a finite or infinite upper critical dimension is probably illustrated most clearly for CA with linear trajectories ($d_w = 1$) with and without impact parameter. In the first case (I) the clusters collide randomly, in the second case (II) the trajectories go through the centers of mass. Fig. 3 compares the numerical results, suggesting that D saturates at the upper critical dimension of the $d_w = 1$ model for model I whereas $D \to \infty$ as $d \to \infty$ for model II. For model II, the transparency condition simply is never satisfied. This condition is required for the meanfield theory to be valid.

The CA model studied so far was for low cluster concentration, precisely for $\rho_0 = N_0/V_0 \to 0$. Experimentally, there have been studies at finite cluster concentrations [15, 16], finding a fractal dimension D which is larger than the value found for $\rho_0 \to 0$. The reason for the increase of D is that at higher concentration the clusters interpenetrate much more. At high concentration, the scaling behavior occurs close to the gel point, when an infinite spanning cluster appears. Thus, CA provides a mechanism to describe kinetic gel formation [17]. This aspect has been studied as a function of the cluster mobility exponent α, with the conclusion that the same qualitative picture based on the Smoluchowski equation that characterises CA for $\rho_0 \to 0$ also applies to CA at high concentration. Of course, the relation between α and ω is different in this case. Of importance is the distinction between finite and infinite gel time, having entirely different scaling properties. In Fig. 4, CA at high concentration in two dimensions is shown.

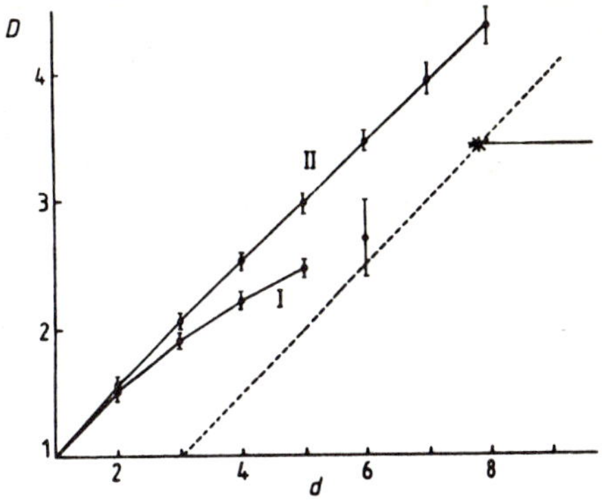

Figure 3 : Fractal dimension D as a function of the spatial dimension d for CA with linear trajectories ($d_w=1$) Model I(II) corresponds to trajectories with (without) impact parameter, leading to a finite (infinite) upper critical dimension.

The aggregation process considered so far is idealised in the sense that two clusters stick together at the first contact. In the experiments with gold colloids [4], it is possible to vary the reactivity, obtaining different fractal properties in the diffusion limited and the reaction limited case. The bonds formed are also irreversible in the reaction limited case. This new model (the $d_w = 0$ version of CA) has been studied theoretically [7] by introducing a sticking probability $0 < p < 1$: when two clusters touch each other a band is formed with probability p. If they don't stick together they continue to diffuse independently, respecting nevertheless the steric constraints. p = 1 (p = 0) corresponds to diffusion limited (chemically limited) aggregation, $d_w = 2$ ($d_w = 0$) ; for $0 \lesssim p \ll 1$ there is a crossover from chemical to diffusive aggregation. Asymptotically, the clusters have the same properties as diffusive CA. p = 0 is best investigated using the hierarchical version. The fractal dimension obtained from simulations, shown in table 1, agree well with the results from the experiments.

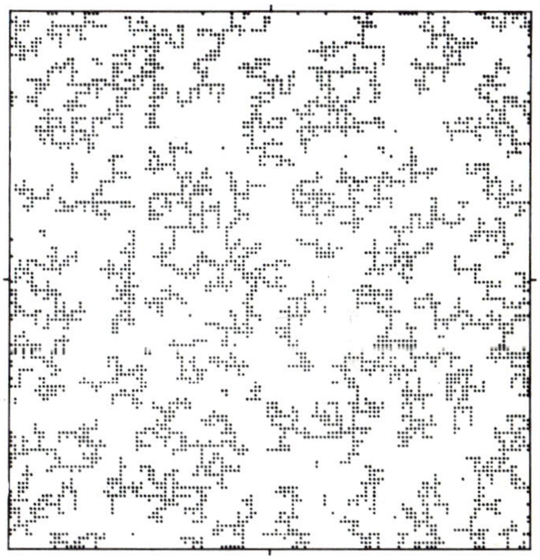

Figure 4 : Simulation of CA in two dimensions at high concentration [17]. The system is close to the gel point.

In the experiment by Allain and Jouhier [15] and by Camoin and Blanc [18], it has been observed that there is a restructuring of the clusters, which changes (compactifies) their appearance, most clearly on the scale of a few particle radii. Notably, the clusters rotate around until they form loops, that strengthen their mechanical stability. This effect has been modelled by letting the clusters in CA rotate around their point of contact until they form a loop [19]. The simulations have been done as follows : when two clusters touch at a point, they are rotated instantly, either in the direction of the smallest, the largest or both (at random) angles, until a loop forms (sometimes two rotations are required). In the hierarchical version simulated, the choice of now one rotes does not change the results. The fractal dimension D in two dimensions is slightly larger with restructuring (1. 49 instead of 1.42) though one cannot exclude that they are asymptotically the same, as the average angle of rotation may tend to zero for larger clusters. The very pronounced difference of the clusters can be seen in Fig. 5 where the particles form practically a crystalline structure on a very short scale.

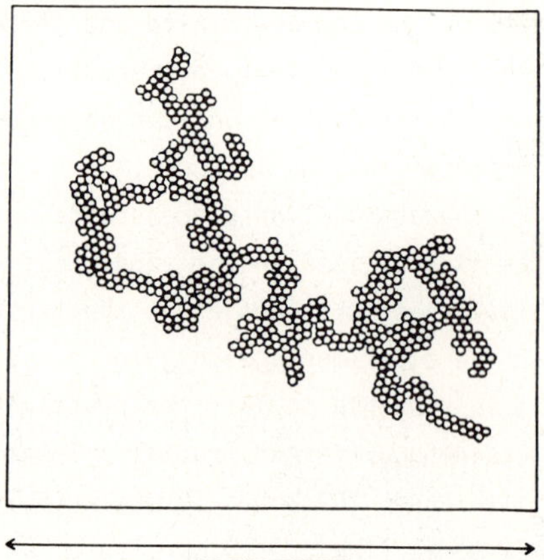

Figure 5 : Cluster from hierarchical CA, with 512 particles where rotation of the aggregating clusters, to form loops, makes the structures very different on a short scale [19].

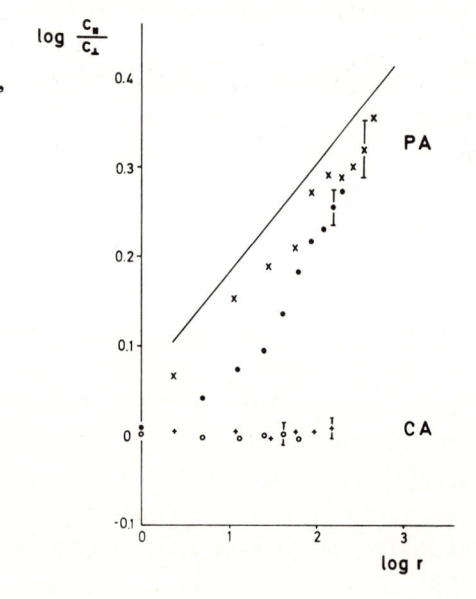

While the fractal concept has been applied to many growth processes with success, no relation between the growth mechanism and the internal structure has been established so far. Particle and cluster aggregation do grow in a very different way : particle aggregation (PA) has a distinct center (the seed particle), cluster aggregation does not (see footnote number 10 in ref. 10). In order to characterise the internal structure and quantify the distinction between these growth models, angular density correlations were considered [20]. They show, that PA grows out of its center, whereas CA is isotropic. In the scaling region deep inside the cluster, the correlations parallel and perpendicular to the direction of growth scale with different exponents for PA. The difference is small, $\Delta A = 0.16 \pm .05$ which explains,

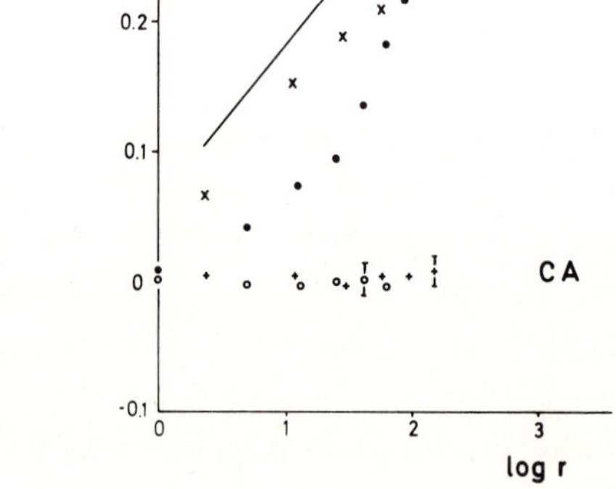

Figure 6 : Parallel ($c_{\parallel}$) and perpendicular ($c_{\perp}$) density correlation inside clusters for PA and CA. The ratio $c_{\parallel}/c_{\perp}$ shows that, for PA only, the correlations scale differently in these directions ; this leads to corrections to scaling.

why the PA clusters appear isotropic on a small scale. The clusters for CA are fully isotropic. In Fig. 6, the ratio of correlations parallel and perpendicular to the center of the cluster is shown, illustrating a marked difference for PA and CA. On a lattice, there are additionally anisotropies in the growth due to preferred (axial) directions. This, however only changes the amplitudes of the correlation functions.

We acknowledge the collaboration and discussions with H. Herrmann and P. Meakin. This work has been supported by an ATP of the CNRS and by the CCVR, Palaiseau.

References

1 T.A. Witten and L.M. Sander : Phys. Lett. 47, 1400 (1981)
2 L. Niemeyer, L. Pietronero and N.J. Wiesman : Phys. Rev. Lett. 52, 1033 (1984)
M. Matsushita, M. Sano, Y. Hayakawa, N. Honjo and Y. Sawada : Phys. Rev. Lett. 53, 286 (1984)
R. Brady and R. Ball : Nature 309, 225 (1984)
3 P. Meakin : Phys. Rev. Lett. 51, 1119 (1983)
M. Kolb, R. Botet and R. Jullien : Phys. Rev. Lett. 51, 1123 (1983)
4 S.R. Forrest and T.A. Witten : J. Phys. A12, 1109 (1979)
D.A. Weitz and M. Oliveria : Phys. Rev. Lett. 52, 1433 (1984)
D.A. Weitz, M.Y. Lin and C.J. Sandroff : Surf. Sci. to appear (1985)
5 R. Jullien, M. Kolb and R. Botet : J. Physique Lett. 45, L211 (1984)
6 R. Jullien : J. Phys. A17, L771 (1984)
7 R. Jullien and M. Kolb : J. Phys. A17, L639 (1984)
M. Kolb and R. Jullien, J. Physique Lett., 45, L977 (1984)
8 M. Kolb : Phys. Rev. Lett. 53, 1653 (1984)
9 R. Botet and R. Jullien : J. Phys. A17, 2517 (1984)
10 R. Botet, R. Jullien and M. Kolb : Phys. Rev. A30, 2150 (1984)
11 R. Ball and T.A. Witten : Conference on Fractals ; Gaithersburg Md.(Nov.1983)
12 R. Botet, R. Jullien and M. Kolb : Phys. Rev. A30, 2150 (1984)
13 R. Botet : J. Phys. A, to appear
14 R. Ball and R. Jullien, J. de Physique Lett. 45, L1031 (1984)
15 C. Allain and B. Jouhier : J. de Physique Lett. 44, L421 (1983)
16 P. Richetti, J. Prost and P. Barois : J. de Physique Lett. 45, L1137 (1984)
17 M. Kolb and H. Herrmann : J. Phys. A Lett., to appear
18 C. Camoin and R. Blanc : J. de Physique Lettres 46, L67 (1985)
19 P. Meakin and R. Jullien, submitted for publication
20 M. Kolb, submitted for publication

Dynamic Scaling in Aggregation Phenomena

Fereydoon Family and Tamas Vicsek
Department of Physics, Emory University, Atlanta, GA 30322, USA

The dynamics of the diffusion-limited cluster-cluster aggregation model (CCA) is reviewed through the temporal evolution of the cluster size distribution $n_s(t)$, which is the number of clusters of size s at time t. In recent studies it was shown that the results of the Monte-Carlo simulations can be well represented by a dynamic scaling function of the form $n_s(t) \sim s^{-2} f(s/t^z)$ where f(x) is a scaling function which depends on the spatial dimension and the microscopic details of the aggregation process. Two basic factors which affect the aggregation process are the mobility and the reactivity of the aggregating clusters. The diffusion constant of a cluster of size s is assumed to be proportional to s^γ. Depending on such factors as the chemical reactivity, kinetic energy, mass, etc. of the aggregates, the coagulation of two clusters may or may not take place. These effects are simulated by assuming that the probability that two clusters of sizes i and j irreversibly stick together is proportional to $(ij)^\sigma$. The overall behavior of $n_s(t)$ and its moments have been determined for a set of values of γ and σ. We find that the results are consistent with the scaling theory, and the exponents in $n_s(t)$ depend continuosly on γ and σ. Moreover, there is a critical value of γ for a given σ, and vica versa, at which the shape of the cluster size distribution crosses over from a monotonically decreasing function to a bell-shaped curve. This phenomenon, which is consistent with experiments, can be described by the above scaling form for $n_s(t)$ with appropriate forms for f(x) in the two regimes.

We also investigated the CCA model, under conditions which for long times lead to steady-state coagulation. Single particles are added to the system at a constant rate and the larger clusters appearing as a result of the aggregation process are removed according to various rules. Our results indicate scaling as a function of the feed rate. The data obtained for steady-state coagulation are used to discuss the upper critical dimension for the kinetics of the CCA model.

1. INTRODUCTION

The main process in many phenomena of practical importance in physics, chemistry, biology, medicine and engineering is the formation of clusters by aggregation of particles or the clusters themselves [1]. A widely used method of describing such systems is the determination of the cluster size distribution function [2,3]. This quantity describes the dependence of the number of clusters on their mass, and is measured in experiments and is the focus of a kinetic equation approach to coagulation.

A model which embodies many of the physical features of an aggregating ensemble of clusters and is well suited for the investigation of the dynamic cluster size distribution is the cluster-cluster aggregation (CCA) model, recently introduced by MEAKIN [4] and independently by KOLB et al. [5]. In this model, initially N_0 particles are randomly distributed on a periodic d-dimensional lattice.

Particles are then picked at random, and moved by one lattice spacing in one of the z equally probable directions, where z is the coordination number of the lattice. If a particle becomes the nearest-neighbor of another particle, they are permanently bound together and form a two-particle cluster. These clusters and the single particles are then picked at random, moved by one lattice spacing, and joined together if they contact another cluster or a single particle. In this manner, the clusters grow larger and larger until only one large cluster remains (in a finite system). The formation of this large cluster can be interpreted as the onset of gelation. The gel time, t_{gel}, at which gelation phenomena occurs, depends on the initial concentration of the monomers [6]. For $t \ll t_{gel}$ CCA describes the flocculation process. In particular, in the low-density colloidal aggregation the gel time tends to infinity, and CCA describes the flocculation phenomena for finite times.

In actual experimental situations, various factors influence the aggregation process. One important physical parameter affecting the aggregation phenomena is the cluster mobility [7,8]. Other factors such as chemical reactivity, kinetic energy, mass, etc. of the clusters also influence the aggregation process. Depending on these parameters, the coagulation of two clusters may or may not take place during a collision. In the simulations, the cluster mobility can be taken into account through a mass-dependent diffusion coefficient, and the chemical reactivity can be accounted for by the sticking probability of two clusters [9].

In this article we review some of our recent results on dynamics of aggregation through the time-dependent cluster size distribution function. In the second section we describe a dynamic scaling theory, and its applications for the cluster size distribution and its moments [3]. In the third part of the paper we discuss the results of the simulations of the cluster-cluster aggregation model with a mass dependent cluster mobility [7] and compare them with the predictions of the Smoluchowski equation [10] and the dynamic scaling theory. We define a chemically-controlled coagulation model, and discuss its dynamic scaling properties in section four. Aggregation is generally considered as a process with a permanent evolution in time. There are important classes of coagulation phenomena in which a steady-state cluster size distribution develops in the system. In section five we investigate a steady-state model in which single particles are fed into the system and the larger clusters are removed according to some rules.

2. DYNAMIC SCALING

The dynamic properties of cluster-cluster aggregation can be investigated by determining the cluster size distribution function $n_s(t)$ and its moments. Very recently, a dynamic scaling description was introduced [3,8], for the cluster size distribution in CCA. It was shown [3] that the Monte-Carlo simulations can be well represented by a dynamic scaling for $n_s(t)$ of the form

$$n_s(t) \sim s^{-2} f(s/t^z) \qquad (1)$$

where f(x) is a scaling function which depends on the dimension and the microscopic properties of the aggregating particles and clusters. Two important factors which affect the dynamics of the aggregation process are the cluster mobility and the bond formation probability of the clusters. In order to account for these factors, we have assumed that the diffusion coefficient D_s of an s-site cluster is given by

$$D_s = D_o s^{\gamma} \qquad (2)$$

and the bond formation probability $P_{ij}(\sigma)$ which is the probability that a cluster of size i sticks to a cluster of size j is

$$P_{ij}(\sigma) = P_o (ij)^{\sigma} \qquad (3)$$

There are two regimes, depending on the values of γ and σ and correspondingly in the scaling form (1) the following scaling functions appear

$$f(x)=x^2\, g(x) \qquad (\gamma<\gamma_c \text{ or } \sigma<\sigma_c) \tag{4a}$$

with $g(x)<<1$ for $x<<1$ and $g(x)<<1$ for $x>>1$, and

$$f(x)\sim\begin{cases} x^{\delta} & x<<1 \\ <<1 & x>>1 \end{cases} \qquad (\gamma>\gamma_c \text{ or } \sigma>\sigma_c) \tag{4b}$$

According to (4), for $s/t^z<<1$,

$$n_s(t) \sim t^{-w}s^{-\tau} \qquad \text{with} \qquad w = (2-\tau)z, \tag{5}$$

where $w = z\delta$ and $\delta = 2-\tau$.

An immediate consequence of the scaling form (1) is that the mean cluster size $S(t)=\Sigma s^2 n_s(t)$ diverges as $t\to\infty$. Expressing $S(t)$ in terms of $n_s(t)$ and using (1), we find [3]

$$S(t) \sim t^z \tag{6}$$

Another consequence of the scaling form (1) is that the total number of clusters in a unit volume $n(t)=\Sigma n_s(t)$ also scales with time. Using expression (1) we find [11]

$$n(t) \sim \begin{cases} t^{-z} & \tau>1 \\ t^{-w} & \tau<1 \end{cases} \tag{7}$$

Therefore, the time-dependence of the total number of clusters is determined by the value of τ. The cluster size distribution and the exponents z, τ and w have been measured [3,7,9] by Monte-Carlo simulations in d=1-3 dimensions for various values of γ and σ and the results agree with the scaling theory.

The cluster-size distribution is expected to satisfy (1) in the flocculation regime, i.e. for $t<<t_{gel}$. However, with the initial density $\rho=N_0/L^d$ going to zero, $t_{gel}\to\infty$, and the scaling theory holds for finite times.

3. CLUSTER MOBILITY

Next we discuss the effect of a mass dependent diffusion constant D_s of the form given in (2) on the cluster size distribution. For $\gamma>>0$ the large clusters move much faster, and relatively many small clusters remain intact during the process. If $\gamma<<0$ the speed of the small clusters is higher, therefore, these clusters die out by joining together and building up large clusters. At a value of $\gamma=\gamma_c$ the behavior of the cluster size distribution changes qualitatively and becomes nonmonotonic. Figure 1 shows cluster size distribution functions obtained from some of our two-dimensional simulations of cluster-cluster aggregation, with several values for the diffusivity exponent γ. The characteristic shapes of the cluster size distribution function for $\gamma<\gamma_c$ and $\gamma>\gamma_c$ are different. There are, therefore, two main regimes, and the results can be represented by the scaling form (1) with (4a) and (4b). The exponents τ and z and w change continuously as γ decreases. The dependence of z on γ obtained from the simulations agree well with the Smoluchowski equation prediction in d=3.

One way of testing the dynamic scaling is to plot $s^2 n_s(t)$ versus s/t^z. If eq. (1) is valid, all results for a given γ must fall on one curve which is the scaling function $f(x)$. Plots of $n_s(t)$ for different times have been shown to scale in this way into one universal scaling function [7].

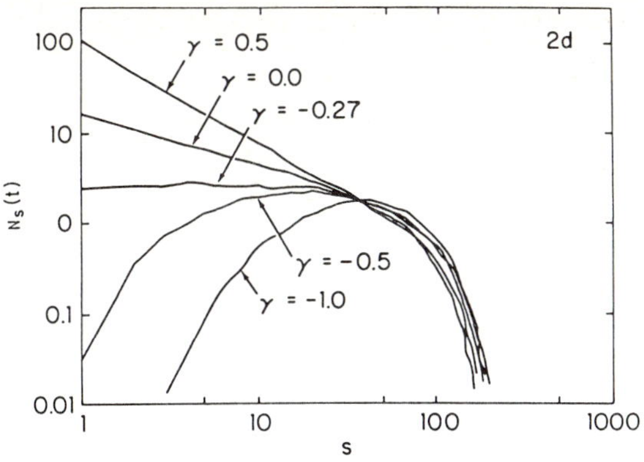

Figure 1: Cluster size distribution functions obtained from two-dimensional simulations of cluster-cluster aggregation on a 500x500 cell with a density $\rho=0.05$ for several values of the diffusivity exponent γ. As γ is decreasing, at a critical value of the diffusivity exponent $\gamma_c \simeq -1/4$ the monotonic decay of the cluster size distribution crosses over into a different, bell-shape behavior.

4. CHEMICALLY-CONTROLLED COAGULATION

In order to simulate an aggregation process in which the chemical bond formation between the clusters has a certain probability, we assume that the probability that a cluster of size i is permanently joined to a cluster of size j to form a cluster of size i+j is given by (3). The rules for joining the particles together with a mass-dependent sticking probability are as follows: (1) Two clusters "touching" each other (via nearest neighbor occupancy) are "merged", if a random number x evenly distributed between 0 and 1 is less than $P_{ij}(\sigma)$. (2) If there is more than one contact point between the two clusters, they are tested independently. Sticking may occur with a probability $P_{ij}(\sigma)$ at each of the contact points. (3) If two "touching" clusters do not stick, they are allowed to remain in contact. (4) If a randomly chosen step for one cluster would cause it to overlap with a cluster it is already contacting (but not joined to), the move is not permitted. However, the time is incremented, and a new attempt is made to stick the contacting clusters together by testing a random number against $P_{ij}(\sigma)$ at each contact point. (5) If a cluster moves from one contacting position to another contacting position with the same cluster, a new attempt is made to join the clusters together.

In order to investigate the effects of a mass independent sticking probability ($\sigma=0$) on the dynamics of the cluster-cluster aggregation, we have measured $N_s(t)$ (where $N_s(t)=L^d n_s(t)$ is the number of clusters of size s in the L^d system) and its moments in two and three dimensions for several values of P_o and γ. The results of the two-dimensional simulations show that initially N(t) decays slowly for small P_o, but asymptotically it decays with the same exponent independent of P_o. The approach to the asymptotic behavior is reached considerably slower at smaller values of P_o. Similar behavior is observed in d=3, but for small P_o much longer times are required for reaching the asymptotic limit.

The above results show that even though a constant sticking probability does not change the scaling behavior of N(t), the initial stages of the aggregation are chemically-controlled. This is consistent with the results of the studies of the effects of sticking probability on the fractal dimension [12,13], which showed that the fractal dimension [14] in chemically controlled aggregation is higher than in diffusion-limited CCA (In d=3, D=2.0 instead of 1.8 and in d=2, D=1.55 instead of 1.45). In the early stages, the number of clusters decays much slower than in the case $P_o=1$. The reason is that initially there are only a small number of chances for two given clusters to meet, and since $P_o \leq 1$, there is a correspondingly small probability for them to join together. However, as the time increases, the number of possible contacts between the two clusters, and the probability that they join, increases. That is, after a long time the probability that two clusters which have approached each other will join is essentially unity, and there is a crossover from chemically-controlled to diffusion-limited aggregation.

We have also studied the effects of size-dependent sticking probability on $n_s(t)$ and its moments by assuming that the diffusion coefficient and the bond probability have the forms given in (2) and (3), respectively, with various values of σ and γ. The results indicate that in analogy with the effects of γ on $n_s(t)$, there is a critical value of σ, say σ_c, at which the shape of $n_s(t)$ changes from a monotonically decreasing function to a bell-shaped curve. From the simulations, it has been found that for $\gamma=0$, $\sigma_c=-0.6$ in d=2 and $\sigma_c=-0.8$ in d=3, respectively.

From the data for $N_s(t)$, $N(t)$ and $S(t)$, the exponents τ, w and z for various values of σ and γ have been determined. Within the errors involved in the calculations, the exponents satisfy the scaling relation (5).

In order to describe the chemically-controlled coagulation process within the Smoluchowski kinetic equation approach, one must relate the kernel K_{ij} to σ. If we assume that once an i-mer touches a j-mer they stick together with a probability $P_{ij}(\sigma)\sim(ij)^{\sigma}$, then the kernel in the Smoluchowski equation becomes $K_{ij}\sim P_{ij}(\sigma)R_{ij}^{d-2}D_{ij}$. Therefore, $K_{i\lambda j\lambda}\sim\lambda^{2\omega}$ $K_{ij}\sim$ $\sim\lambda^{2\sigma+(d-2)/D+\gamma}$ K_{ij} and we have $2\omega=2\sigma+\gamma+(d-2)/D$. Note that this relation includes both the effects of cluster diffusivity and the sticking probability, and suggests that ω depends linearly on σ. We have used the relation $z=(1-2\omega)^{-1}$ between z and ω to calculate ω for several values of σ. The dependence of ω on σ is shown in Fig. 2 for d=3. It is clear from these plots that ω does not depend linearly on σ as suggested by the relation given

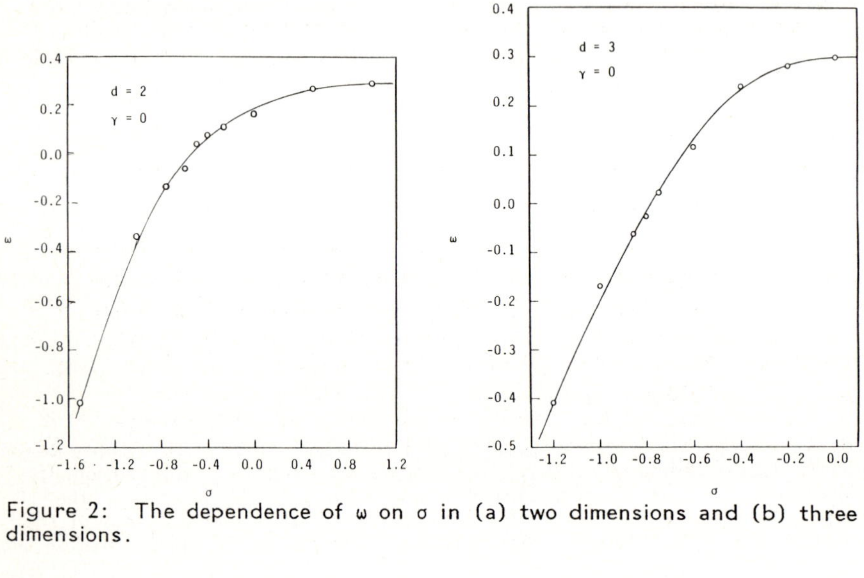

Figure 2: The dependence of ω on σ in (a) two dimensions and (b) three dimensions.

above. This implies that the sticking probability does not enter the kernel in the Smoluchowski equation in a straightforward manner.

4. STEADY-STATE COAGULATION

Cluster-cluster aggregation in its original form describes a process with a permanent evolution in time as the number of clusters in the system is always decreasing. One can modify this model to simulate an important process of both practical and theoretical interest, in which a steady-state cluster size distribution develops in the system. This goal can be achieved by feeding single particles into the system and removing the larger clusters according to some rules [15]. Steady-state conditions are very typical in many applied fields. For example, in the stirred tank reactors for aerosols, often used for modelling chemical reactors in industry, an analogous process takes place, but the particles are removed by letting them flow out from the chamber.

We have simulated the following model. At every unit time k particles are added at k different sites selected randomly. In addition, a cluster is discarded as soon as it becomes larger than a previously fixed number s_r. This is an extreme version of the situation in which the larger clusters leave the system with a higher probability. In this way, both the number of clusters N(t) and the number of particles M(t) in the system go to a constant value (N_∞ and M_∞) for long times. It is easy to show that in the Smoluchowski equation approach, both N_∞ and the characteristic time τ needed for relaxing to the final state scales as $k^{\frac{1}{2}}$. In the case of a more general coagulation equation, $N \sim k^\alpha$ and $\tau \sim k^\beta$ with the exponents α and β satisfying the scaling relationship $\alpha+\beta=1$ [15] independent of K_{ij}. Thus, in the steady-state coagulation there are two exponents which do not depend on the details of the aggregation process. Therefore, perhaps the best way to estimate the critical dimension for CCA kinetics and to investigate the effects of the fractal geometry of the clusters on the time-dependence of the aggregation process is to calculate the exponents α and β from simulations, and compare these results with the predictions of the SE.

Our results for the steady-state value of the number of clusters N_∞ and for the relaxation time τ indicate scaling of these quantities as a function of the feed rate k. For the case when $k/L^d \ll 1$ and the initial number of particles is very small, our data for N(t) can be well represented by the following scaling form for the number of clusters in the system at time t

$$N(t) \sim k^\alpha f(k^\beta t), \qquad (8)$$

where f(x) is a scaling function with $f(x) \sim x$ for $x \ll L^d$ and $f(x)=1$ for $x \gg L^d$. The actual shape of f(x) may depend on the parameters γ or s_r but for a fixed set of these numbers N(t) can be expressed through the scaling form (8).

The values of the exponents α and β can be determined from the slopes of the straight lines drawn through the data on the log-log plots of N_∞ and τ versus k. The numbers we obtained for α and β depend on the dimension of the space in which the cluster-cluster aggregation takes place, but they seem to be insensitive to the mass dependence of the diffusivity or to the parameter s_r. In one dimension we found that $\alpha=0.33\pm0.02$ and $\beta=0.65\pm0.03$. The two-dimensional simulations gave $\alpha=0.40\pm0.04$ and $\beta=0.58\pm0.05$ without and $\alpha=0.52\pm0.04$ and $\beta=0.46\pm0.05$ with the logarithmic corrections taken into account. In three dimensions $\alpha=0.47\pm0.05$ and $\beta=0.54\pm0.05$ were obtained.

Having determined α and β in one to three dimensions, we are able to discuss the relevance of the Smoluchowski equation to the cluster-cluster aggregation under steady-state conditions. One of the notable features of the numbers obtained for the exponents α and β is that their sum is approximately equal to one $\alpha+\beta \simeq 1.0$ in all dimensions we investigated. On the other hand, from the scaling properties of the original SE it follows that α and β should be equal to 1/2 in all dimensions. This value is in clear contradiction with our simulation results in one dimension, while it is in a reasonable agreement with the results obtained for the two and three-dimensional cases if we assume logarithmic corrections to scaling and indicate that two is the upper critical dimension for the kinetics of the problem.

ACKNOWLEDGEMENTS

This research was supported by grants from NIH Biosciences grant through Emory University and by the National Science Foundation Grant No. DMR-82-08051.

1. See e.g., Kinetics of Aggregation and Gelation, eds. F. Family and D.P. Landau (North-Holland, Amsterdam, 1984).
2. S. K. Friedlander, Smoke, Dust and Haze: Fundamentals of Aerosol Behavior (Wiley, New York, 1977).
3. T. Vicsek and F. Family, Phys. Rev. Lett. 52, 1669 (1984).
4. P. Meakin, Phys. Rev. Lett. 51, 1119 (1983).
5. M. Kolb, R. Botet, and R. Jullien, Phys. Rev. Lett. 51, 1123 (1983).
6. M. Kolb and H. J. Herrmann (unpublished) 1984.
7. P. Meakin, T. Vicsek and F. Family, Phys. Rev. B31, 564 (1985).
8. M. Kolb, Phys. Rev. Lett. 53, 1653 (1984).
9. F. Family, P. Meakin and T. Vicsek, preprint
10. M. von Smoluchowski, Z. Phys. Chem. 92, 129 (1918); Physik Z. 17, 585 (1916).
11. T. Vicsek and F. Family, Ref. 1, pg. 101.
12. P. Meakin and Z. R. Wasserman, Phys. Lett. 103A, 337 (1984).
13. R. Botet, R. Jullien and M. Kolb, Phys. Rev. A30, 2150 (1984).
14. B. B. Mandelbrot, The Fractal Geometry of Nature (Freeman, San Francisco, 1982).
15. T. Vicsek, P. Meakin and F. Family Phys. Rev. A, 1985.

Fractal Growth of Viscous Fingers: New Experiments and Models

Gerard Daccord and Johann Nittmann
Etudes et Fabrication Dowell Schlumberger, F-42003 St. Etienne, France
H.E. Stanley
Center for Polymer Studies and Department of Physics, Boston University, Boston, MA 02215, USA

We report on experiments which demonstrate that the instability pattern obtained when water pushes a more viscous but miscible, non-Newtonian fluid in a Hele-Shaw cell are **fractals**. Reproducible values for the fractal dimension d_f were found, which were interpreted using a modification of the diffusion-limited aggregation model. We also discuss the application of the gradient-governed growth model to two-fluid instability problems.

Introduction

Viscous fingering is the name given to an instability phenomenon which arises when a low viscosity fluid pushes a high viscosity fluid through materials which provide resistance to flow (e.g., a porous medium like rock). The low viscous fluid penetrates the high viscous fluid, producing thereby an instability pattern which resembles a finger of a glove.

The viscous finger instability was first studied by Saffman and Taylor [1] and Chuoke et al [2] in parallel plate flow (Hele Shaw cell). The flow in a Hele Shaw cell is very similar to flow in porous media—for both systems a Darcy-type pressure-flow velocity relation exists:

$$u = -(k/\mu) grad\ p \qquad (1),$$

with u, k, μ, and p representing the flow rate, permeability, viscosity, and pressure, respectively. One can show [3] that $k = b^2/12$ for a Hele Shaw cell, with b being the distance between the two plates.

Early work on viscous fingering has focused on the displacement of oil by water,because of the importance of this problem to the oil industry. In enhanced oil recovery, the oil-containing reservoir is flushed with a cheap fluid (water) to push remaining oil toward the production wells. The objective is to have a uniform displacement front to completely empty the oil reservoir. In general, however, fingering will reduce the area sweep efficiency and very often less than half of the oil is produced.

In order to limit the number of physical variables on which fingering depends, we study the instability under conditions where fingering will be most profound—namely for miscible fluids with a high viscosity difference. Highly ramified structures have indeed been observed for fingering in a porous medium by Habermann [4]. Ramified materials often do show fractal [5] structures, in which case analysis techniques based on the self-similarity principle can be applied.

Experiments

Oil and water is a immiscible fluid system; the finger growth is controlled by the interfacial tension between the two fluids. The pressure discontinuity Δp at the interface is given by

$$\Delta p = s(2/b + 1/R). \tag{2}$$

Here s is the surface tension and R is the curvature of the interface.

Chuoke et al [2] have shown that interfacial tension introduces a critical wavelength λ_{min}. Only pertubations with a wavelength $\lambda > \lambda_{min}$ will grow. In order to reduce this minimum wavelength to the smallest possible value [6], we choose miscible fluids for which no interfacial tension exists. We used water displacing a water-based polymer solution of high molecular weight polysaccharide. The intermolecular diffusion is very slow ($D = 10^{-5} cm^2/sec$) so that appreciable mixing between the two fluids will not occur during the time interval of our experiment.

Polymer solutions are non-Newtonian fluids. In contrast to Newtonian fluids ,where the viscosity is a function of temperature and pressure, the viscosity also depends on the shear rate. Many polymer solutions are shear thinning fluids, i.e., the viscosity decreases with increasing shear rate. The shear thinning character of our fluid is very important,as it enables us to pump the fluid easily in a thin cell without causing bending of the plates. The viscosity of the polymer solution is between 10^3 and 10^4 cP, depending on the flow velocity. We used a wide range of flow rates and polymer concentrations. Two different plate distances b were used: 0.5 and 1.0 mm. The instability structures were random and highly branched for both [6].

Analysis

Visual examination of the instability pattern showed the presence of fingers of many different length scales [6]. We therefore found it appropriate to analyse the structures using the concept of self-similarity or fractals [4]. We first digitized the finger system using a pixel size equal to the characteristic thickness of the viscous fingers. Then we used two methods to determine the fractal dimension of our patterns. We first measured the total exterior perimeter length using a ruler of length $L = 1$ measured in the units of the pixel. Then we repeated the same operation for a range of increasing L up to $L = 20$.

The apparent length of any fractal, in units of L, decrease as L^{-d_f}. We found the quantitative parameter $d_f = 1.40 \pm 0.04$.

As a second method, we placed an imaginary box of edge L on each point of the finger. For each box we found the number of points (total mass) inside the box. As L increases, the mass should increase as L^{d_f}, if the finger has a fractal structure. We find that method 2 gave smaller values (by about 8%) for small clusters as a large proportion of the finger points lie on the perimeter of a pattern. Boxes centered at these points will cover much unoccupied area outside the structure. For large clusters both methods gave an identical fractal dimension.

Statististical Model: Diffusion-Limited Aggregation

In Eq. (2) the velocity u is defined as a gradient of a potential ϕ ($\phi = -p$). Together with the equation for the conservation of mass for an incompressible fluid, $div\ u = 0$, we obtain the following equation for the potential in the Hele Shaw cell

$$div\left[(1/\mu) grad\phi\right] = 0. \tag{3}$$

If we assume that the less viscous fluid has neglibible viscosity compared to the high viscous fluid, we can approximate the potential within the low viscous fluid as being constant. In this case, we only need to solve Eq. (3) inside the domain of the high viscous fluid and for which we obtain a Laplace Equation for the potential

$$div\ grad\phi = 0. \tag{4}$$

As boundary conditions, we choose $\phi = \phi_1$ $(constant)$ at the fluid interface and $\phi = \phi_0$ at the outlet.

With these boundary conditions, Eq. (4) is identical to the Laplace equation describing the probability $u(\underline{r}, t)$ of a random walker to be at the point $\underline{r}$ at the time t [7,8]. This is similar to diffusion-limited aggregation (DLA) [8,9]. Hence we propose the following quantitative model for viscous fingering in a rectangular Hele Shaw cell:

Assume a rectangular domain of width $w = 30$ units and length $l = 100$ units. Random walkers are then launched randomly from the left boundary. A seed particle is placed in the middle of the right end boundary. If one of the random walkers hits the seed, it sticks, whereas if it hits any of the four walls it "dies" and a new walker is released. The width w was chosen in accordance to the experimental value for $R = W_{HS}/W_f$ (ratio of the Hele Shaw cell width to characteristic finger thickness).

Two striking features of the actual viscous fingering experiments are mirrored by our statistical method. The first one is the tendency of growing side branches to die out after a relatively short period of growth. The second is the fact that the entire viscous finger strongly avoids coming even moderately close to the wall. Both features are further discussed in Nittmann, Daccord and Stanley. In addition, Figs. 3 and 4 of Ref. 6 show the excellent agreement in the time evolution of model and experimental "fingers." Calculating d_f as discribed above revealed $d_f = 1.41 \pm 0.04$, a value remarkable close to the experimental value. The difference from the value ~ 1.7 found for DLA in radial simulations is explained by the growth limitation introduced by the two lateral and left end boundaries. Even at initial stages of growth the small cluster cannot capture enough mass,as incoming walkers are strongly influenced by the absorbing walls. Therefore in the **linear** Hele Shaw problems d_f never takes on its bulk value (1.7) before it starts to turn over to its ultimate value of $d_f = 1.0$, but instead takes a smaller value that depends monotonically on W_{HS} (Fig. 6 of Ref. 6).

Statistical Model: Gradient-Governed Growth

In the previous sections ,we have shown that viscous fingers observed on a laboratory scale show a fractal structure if we consider fluids of zero interfacial tension and with a high viscosity ratio. For that case we have seen that Eq. (3) simplifies to Eq. (4) and DLA-type models can simulate the fractal dimension very closely. DLA, however, has the conceptual "drawback" that it forms fractal clusters by aggregating mass from outside rather than growing them from the inside,as observed in fluid flow. In addition, one would wish to model the instability for a wide range of fluids with or without interfacial tension and with a finite viscosity ratio. We therefore suggest a model put forward recently by Sherwood and Nittmann [10]. It is an extension of the work by Niemeyer et al [11] on a dielectric breakdown model and by Meakin [12],who has introduced the term "diffusion-limited growth." DLG is similar to DLA in solving Laplace equations for the potential distribution outside the cluster, not with random walkers but with a successive overrelaxation method. In the gradient-governed growth model (GGG), Eq. (3) is solved in both fluids, and the interface grows proportional to the local pressure gradient. In Refs. 10 and 13 a detailed study is made of the dependence upon the viscosity ratio κ. In the high viscosity ratio case the structure is very close to the fractal structure of Ref. 11. With decreasing κ the fingers become thicker and the ramification is fading. For the case that the displacing fluid is more viscous than the displaced fluid (e.g., oil displacing water) the interface is circular [13].

[1] P. G. Saffmann and G. I. Taylor, Proc. R. Soc. A **245**, 312 (1958).

[2] R. L. Chuoke, P. Van Meurs, and C. J. Van der Poel, J. Petrol. Tech. **11**, 64 (1959).

[3] H. Lamb, **Hydrodynamics** (Cambridge University Press, London, 1932).

[4] B. Habermann, Trans., AIME **219**, 264 (1960).

[5] B. B. Mandelbrot, **The Fractal Geometry of Nature** (Freeman, San Francisco, 1982).

[6] J. Nittmann, G. Daccord, and H. E. Stanley, Nature **314**, 141 (1985).

[7] T. A. Witten and L. M. Sander, Phys. Rev. Lett. **47**, 1499 (1981).

[8] L. Paterson, Phys. Rev. Lett. **52**, 1621 (1984).

[9] L. P. Kadanoff, J. Stat. Phys. (submitted).

[10] J. D. Sherwood and J. Nittmann, J. de Physique (submitted).

[11] L. Niemeyer, L. Pietronero, and H. J. Wiesmann, Phys. Rev. Lett. **52**, 1033 (1984).

[12] P. Meakin, preprint.

[13] G. Daccord, J. Nittmann, and H. E. Stanley, preprint.

Fingering Patterns in Hele-Shaw Flow

J.V. Maher
Department of Physics and Astronomy, University of Pittsburgh, Pittsburgh, PA 15260, USA

The formation of fingers through the Saffman-Taylor instability in Hele-Shaw flow presents one of the simplest of pattern formation problems. The flow is governed by the Poisson equation, and thus should have a close formal connection[1,2] through its similarity to the random walk problem to diffusion-limited aggregation (DLA)[3]. This poses the fascinating simulation problem of how much and which physical properties must be added to the essentially fractal DLA problem to mock up the interfacial conditions which produce the smooth fingering patterns seen in the Saffman-Taylor flow[4,5,6]. The experimental work I will present in this paper[7] exploits the well-established critical properties of binary mixtures to vary the crucial viscosity contrast parameter more precisely and over a greater range than ever before. In addition, the temporal development of the non-linear pattern is determined more accurately than in most earlier work.

Hele-Shaw flow is flow between parallel plates whose separation distance, b, is small compared to any other length in the problem. Hill[8] and Saffman and Taylor[9] have demonstrated that the interface between two liquids in a Hele-Shaw cell becomes unstable when the interface between fluids 1 and 2 advances toward fluid 1 at a speed V such that

$$(\rho_1 - \rho_2)\, g + [12\, (\mu_1 - \mu_2)/b^2]\, V \geqq 0 \tag{1}$$

where μ_i, ρ_i are the shear viscosity and density of the ith liquid. Thus when gravity tends to stabilize the flow ($\rho_1 < \rho_2$), the instability arises only if $\mu_1 > \mu_2$, in which case the critical velocity, V_c, is reduced as the viscosity contrast increases. It is this instability which makes it difficult to vacuum up an oil spill. This instability is also closely related to the fingering instability which complicates recovery from oil wells, but, as Wilkinson discusses elsewhere in this conference[10], the boundary conditions are significantly different in a porous medium. Equation 1 also shows that the instability can be gravity-driven if $\rho_2 < \rho_1$, but it is distinctly different from the Rayleigh-Taylor instability because the Hele-Shaw flow is a Darcy flow, and thus lacks the inertial terms which make the Rayleigh-Taylor instability so violent.

Chuoke et al.[11] included interfacial tension, σ, in a linear stability analysis and concluded that the instability was present for all wavelengths greater than

$$\lambda_c = 2\pi\, [\sigma b^2/12(\mu_1 - \mu_2)(V - V_c)\,]^{1/2} \tag{2}$$

This analysis also indicated a fastest growing wavelength $\simeq \sqrt{3}\,\lambda_c$ and experiments tend to show this wavelength during the early stages of the growth of the fingering pattern[12]. Some experiments, including the original work of Saffman and Taylor[9], suggest that the system reaches a long-time steady-state pattern with only one finger of the less viscous liquid penetrating the other. Theoretical arguments have been raised questioning the stability of any such single-finger solution[13], but at any rate the early experiments clearly show a finger-amalgamation/competition mechanism with the late-time pattern exhibiting a significantly larger average wavelength than does the early-time pattern.

Recently Tryggvason and Aref [14] (TA) have performed vortex sheet calculations to follow the dynamics of this instability far into the non-linear regime. They cast the problem in terms of two control parameters, a dimensionless viscosity contrast

$$A = (\mu_2 - \mu_1)/(\mu_2 + \mu_1) \tag{3}$$

and a dimensionless surface tension

$$B = \sigma b^2/6U^* W^2(\mu_2 + \mu_1) \tag{4}$$

where W is the width of the Hele-Shaw cell and U^* is a characteristic velocity

$$U^* = \left| \, [(\mu_1 - \mu_2)V + 1/12(\rho_1 - \rho_2)gb^2]/(\mu_1 + \mu_2) \, \right| . \tag{5}$$

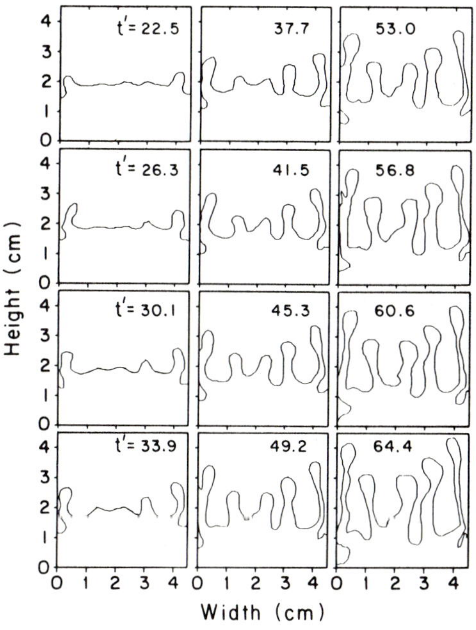

Figure 1

A dimensionless time can be defined as

$$t' = U^* t/WB^{1/2} \tag{6}$$

($B^{1/2}$ is included to scale surface tension out of the Hele-Shaw equations [14]). Similarly, the length of the mixing zone (θ, the distance along the direction of flow between the tips of the longest fingers in each direction) can be made dimensionless as

$$\theta' = \theta/WB^{1/2} . \tag{7}$$

The TA results find a reduction in the number of fingers with time only when A is large. For large A they find some tendency for fingers to amalgamate but, more importantly, long-range interactions along the interface act in such a way that, once one finger gets ahead of the others, the other nearby fingers grow still less rapidly. Neither of these effects occurs for small A in the TA calculations. Most available data [12,13] correspond to A ~ 1 (values of A near 1/2 have usually been achieved by using miscible liquids, in which case there is a temporally expanding diffusion zone instead of a sharp interface).

I am reporting measurements of viscous fingering for a system of immiscible liquids with A ~ 0. At the same time, measurements were also made for a system with A ~ 1 to allow direct comparison. The A ~ 0 system used the binary-liquid mixture isobutyric acid + water at critical composition. The important properties of this system are well-known [16] power laws in the reduced temperature $\varepsilon = (T_c - T)/T_c$ where T is the temperature and T_c is the system's critical temperature (26.12°C). Using the published values [16] of $\rho_2 - \rho_1$ and assuming quite reasonably that kinematic viscosity is the same in both phases, one finds that $A = 0.053\varepsilon^{\beta}$ and $B = 0.024\varepsilon^{2\nu-\beta}$ where $\beta = 0.31$ and $\nu = 0.61$. The temperature of the system was controlled to ± 1mK and measurements were made at several temperatures between 21.09°C (A = 0.015) and 26.07°C (A = 0.004). The Hele-Shaw cell containing the binary mixture was a square of side length 45 mm with a 1 mm gap. A measurement consisted of preparing the system at equilibrium with a flat, gravity-stabilized meniscus, then inverting the cell and photographing the interface as the non-linear pattern developed. A digital pattern of the interface was then formed for each picture by running the cursor of a computer-interfaced digitizer over the photograph. A selection from a time series of such patterns is shown in Figure 1 for a measurement with A = 0.015 and B = 6.6 x 10^{-4}. The patterns in Figure 1 show five fingers; by varying the temperature, measurements were made for patterns with as many as fifteen fingers. The variation of finger width with temperature followed the $\varepsilon^{1/2}$ dependence expected from the arguments above. Figure 2 shows patterns for several different temperatures. Most features of the pattern formation in this liquid mixture even survive the beginning of breakdown in the Hele-Shaw assumptions: as $T \to T_c$ the characteristic wavelength becomes progressively smaller, and eventually becomes comparable to b, at which point a gravitationally unstable Saffman-Taylor instability should begin to change into a Rayleigh-Taylor case. For the A = 0.0063 and A = 0.0039 cases in Figure 2 the wavelengths are respectively 3.5b and 3.0b, and the beginning of Hele-Shaw breakdown manifests itself very gently with the formation of two contact lines, one running along each face of the cell. Nevertheless, these fingering patterns show a flow which exhibits all the characteristics discussed above for the simple Hele-Shaw case, and also follow the power law to be discussed below.

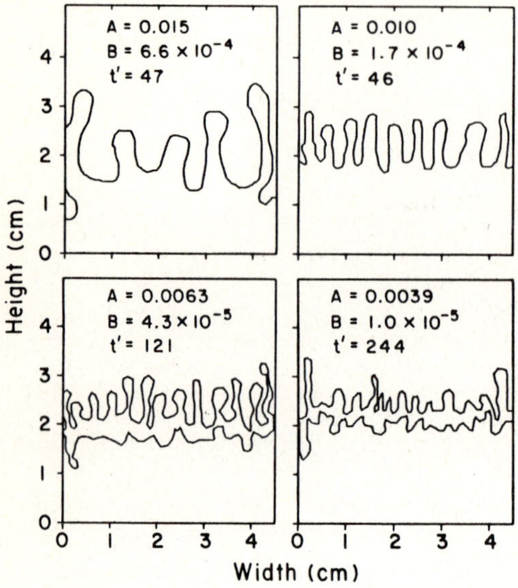

Figure 2

The A ~ 1 system advanced water against paraffin oil in a Hele-Shaw cell of length 1 m, width 36 cm and gap 0.3 cm. The interface advance was forced by applying a pressure gradient; data were acquired photographically and processed as described above. Figure 3 shows a selection of frames from a time series of patterns for this system with A = 0.93 and B = 8.3 x 10^{-4}. It should be noted that, while the A~1 system is gravitationally stable and driven unstable by application of a pressure gradient, the A ~ 0 system is gravitationally unstable. Since it is experimentally quite difficult to apply an external pressure to the binary liquid mixture while maintaining the requisite temperature control, the comparison of A ~ 0 with A ~ 1 depends on there being no essential difference between the two driving mechanisms. Indeed, pressure gradient and gravitational field play equivalent roles in the Hele-Shaw equations and in the characteristic velocity, U^*, defined above.

The most dramatic differences between high and low-A systems can be seen by comparing Figures 1 and 3. The low-A system has never been observed to change the number of fingers during the development of the fingering pattern from the linear regime. Additionally, rapid growth of one finger in the low-A system does not appear to affect its neighbors so, despite some apparent randomness in the length and width of fingers, there is no systematic impoverishment of fingers in the neighborhood of a rapidly developing finger. Figure 3 shows a high-A case where two fingers grow and the others experience only a very stunted growth. Other high-A measurements (from this experiment but not shown in the figure) have occasionally shown a finger-amalgamation mechanism, and always show a stultification of growth of uncompetitive fingers at dimensionless times very small compared to the largest times, at which the low-A system has been observed. Since this experiment is designed with wide cells to allow the early-time formation of many fingers and careful observation of these patterns as they develop out of the linear regime, it is not able to answer whether eventually there are steady-state patterns with one or many fingers. The strict

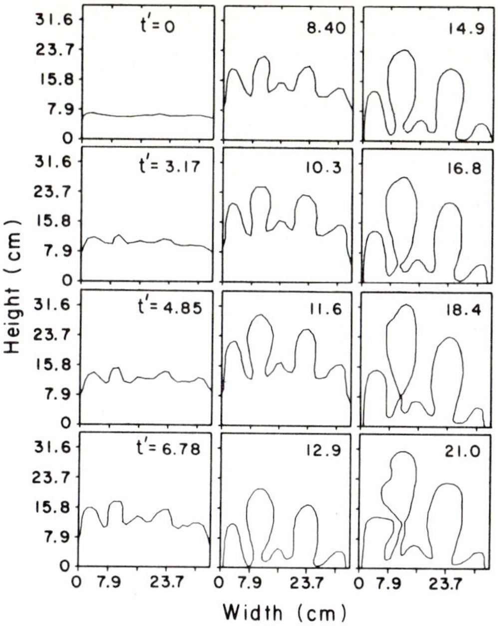

Figure 3

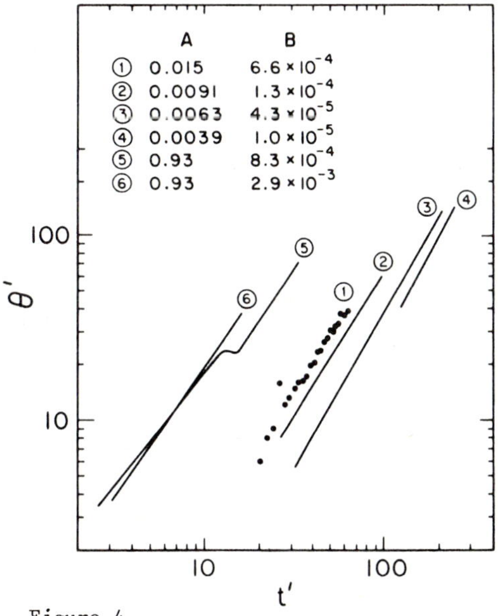

Figure 4

symmetry of the flow only occurs for A = 0 and there is no other value of A for which there should be a change of character, so one might expect that all $A \neq 0$ flows should eventually show the same late time characteristics. This poses an interesting question as to why flows at $0 < A \ll 1$ show no tendency to reduce their number of fingers. Since this tendency appears in the TA calculations along with the data, the physics is clearly contained in the Hele-Shaw equations even though we cannot at present identify the source.

The dimensionless length of the mixing region, θ', shows a remarkably regular dependence on dimensionless time for all measurements. If one ignores a somewhat noisy early period during which the linear pattern establishes itself, $\theta' \sim (t')^{1.6}$ from the emergence of the linear pattern through to the end of the measurement. This can be seen in Figure 4 where straight lines have been drawn through the data of log θ' vs log t' for several measurements and the individual data are shown for one additional measurement. There is some scatter in the slopes of these lines about their average value of ~ 1.6, but this does not seem large when one considers that the uncertainty in any one slope is relatively large (the θ' measurements span less than two decades). One of the paraffin oil + water measurements ($A = 0.93$, $B = 8.3 \times 10^{-4}$) shows an abrupt shift in the θ' vs t' curve at a rather late time. This shift results from the water abruptly displacing paraffin oil along the teflon spacers at the side of the cell after a long period of stasis at the side wall. No such shift ever appeared in any measurement with the binary liquid system; the two phases of the binary liquid systems have such similar composition that their wetting properties at the walls are very similar and no sticking was ever observed. Figure 1 does show a slightly earlier growth of fingers near the side walls for the binary liquid mixture. This tendency was seen for all the A ~ 0 measurements and has been reported in previous A ~ 1 measurements [15]. It is difficult to assess whether or not this wall effect should always be present; it is smaller in the present A ~ 1 measurements than in Ref. 15 and smaller still in the A ~ 0 patterns of Fig. 2. A related question which has recently been addressed [17] involves the appropriate pressure drop across a curved interface, but even Ref. 17 ignores variations in local interfacial curvature due to variations of the contact angle with velocity.

The Power law $\theta' \sim (t')^{1.6}$ corresponds to an acceleration of the length of the mixing zone $(d^2\theta'/dt'^2)$ which varies roughly as $(t')^{-0.4}$. Presumably in a sufficiently long cell a terminal velocity would be reached, but it is interesting to note that through rather advanced stages of non-linear growth and for the wide range of viscosity contrasts reported herein, the mixing zone acceleration shows such a regular time-dependence. The TA predictions for θ' vs t' are only qualitatively similar to this observation; their $A = 1$ calculation falls reasonably well on a power law with $\theta' \sim t'^{1.85}$ while their $A = 0$ calculation is essentially identical to the $A = 1$ at early times $t' < 25$ and subsequently shows a much slower growth $(\theta' \sim t^{1.3})$. Not only is no such difference between $A \sim 1$ and $A \sim 0$ seen in this experiment, but the $A \sim 1$ fingers are observed to develop at much earlier dimensionless time than are the $A \sim 0$ patterns.

A dimensionless stretching of the interface can be defined, following TA, as $L' = (L - L_o)/L_o$ where L is the length of the interface and L_o is its initial value. For smooth fingers at very late times L' should vary linearly with θ', but this is not the case through the stages of pattern formation discussed here. Instead L' grows faster than θ'. In the $A \sim 0$ case this can be partly understood as a tendency for a few fingers to grow faster than others as the patterns first form, but for the others to catch up at later times. These $A \sim 0$ cases all show a very similar dependence of L' on θ' which can be reasonably well expressed as $L' \sim (\theta')^{2.2}$. The $A \sim 1$ data show a less dramatic growth of L' with θ' along with a greater variation in this dependence, but they do appear to follow a power law with an exponent averaging ~ 1.6. Figure 5 shows L' vs θ' for several cases. This faster growth for L' than for θ' in the presence of stultification of some fingers and a secular reduction in the number of important fingers is very difficult to understand and appears to arise from an increase of complexity of the shape of the individual fingers. Nittmann et al. [18] have reported observing fractal behavior in the shape of viscous fingers formed between two miscible liquids, one of which was non-Newtonian. It is possible (but not at all clear) that the interfacial stretching in the present case shows a last vestige of a fractal behavior which is smoothed by surface tension. Similarly the growth law $\theta' \sim (t')^{1.6}$ is reminiscent of the fractal growth rate calculated by Witten and Sanders[3] for diffusion-limited aggregation.

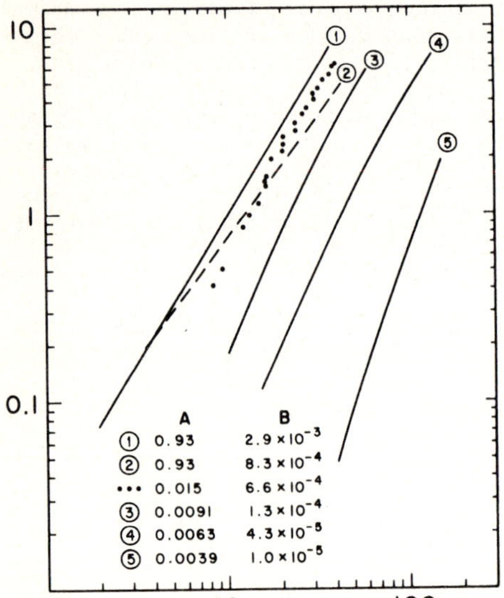

Figure 5

It would be interesting to understand the role of the shear-dependent viscosity of the polymer solution of Nittmann et al. [18] in producing a fractal pattern. Clearly, some miscible liquids produce smooth fingering patterns [12]. We plan to vary the capillary length and the diffusion constant for a miscible pair of Newtonian fluids by extending our isobutyric acid + water measurements to observe flow at temperatures in the one-phase region of mixtures prepared at different viscosity contrasts in the two-phase region.

Further analysis of the present data will concentrate upon distributions of curvature, finger width and finger length and their dependence on t', A and B. However, the results I have presented in this paper show dramatic differences in the development of non-linear fingering patterns at different values of A and a strikingly simple power-law dependence on time of finger length for all values of the control parameters. Specifically these results are: 1) neighboring fingers' growth rates do not affect each other noticeably for A ~ 0 but do for A ~1; 2) fingers do not amalgamate when A ~ 0 but they occasionally do when A ~ 1; 3) the dimensionless interpenetration zone of the liquids follows the power law $\theta' \sim (t')^{1.6}$ at late times for all observed values of A and B; and finally, 4) the dimensionless stretching of the interface, L', grows faster than θ' for all values of A. Reproducing these dynamical systematics may provide a tractable case to bridge the gap from fractal-growth simulations to simulations of smooth growth.

This work is supported by the U.S. D.O.E. under grant #DE-F602-84ER45131.

REFERENCES

1. L. Paterson, Phys. Rev. Lett. 52, 1625 (1984).
2. L.P. Kadanoff, J. Stat. Phys, in press. C. Tang, Phys. Rev. A 31, 1977 (1985).
3. T.A. Witten and L.M. Sander, Phys. Rev. Lett. 47, 1499 (1981).
4. S. Liang and L.P. Kadanoff, Phys. Rev. A 31, 2628 (1985).
5. T. Vicsek, Phys. Rev. Lett. 53, 2281 (1984).
6. D. Jasnow, private communication.
7. J.V. Maher, Phys. Rev. Lett. 54, 1498 (1985).
8. S. Hill, Chem. Eng. Sci 1, 247 (1952).
9. P.G. Saffman and G.I. Taylor, Proc. Roy. Soc. London A245, 312 (1958).
10. D. Wilkinson, paper presented at this conference.
11. R.L. Chuoke, P. van Meurs and C. van der Poel, J. Petrol. Tech. 11, 64 (1959).
12. R.A. Wooding and H.J. Morel-Seytoux, Ann. Rev. Fluid Mech. 8, 233 (1976); and references therein.
13. J.W. McLean and P.G. Saffman, J. Fluid Mech 102, 455 (1981).
14. G. Tryggvason and H. Aref, J. Fluid Mech 136, 1 (1983), and preprint.
15. C.-W. Park, S. Gorell and G. M. Homsy, J. Fluid Mech 141, 275 (1984); I. White, P.M. Colombera and J.R. Philip, Soil Sci. Soc. Am. J. 41, 483 (1977).
16. B. Chu, F.J. Schoenes and W.P. Kao, J. Am. Chem. Soc. 90, 3042 (1968).
17. C.-W. Park and G. M. Homsy, Bull. Am. Phys. Soc. 29, 1557 (1984) and J. Fluid Mech 139, 291 (1984).
18. J. Nittmann, G. Daccord and H.E. Stanley, Nature (London) 314, 141 (1985).

Part VIII **Flow and Diffusion in Porous Media**

Two-Component Transport Properties in Heterogeneous Porous Media

Groupe Poreux PC*
Laboratoire d'Hydrodynamique et de Mécanique Physique, E.S.P.C.I.,
10, rue Vauquelin, F-75231 Paris Cedex 05, France

1. INTRODUCTION

Recent developments in the study of porous media (POM) emphasize the correspondence between geometrical features (percolation, aggregates, fractal and, more generally, multiple scale structures) and the physical properties associated with them. This approach makes use of recent cooperative research works undertaken on the physics of disordered media : in particular in France a CNRS research program on MIAM (MIlieux Aléatoires Macroscopiques) associates scientists from fundamental and applied research laboratories on comparative studies on electrical and mechanical study of composites, visco-elasticity of gels, concentrated suspensions and transport in POM (porous media), which are essential problems discussed in the present book.

In the present article, we will discuss physical (essentially transport) properties of fluids in POM. This has been a study of long lasting interest, in particular because of their large range of applications : until recently these problems had been essentially approached by looking for homogenized properties (electrical conductance, permeability, diffusion...). They have been revisited in the last few years by applying to them scaling concepts present within the geometrical description of disordered systems. These concepts apply best to POM structures having a multiplicity of scales, like fractal structures which have a symmetry of dilation, or near a permeability threshold which corresponds to a percolation problem. The most natural example is that of a multiply connected or percolating POM saturated with one single phase. Let us analyse briefly, in this monophasic case, the basic assumptions of the classical theory and their possible

+C. Baudet, E. Charlaix, E. Clément, E. Guyon, J.P. Hulin, C. Leroy

shortcomings. In his historical work, Henri Darcy expressed the proportionality between the flow rate per unit area $\vec{Q}$ of a fluid saturating a porous medium and an applied pressure gradient $\vec{\nabla}p$

$$\vec{Q} = -(K/\eta)\ \vec{\nabla}p \tag{1.1}$$

The permeability K (length^2) is a function of the pore geometry. The properties of the fluid enter the equation through the viscosity η.

This classical description is based on two major assumptions :

1 - existence of a length scale l_o large, on one hand, with respect to the pore size and the characteristic scale of microscopic variations and, on the other hand, small with respect to the macroscopic features of the medium structure (fractures, layering...).

Then, one can define significant macroscopic variable by averaging the microscopic quantities on a "representative elementary volume" (REV) of size l_o.

2 - linear variation of the microscopic velocities with the stresses applied to the fluids.

In many cases, Darcy's law is not valid, even in cases where the flow is laminar and stationary in the pores :

- non-linear effects take place when $(\vec{v}.\vec{\nabla})\vec{v}$ terms become significant in the Navier-Stokes equation due to fast local changes of the velocity field $\vec{v}$;

- heterogeneous materials with a large range of characteristic sizes ranging from small fissures or grains to large fractures, preventing the definition of a suitable averaging length (in some cases, a "fractal" type description may be suitable);

- compact rocks with a low fracture density can have a small (critical) permeability. The fractured system has a structure similar to an infinite percolation cluster, and permeability fluctuations appear even on length scales much larger than the fracture individual size.

In such cases, a statistical disordered physics approach is better adapted than the classical ones where local averagings are made. Paradoxically, this monophasic flow problem has not been studies in this scaling spirit as much as problems of two-phase flows, and we will address essentially this last class of systems by considering first the problem of immiscible fluid phases in a POM which leads percolation and fractal structures (chapter 2). In chapter 3 we will extend the result to two miscible fluids : in the dilute limit, the problem is that of the diffusion (in the absence of flow) or dispersion (if a flow field is present) of a dye, or small particles diluted in a fluid phase contained in the POM. Transport on percolating or fractal lattices shed light on this class of problem.

The problems of monophasic flows through multiple scale materials are discussed in a twin paper [1.4 chapter 2].

2. DIPHASIC FLOW IN POROUS MEDIA

2.1. Classical description

The classical approach of diphasic flows assumes a good local homogeneity of the POM and introduces new macroscopic variables : the fluid saturations S (S_w and S_o are, for instance, the local average percentages of pore volume occupied by water and oil : $S_w + S_o = 1$). The usual generalization of the Darcy law estimates the flow rates of the two fluids by the equations :

$$\vec{Q}_{ws} = -(K\, k_{wr}/\eta_w)\vec{\nabla}p \qquad (2.1)$$

$$\vec{Q}_{os} = -(K\, k_{or}/\eta_o)\vec{\nabla}p \qquad (2.2)$$

K is the monophasic permeability value. The two dimensionless coefficients k_{wr} and k_{or} are the relative permeabilities for oil and water : they depend on the medium characteristics and on the saturation values.

The figure shows typical variations of k_{wr} and k_{or} with S_w. The relative permeability for each fluid goes to 0 below a threshold saturation value for which this fluid is no more continuous but trapped as isolated pockets inside the rock. These values are respectively called the irreducible oil and water saturations (s_{wi} and S_{oi}).

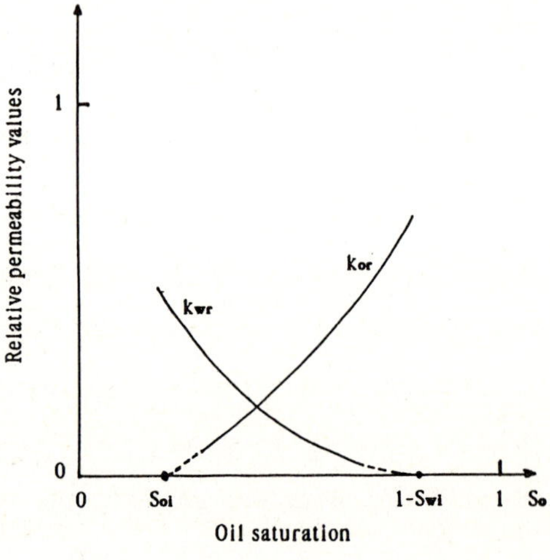

In addition to the difficulties with the application of Darcy's law already discussed in 1, the relative permeability approach suffers additional ones associated with the presence of two fluids :

- near the threshold saturation value for the permeability to a given fluid, its distribution will be very inhomogeneous and similar to the structure of the infinite cluster in a percolation problem. Thus, even if the pore structure is homogeneous and well connected, large fluctuations of S_w occur over a wide range of length scales and prevent the determination of significant local averages;

- it is possible to apply different pressure gradients $\vec{\nabla}p_w$ and $\vec{\nabla}p_o$ to the two phases. In addition to the diagonal coupling terms of (2.1) and (2.2), non-diagonal contributions should enter, describing the mobilization of one phase under the pressure gradient applied to the second one. They are usually neglected;

- the relative permeabilities depend on the history of the invasion process (nature of the original wetting phase, invasion fluid velocity, number of invasion cycles). Different results are obtained when the non-wetting fluid diplaces a wetting fluid already filling up the pore space (drainage) or when the wetting fluid displaces the non-wetting one (imbibition);

- capillary forces create a pressure difference Δp between the two sides of the interface menisci separating the two fluids inside the medium; if r is the radius of a given channel, θ the wetting angle and γ the interfacial tension

$$\Delta p = 2\gamma \cos\theta / r \tag{2.3}$$

At low flow velocities, the capillary effects are dominant and equations (2.1,2) must be modified to take into account the capillary pressure difference (2.3) between the two fluids.

The ratio between the capillary effects and the viscous pressure drops in a given fluid is characterized by the capillary number Ca. $Ca = \eta v/\gamma$ (2.4) where v is a characteristic mean velocity. Ca is a ratio of the viscous and capillary pressure drops on a length scale ℓ representing both the channel width and its length. One has indeed :

$$\frac{\text{viscous pressure drop}}{\text{capillary pressure drop}} \simeq \frac{\eta|\Delta\vec{v}|\ell}{\gamma/\ell} \simeq \frac{\eta v}{\gamma}\frac{\ell^2}{R^2} \simeq Ca\,\frac{\ell^2}{R^2} \tag{2.5}$$

R is another length scale characterizing the velocity gradients and $\ell/R \sim 10^{-1}$ to 10^{-2} for granular systems.

Capillary forces are introduced as corrections in classical models, but in a way that does not account for the essential role they play in slow displacement processes. The classical model of Welge, Buckley and Leverett (B.L.W.) describes the fluid invasion process by assuming that the relative permeability description is valid, and that the k_r are unique functions of the S_w. Values of k_r cannot be determined directly for actual samples of rocks, because generally local saturation measurements cannot be performed economically. The Welge method allows to deduce these values from time variations of the composition of a mixture flowing out of a sample during a water injection at a constant flow rate.

The B.L.W. model predicts the formation of a shock type discontinuity in the invasion front. The displacement of the front results from the mutual replacement of one phase by the other, which is no larger possible beyond both saturation thresholds. This non-monotonic variation of the mutual permeability is responsible for the shock. The shock front gets smeared out by capillary effects. De Gennes[2.3] has reconsidered the B.L.W. model, taking into account the critical variation of the k_r's near the irreducible saturation thresholds, and has found that the diffusion behavior should be "hypodiffusive" (there is a cut-off of the smearing of the front at both ends).

The relative permeability values obtained by the B.L.W. approach are the basis for most practical applications. They rely on assumptions which are not verified for heterogeneous rocks and near-irreducible saturation values.

2.2. Statistical and microscopic approach to diphasic flows

The classical results described above represent a mean field type approximation of the problem. Observations in porous media, in particular near relative permeability thresholds, can only be analysed by a statistical approach[2.4].

2.2.1. Why and when is a microscopic approach necessary ?

- When two non-miscible fluids coexist within a rock, the structure of the mixture depends very much on the invasion process. For instance, let us inject water into a rock sample originally saturated with oil. The residual oil content gets smaller when the injection flow-rate increases. However, as soon as the continuity of the oil phase has been broken, one needs to increase enormously the water flow-rate to mobilize again the oil. This is due to the large capillary forces preventing the motion of isolated blobs. The only driving force is the viscous pressure drop acting between the extremities of the blobs and which must overcome the capillary forces ($U > \gamma/\eta$). On the other hand, in the initial part of the injection, when the displaced fluid path is continuous, small pressure gradients could set this fluid in motion. Clearly, a prediction of the residual saturation values and of the blob size distribution requires a good modelization at the pore level.

- A relative permeability approach can be satisfactory far from the irreducible saturation thresholds. Close to them, large fluctuations of the fluid distribution occur, requiring a proper description. A similar problem occurs in the front part of the invasion profile, where the shock front predicted by the classical theory is broadened by local heterogeneities.

- Invasion front instabilities : the classical Saffman-Taylor approach predicts, in porous media (and in Hele-Shaw cells), the onset of invasion front instabilities when a viscous fluid is displaced by a less viscous one. In this theory, the overall macroscopic invasion front curvature value is used to evaluate the capillary pressures determining the instability wavelength; the local curvature in the pore throats (or between the plates of the Hele-Shaw cell) is neglected in this approach. The instability patterns observed experimentally have complex shapes, showing again the importance of local fluid structures. At high enough velocities, the invasion front may have a structure close to that observed in diffusion-limited aggregation(2.5).

2.2.2. Why can one think of applying percolation to flow through porous media?

The percolation approach is best adapted to the description of the slow invasion of a porous material by a fluid less wetting than the one already present in the POM (drainage). In this case, the transport of fluid by wall films is avoided. Adapting percolation theory to those problems is suggested by several threshold effects in porous media.

- Permeability threshold with saturation : a minimum percentage of the pores must be occupied by a given fluid to allow the existence of a continous path for its flow through the porous material.

- Pressure threshold for the invasion of a porous material : such thresholds exist for the invasion of a porous material by a non-wetting fluid (drainage). In porosimetry, only pores located close to the surface are accessible to non-wetting Hg at low pressures. Above a pressure threshold value, a continuous path of accessible bonds appears across the material, after which the mercury saturation increases very fast.

In these problems appears the very important notion of accessible pore throats. A meniscus separating locally the two fluids can pass through a given channel connecting two pores only if the pressure difference Δp applied between the two fluids is sufficient to overcome the value of the capillary pressure when the meniscus reaches the narrowest section (of radius r_c) of the channel. The threshold pressure Δp_c is given by :

$$\Delta p_c = 2\gamma \cos\theta / r_c \qquad (2.6)$$

2.2.3. *Invasion percolation*[2.6]

A a given injection pressure, the Hg cluster fills up all pores connected to the injection surface by a path,such that the channel width is always larger than r_c. The statistics is the same as for classical bond percolation : conducting bonds correspond to channels with a width $r > r_c$; a cluster is limited by bonds such that $r < r_c$. Some differences appear between this "invasion" percolation and classical percolation :

- only pores connected to the sample injection surface can be invaded;

- when the medium is initially filled with fluid instead of being evacuated, as in porosimetry, trapped pockets of displaced fluid appear, and lead to a larger value of the final residual fluid saturation. At the injection pressure threshold value (breakthrough), the injected fluid volume resembles an infinite percolation cluster, and should have a fractal structure extending through the whole sample volume (although boundary effects may have to be taken into account).

2.2.4. *Validity of the invasion percolation model*

The invasion percolation model corresponds to slow invasion by a fluid with no wettability at all or with well-defined low one, under zero gravity. Let us analyze the effects of gravity and capillarity :

i) gravity. We consider two fluids with different densities in a vertical geometry. To be more specific, let us discuss the drainage obtained by injecting Hg (or Wood metal) from the bottom in a vertical porous glass sample. The effective injection pressure (Δp) decreases from bottom to top, due to the hydrostatic pressure gradient (Ca = 0). This is roughly equivalent to a linear decrease of the percolation parameter with height; this causes both a progressive variation of the fluid saturation in the injection cluster and a smearing of the upper critical front. The magnitude of the gravity effect is measured by the Bond number

$$Bo = \frac{(\Delta \rho) g \ell^2}{\gamma} \sim \frac{\text{hydrostatic pressure drop over distance } \ell}{\text{capillary pressure drop for a channel of width } \ell} \tag{2.7}$$

(again, a dimensionless ratio of lengths should be introduced in a detailed model).

The saturation profile concentration should vary across a height W : only the largest channels of the distribution are invaded in the upper part of the front; the menisci are blocked inside the smallest channels in the lower part. Dimensionally, $W \sim \ell/Bo$ (an additional factor should enter, taking into account the width of the pore-size distribution).

The width of the critical region at the top part of the injected fluid has been estimated in (2.7) by considering that the correlation length, instead of becoming infinite when the effective percolation parameter goes to the critical value, has an upper limit due to the continuous variation of the percolation parameter with height. A simple self-consistent argument then leads to the following value for the upper length extension of the fractal structure of the invasion front around threshold

$$\xi \sim Bo^{-\nu/(1+\nu)} \simeq Bo^{-0.5} \text{ in 3D} \tag{2.8}$$

A similar argument was developed in the (apparently) different problem of the fractal structure of a classical diffusion front(2.8). The authors assume that the diffusion of particles connected to an injection surface by nearest neighbor interactions is characterized by the average classical concentration c profile (c plays the role of the decreasing percolation parameter p or height h in the present problem). By numerically solving a 2D percolation problem with a spatially variable percolation parameter, they estimate a fractal dimension $d_f = 1.76$ for the diffusion front (the "hull" of the 2D percolation cluster). The structure is fractal up to a length scale of the order of the diffusion length.

ii) capillarity. Another perturbation is associated with the viscous friction effects that become important at higher flow velocities. If the invading fluid is more viscous than the displaced one, the width of the invasion front decreases when flow velocity increases. In some sense, the capillary number Ca plays a role similar to that of Bo.

In addition, if the invading fluid is the less viscous one, viscosity-driven front instabilities appear. As stated above, due to the randomness of the medium, the front structure differs from the classical Saffman-Taylor description and resembles, in some ways, aggregates obtained by a diffusion-limited process. (In this case, the structure will also be fractal, but with a lower dimension). These problems are discussed in the communication by Lenormand(2.5).

Finally, when the invading fluid is the wetting one (imbibition processes), the transport of fluid through wall films plays an essential role. When this process is dominant, pores get filled at random inside the medium during the injection; the process resembles more closely site percolation(2.9).

2.2.5. Experimental attempts to test the invasion percolation model

Experimentally attainable predictions of the invasion percolation model cover three main areas :

- existence of a fractal structure of the invasion front, particularly in zero gravity conditions;
- variation of the saturation value with the distance to the injection surface;
- structure and size of the residual blobs of fluid trapped behind during the invasion.

Fractal structure of the invasion front. Present invasion percolation numerical simulations do not take gravity into account. They predict a fractal structure for the infinite cluster with fractal dimensions of 2.52 (3D), 1.89 (2D, no trapping) and 1.83 (2D, trapping)[2.6]. These values are obtained by measuring the number n(r) of invaded sites located within a radius r from a given (also invaded) site. Fitting n(r) with the relation $n(r) \sim r^{d_f}$ gives the fractal dimension d_f. These numerical simulations have up to now only been verified fully on horizontal 2D models; in this case, invading with air acting as a non-wetting fluid a 2D model originally filled with oil, one finds $d_f = 1.82 \pm 0.01$[2.10].

In three dimensions, a particularly convenient experimental procedure is the injection of liquid Wood metal into the porous material. Wood metal is a low melting point alloy which, like mercury, does not wet at all most materials. In our experiments, we inject very slowly Wood metal into an unconsolidated crushed glass volume. The grain size distribution is kept as narrow as possible (225-250 μm) in order to reduce stratification effects (however, the distribution of the pore sizes remains very broad, due to the irregular shape of the grains). After the injection, the Wood metal is slowly solidified, and the crushed glass is taken away from the non-invaded part. The plate (a) shows a Wood metal sample. The value of $B_0 \sim 10^{-2}$ is large on this experience and consequently, the thickness (given by (2.8)) of the zone where the invasion front can be considered as fractal should be quite narrow ($\sim$ 1 mm, i.e. a few pore sizes), whereas the variation of saturation with height takes place over a pair of centimeters.

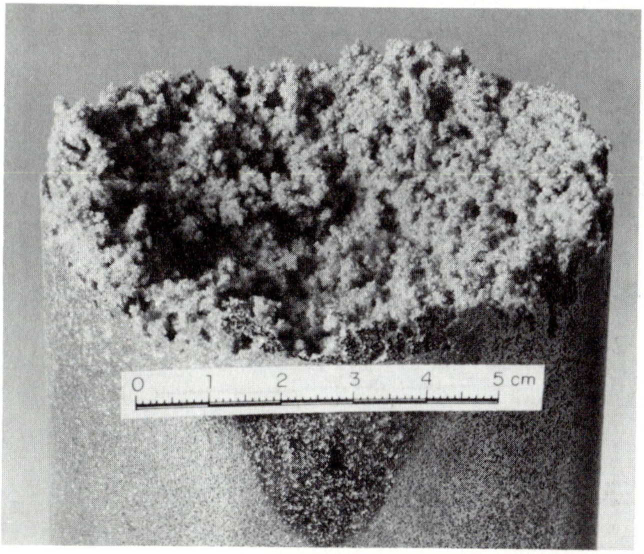

Plate(a)

As in the 2D diffusion front simulation(2.8), the length scale over which the fractal description is valid will be reduced and smaller than the overall invasion front thickness. In addition, in this 3D problem, one cannot define an external frontier or "hull" as in 2D and no prediction for the fractal dimension of an invasion front under gravity is presently available. Therefore it seems that the most reliable information can be extracted from horizontal cuts of the Wood metal block performed perpendicularly to the flow; in these planes, the equivalent percolation parameter is constant,and we shall assume that the structure of the Wood metal invaded areas corresponds to a two-dimensional percolation problem. The percentage of the area occupied by Wood metal will decrease as the level of the cut plane increases. When this percentage corresponds to the 2D percolation threshold, the extension of the fractal structure should be maximum (and probably should exceed the extension parallel to the flow of the fractal zone of the 3D front).

Plate (b) shows such a cross-section for an invasion structure similar to that of plate (a). In order to estimate a fractal dimension of the section, we first digitize the picture by covering it with a grid of mesh size close to the mean grain diameter. Each cell containing some Wood metal is defined as "occupied".

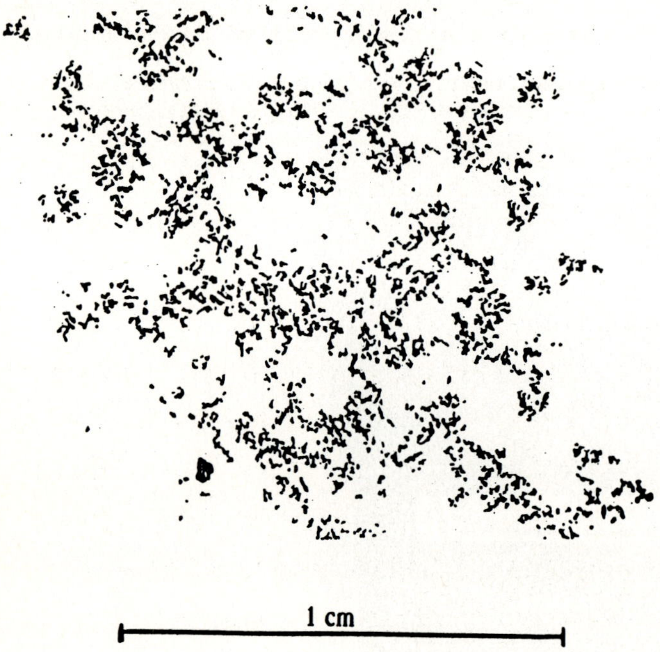

Plate(b)

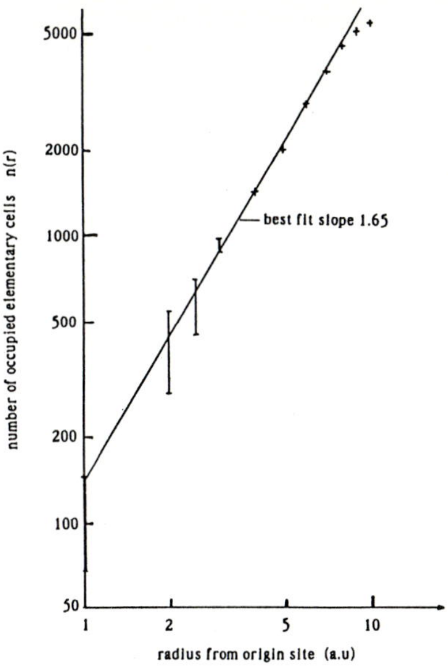

Fig.2

We have obtained a first estimation of the fractal dimension by counting the number N of "occupied sites" within a square of side R. For a fractal structure of dimension d_f, N should increase as R^{d_f}.
We obtain from the above curve $d_f = 1.65 \pm 0.05$, a value somewhat lower than that obtained for a 2D classical percolation cluster using a similar definition for the fractal dimension.

Another invasion surface fractal dimension value has also been determined on the same samples by an electrochemical method developed by le Méhauté[2.11] : the value for the 3D invasion surface is close to 2.3. Let us note, however, that this latter measurement is an average over different heights, i.e. over different values of the 2D percolation parameter.

Independent measurements have been reported[2.12] on invasion surfaces obtained by emptying, under gravity, tubes packed with crushed glass and filled with a matched index liquid. An optical determination of the profile of the invasion surface projected onto an observation plane gives $d_f > 2.3$.

Thus, although invasion fronts seem indeed to have a fractal character, more theoretical work is necessary to predict the relations between the various experimental dimension values.

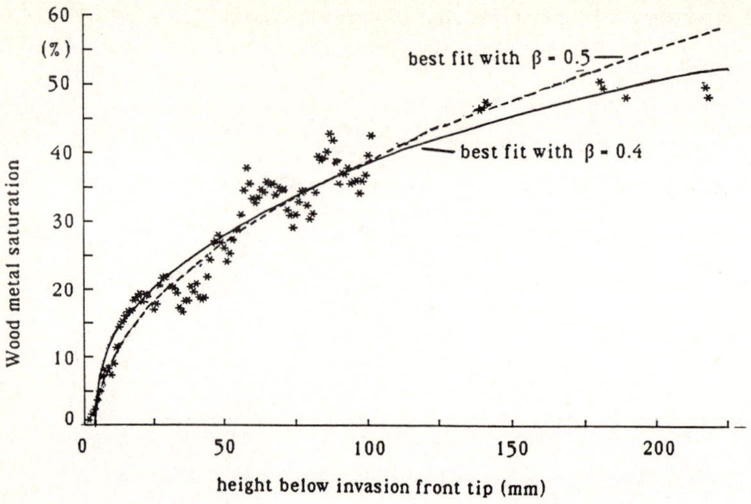

Fig.3

Variation of the saturation value in the invasion front

Figure 3 shows the variation of the relative saturation of Wood metal with the distance from the front obtained for a plane cut parallel to the flow. Theoretically, one expects the saturation to increase as z^{β} with the distance from the leading edge; β is the classical percolation exponent corresponding to the variation of the number of sites on the infinite percolation cluster. The lines drawn on the figure correspond to best fit theoretical variations obtained with $\beta = 0.4$ (3D percolation) and $\beta = 0.5$ (mean field approximation). The results are consistent with the theoretical predictions, but are not precise enough for a good evaluation of β.

Structure and size of the residual blobs of trapped fluid

Lattice percolation models predict a variation of the number n(s) of clusters containing s sites as :

$$n(s) \propto s^{-\tau} \qquad (2.9)$$

with $\tau \sim 2.17$. Experimental verifications have been performed[2.13,2.14] on real rocks by injecting a polymerisable resin into compacted sands and displacing it thereafter by another fluid. The porous medium is destroyed after the polymerisation, and the number and size of the residual blobs are analysed. The results are in reasonable agreement with classical percolation predictions.

In short, comparisons of the invasion percolation predictions with the experimental results are qualitatively encouraging,but precise quantitative verifications are still lacking.

3. DISPERSION

This chapter deals with the behavior of a tracer in solution in a liquid flowing through a porous media, adding the effects of molecular diffusion (D_m) and of convection through the random geometry. In ideal cases (e.g. chromatography), one looks for a macroscopic dispersion law :

$$\partial c/\partial t + |\vec{U}|\,\partial c/\partial x_{/\!/} = D_{/\!/}\partial^2 c/\partial x_{/\!/}^2 + D_{\perp}\partial^2 c/\partial x_{\perp}^2 \qquad (3.1)$$

where c is an average of the tracer concentration over a REV. $D_{/\!/}$ and $D_{\perp}$ are dispersion coefficients giving the spreading along (//) or perpendicular ($\perp$) to the average flow and $\vec{U}$ is the average velocity.

3.1. Well-connected materials

Random arrays of monodisperse spheres or of crushed glass give results which agree with eq. (3.1). However the value of $D_{/\!/}$ depends very much on the flow velocity. In addition, eq. (3.1) does not apply to heterogeneous geometries. Even in the simplest case of a packed bed of spheres with 2 different radii, Lemaitre[1.2] has found that the dispersion of a step function variation of tracer injected in the flow departed strongly from the expected sigmoid variation. This is likely to be due to the different classes of pore sizes and of the current distribution. We may expect multiple scale and weakly connected media to lead to much stronger effects. However, it is likely that, at times long compared with the slow diffusion processes, normal dispersion laws will be recovered. Let us analyse the different mechanisms controlling the dispersion.

3.1.1. Molecular diffusion

At extremely low or zero flow-rate, molecular diffusion is the dominant factor. $D_{/\!/}$ is equal to a value D_m corresponding to the bulk molecular diffusion coefficient for the fluid multiplied by a geometric factor.

At higher flow-rates, fluid flow plays a key role : one has a competition between the dispersive effect of the velocity gradients in the flow,and the various processes homogeneizing the tracer concentration across the channels. The order of magnitude of $D_{/\!/}$ is :

$$D_{//} = v^2\tau$$

v and τ are the characteristic flow velocity and homogeneisation time.

3.1.2. Hydrodynamic dispersion

In this mechanism described by Taylor[3.1] τ is the characteristic time (a^2/D_m) associated with the molecular diffusion of the tracer across the width a of the channel. Aris has extended the theory down to very low velocities where molecular diffusion is dominant and found :

$$D_{//} = D_m + U^2\, a^2/D_m \tag{3.3}$$

This mechanism is valid at moderate velocities, such that a^2/D_m is shorter than the time d/U necessary to go from one pore to another through a connecting channel of length d.

3.1.3. Geometric dispersion

At large Peclet numbers, the characteristic time is a convective one, and $\tau \sim d/U$ (if one assumes that the tracer gets homogeneized at the pores) and :

$$D_{//} \sim Ud \tag{3.4}$$

One observes indeed experimentally a linear increase of $D_{//}$ with $|U|$ at large applied velocities U.

In fact, this linear variation is not obvious as already indicated in the original treatment by Saffman[3.3] and several factors will perturb it.

3.1.4. Influence of low velocity and nearly stagnant regions

Several types can be encountered :

- slow fluid regions along the solid walls of long enough tubes;
- "equipotential" channels across the flow field.

If D_m was strictly 0, the tracer would not be able to move out of these regions and the residence time would diverge when the local velocity goes to zero; due to molecular diffusion, its upper limit is of the order of

$$\tau_M \sim \ell^2/D_m \tag{3.5}$$

ℓ is a characteristic length scale across the stagnation line (solid wall) or along it (equipotential channel).

Further computations show that the second moment of the transit time distribution contains a diverging terms $\mathrm{Log}(\tau_{..}/\tau)$ where $\tau \sim d/U$ is the characteristic time corresponding to high velocity regions. This term appears during the integration performed either on the volume or on the various possible orientations for the channels. The divergence is only logarithmic,because the fraction of the total volume corresponding to equipotential channels or to zero velocity regions at the walls is infinitely small.

When finite-size dead arms are present, the second moment of the transit time distribution contains a term linear in τ_M which increases much more the dispersion. In the case of equipotential channels $\ell \sim d$ and we can write by using eq. (3.5).

$$\text{Log } \tau_M/\tau \sim \text{Log } Ud/D_m \sim \text{Log } P_{em} \qquad (3.6)$$

This leads to a non-analytical dispersion coefficient.

$$D_{/\!/} = Ud \text{ Log } P_{em} \qquad (3.7)$$

We have also considered (see ref. (1.4) appendix) the similar contribution of stagnation points. This applies better to unconsolidated porous geometries. We analyse the slow flow of fluid along a set of independent spherical objects in dilute fixed beds. Like in Saffman's problem with a Poiseuille flow in a channel, the contribution of the slow fluid around a sphere leads to a t^{-3} distribution of the residence time E(t) which has a cut-off due to transverse diffusion.

It also contributes to a log correction to the dispersion. More precisely, the divergence of the second moment of E(t) is due to the flow close to the stagnation points,and should be found for other geometries of obstacles(3.4) : for a random assembly of such objects, the long-time behavior of the dispersion is due to the exploration by the flow of a number of stagnant zones,between which the particles circulate in the main flow field.

3.2. Multiple scale and dead end effects

As stated above, one expects much stronger anomalies as in well-connected media. One does not even expect equation (3.1) to hold in finite size samples.

3.2.1. *Multiple scales*

On a fractal lattice, as might be obtained with an Apollonian filling of spheres or with a percolating one below ξ, we expect molecular diffusion to behave like the anomalous random walk of the ant (with $< R^2 > \sim t^{dw}$), a physical unit time step τ being a molecular diffusion time along the shortest bonds. Mitescu and Roussenq

(3.5) are presently considering the related problem of an ant whose walk is linearly biased by the local current distribution. In the limit of low bias (low Pe) we can consider τ to be constant. However, in the opposit limit (large Pe) the distribution of τ's should be taken as proportional to that of the $(\text{currents})^{-1}$. The problem reduces to that of the distribution of passage times through the fractal structure. We have conjectured[1.1] that, for a size L, geometric dispersion should scale as UL, as also found in recent simulations by Redner, Koplik and Wilkinson. There might also be a slight dependence U(L) of the average current due to short distance effect from the injection wall.

The form U(L)L suggests an analogy with pair turbulent diffusion, which describes the rate of separation of points distant of L, in the inertial (fractal) range : the random velocity field is frozen in a porous medium, unlike the turbulent case, but a lagrangian particle experiences (in an ergodic fashion?) the randomness of the field while moving through it. Above the fractal range of percolating lattices, we expect a normal dispersion. However the characteristic time to reach this regime should be controlled by the slower particles in the fractal rather than by the fast ones, and the effect of the multiple scale geometry may be felt at distances much larger than ξ.

3.2.2. *Effect of the presence of dead arms*

Another contribution taking place in a weakly connected (percolating like) material is the existence of dead arms. De Gennes[3.6] has considered the problem and shown that, near p_c, it should give a singular contribution

$$D_{/\!/} = U^2 \tau_\xi \qquad (3.8)$$

where $\tau_\xi = \xi^2/D_a$ is the ant diffusion time over a length ξ and U an average velocity taken on all the percolating fluid phase. One gets

$$D_{/\!/} = \Delta p^{\mu-\beta-2\nu} \qquad (3.9)$$

with $\mu-\beta-2\nu = -0.5$ (3D) or $= -1.4$ (2D).

On the other hand, no such singular behavior is found on $D_\perp$.

Sahimi[3.7] has obtained numerically a sharp increase of the ratio $D_{/\!/}/D_\perp$ in a percolating lattice near p_c. However his calculation does not include the effect of dead arms, and the increase of $D_{/\!/}$ is probably due to the trapping in nearly stagnant zones those are found in blobs in his description of the medium as a system of nodes, blobs and links.

The experiment of Gaudet[3.8] is a beautiful example of the effect of heterogeneity of fluid distribution on dispersion. In a porous medium fully saturated with water, dispersion behaves as described by eq. (3.1). However, for an unsaturated medium partly filled with gas, the distribution becomes dramatically altered.

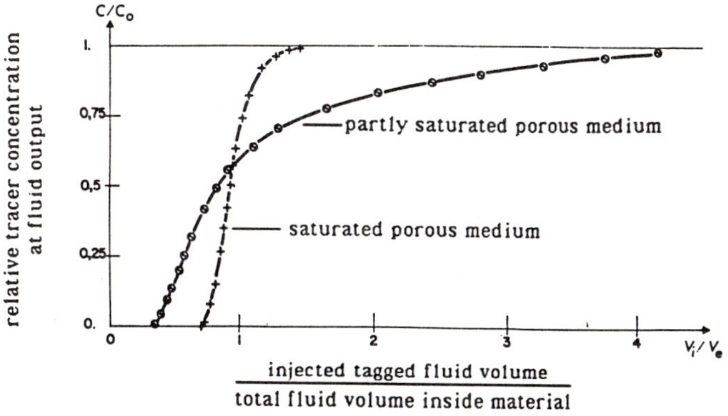

Fig.4

The first exit time (measured from the volume of tagged fluid injected into the medium divided by the total volume of liquid Ve in the porous medium) becomes shorter than in a well-connected medium because only a fraction of the pore liquid is mobile (on the backbone). The tail in the long-time behavior is due to trapping in stagnant regions and dead arms.

3.2.3. Experimental analysis of anomalous dispersion in 2D micromodels

Together with Lenormand and Zarcone, we study the various contributions on 2D micromodels qualitatively (using dyes as tracers) and quantitatively (injection of salt water and conductivity detection between closely spaced parallel lines). Our preliminary experiments support the above description :

a) They emphasize the extreme dye concentration heterogeneity in the path lines of the dye and the weak mixing between them. Those features control the fast release of the dye. The problem of the distribution of the dye-carrying lines appears to be closely connected with the notion of mechanical backbone (or shortest path).

b) Long after the injection of a dye step function, some dye can only be found in the stagnant regions. The image of the dye distribution looks like a complement of that observed shortly after the injection (then very little dye was present in the stagnant zones).

Features a) and b) controlled the extreme parts of the dispersion curve of Gaudet,but the critical behavior for small residual saturation is yet to be studied.

To conclude, we wish to stress that the anomalous features of dispersion in heterogeneous geometries provide a frame of description for two major features in hydrogeology :

i) the dispersion coefficient of pollutants is usually found to vary with the distance from the source ;

ii) anomalously fast as well as very slow release times are found. These features are qualitatively associated with the multiplicity of geological scales and dead arms.

We participate in an extended program of study of permeability and dispersion on granitic sites (in Massif Central) for which the distributions of fractures are documented. It is an open question to see how much the approaches sketched in this review will be of use in a full scale experiment.

Next to real site experiments, it is also highly desirable to "design" model materials to probe the different mechanisms. "Ridgefield sandstone" obtained by sintering monodisperse glass spheres, is such a system. In addition to a detailed study of permeability, conductivity...[1.2], preliminary studies of dispersion have been made on the same material.

Experimental studies on other model materials will be needed to include the effects of multiple scales and dead arms.

REFERENCES

1.1. The present review extends and complements the article by E. Guyon, J.P. Hulin and R. Lenormand (in French) in Ann. des Mines, special issue "Ecoulements dans les milieux fissurés" 191, 5.6 p. 17 (1984) where a larger number of references can be found.

1.2. P.Z. Wong, J. Koplik and J.P. Tomanic, to appear in Phys. Rev. B.

1.3. R. Lemaitre, thèse Université de Rennes (1985).

1.4. Groupe Poreux P.C., to appear in the proceedings of the NATO conference on "Scaling phenomena in disordered systems", Geilo Norway 10-21 April 1985, R. Pynn Editor, Plenum (London).

2.1. F.A.L. Dullien, "Porous media, fluid transport and pore structure", Academic Press, New York (1979).

2.2. J. Bear, "Dynamics of flow in porous media", American Elsevier, New York (1972).

2.3. P.G. de Gennes, J. Fluid Mech. 136, 189 (1983).

2.4. P.G. de Gennes and E. Guyon, J. Meca. 17, 403 (1978).

2.5. R. Lenormand and C. Zarcone, to be published in J. Phys. Chem. Hydr. (1985).

2.6. R. Lenormand and S. Bories, C.R. Acad. Sc. Paris 291 B, 279 (1980).
D. Wilkinson and S.F. Willemsen, J. Phys. A. 16, 3365 (1983).

2.7. D. Wilkinson, Phys. Rev. A 30, 520 (1984).

2.8. B. Sapoval, M. Rosso and S.F. Gouyet, to be published in J. de Phys. Lett.

2.9. R. Lenormand and C. Zarcone, SPE paper n° 13264 (1984).

2.10 R. Lenormand and C. Zarcone, to be published in Phys. Rev. Lett. (1985).

2.11 A. Le Méhauté, Solid State Ionics, 9-10, 17-30 (1983).

2.12 B. Legait and C. Jacquin, see ref. 1.1., p. 57.

2.13 R.G. Larson, H.T. Davis, L.E. Scriven, Chem. Eng. Sci. 36, 75 (1981).

2.14 R.L. Robinson and R.F. Hareng, "Experimental studies of residual saturation in unconsolidated sand", Jersey Production Research Company Report - Production division (1962).

3.1. G. Taylor, Proc. Roy. Soc. Ser. A 225, 473-77 (1954).

3.2. R. Aris, Proc. Roy. Soc. Ser. A 235, 67 (1956).

3.3. P.G. Saffman, J. Fl. Mech. 6, 321 (1959).

3.4. Y. Pomeau, priv. Comm.

3.5. C.D. Mitescu and J. Roussenq work in progress.

3.6. P.G. de Gennes, J. Fl. Mech. 136, 189 (1983).

3.7. M. Sahimi, H.T. Davis and L.E. Scriven, Chem. Eng. Com. 23, 329 (1983).

3.8. J.P. Gaudet, thèse Grenoble (1978).

Multiphase Flow in Porous Media

David Wilkinson
Schlumberger-Doll Research, Old Quarry Road, Ridgefield, CT 06877, USA

We discuss in detail the observable consequences of percolation models of immiscible displacement in porous media, with emphasis on the critical behavior. At the microscopic level these include the fractal nature of the non-wetting fluid configuration in drainage, and the size distribution of the residual non-wetting clusters in imbibition. At the macroscopic level it is suggested that percolation ideas are consistent with the usual multiphase Darcy equations, and critical behaviors of the relative permeability and capillary pressure curves are obtained. Using these results we derive the shape of the saturation profiles in the presence of buoyancy or viscous pressure gradients.

1. Introduction

In this paper I consider the process of the displacement of one fluid from a porous medium by another, immiscible, fluid. A very nice introduction to this topic has been given by Jean-Pierre Hulin in the previous lecture, so I will describe the system very briefly. Consider a porous medium which is filled completely (or perhaps at high saturation) with one fluid of density ρ_1 and viscosity μ_1 which is then flooded with a second fluid of density ρ_2 and viscosity μ_2 at some Darcy velocity (volume flow per unit area) v. Crucial to the nature of the displacement process are the wetting characteristics, and the interfacial tension γ between the two fluids. The fluid in which the contact angle between the fluid-fluid interface and the solid is less than 90 degrees is termed the wetting fluid, and the other the non-wetting fluid. If the contact angle is zero then we have perfect wetting. When the displacing fluid is the wetting fluid the process is called imbibition, and when the displacing fluid is the non-wetting fluid the process is called drainage.

Here we consider the case where the viscosities of the two fluids are comparable, but $\mu_2 > \mu_1$, so that the displacement is stable and viscous fingering does not occur. Many authors [1]-[7] have stressed the importance of percolation ideas in immiscible displacement. These concepts are relevant because the pore spaces of realistic porous media are random and multiply-connected. It is also necessary that pressure differences due to buoyancy and viscosity be small compared to typical interfacial pressure differences. The latter are of the order

$$\Delta p_{int} = \frac{\gamma}{R} \quad , \tag{1}$$

where R is a microscopic length, for example a typical grain size. In the case of buoyancy, the change in the pressure difference between the phases (i.e. the change in the capillary pressure) across a grain size R is given by

$$\Delta p_{grav} = \Delta\rho\, g R \quad , \tag{2}$$

where $\Delta\rho$ is the density difference, and g is the acceleration due to gravity. Taking the ratio of (1) and (2) gives

$$\frac{\Delta p_{grav}}{\Delta p_{int}} = \frac{\Delta \rho g R^2}{\gamma} \equiv B \quad . \tag{3}$$

The quantity B is called the Bond number, and represents the local competition between buoyancy and interfacial forces. For the case of viscosity, in the applications we consider here it is the pressure drop in the displacing fluid which is important, so the viscous pressure drop across a grain size R may be estimated as

$$\Delta p_{visc} = \frac{\mu_2 v R}{k} \quad , \tag{4}$$

where v is the superficial Darcy velocity of the flood and k is the permeability of the medium. Taking the ratio of (1) and (4) gives

$$\frac{\Delta p_{visc}}{\Delta p_{int}} = \frac{Ca}{K} \tag{5}$$

where

$$Ca = \frac{\mu_2 v}{\gamma} \tag{6}$$

is the capillary number, expressed in terms of the displacing fluid viscosity and the superficial velocity, and

$$K = \frac{k}{R^2} \tag{7}$$

is a geometrical constant. Since the permeability is dominated by the narrow constrictions in the medium, the constant K is typically rather small, of order 10^{-3}.

When the Bond number and the capillary number are small, the system is in local capillary equilibrium. That is, at a given capillary pressure (pressure difference between the non-wetting and wetting phase) .

$$p_{cap} = p_{nw} - p_w \quad , \tag{8}$$

the individual menisci between the phases adopt configurations which are determined only by the local geometry, and are independent of the global pressure gradients. Percolation concepts arise because of instabilities in the capillary equilibrium: there are certain parts of the pore space which fill with the non-wetting fluid (in drainage) or the wetting fluid (in imbibition) not gradually, but suddenly, when the capillary pressure rises above, or drops below, some critical value. Examples of such mechanisms are the Haines jump which occurs in drainage when the capillary pressure becomes large enough for the non-wetting fluid to penetrate a throat, or the snap-off which occurs in imbibition when the capillary pressure drops to the value where a wetting film occupying the walls of a throat becomes unstable and the wetting fluid occupies the entire throat. A detailed discussion of such mechanisms is given in LENORMAND and ZARCONE [1], where it is pointed out that some combinations of such mechanisms are percolation-like, and some are not. Here we will assume the former situation and simply assert that, in both drainage and imbibition, there is one-to-one correspondence between the capillary pressure p_{cap} and the fraction p of pores (or throats) which could be occupied by the non-wetting fluid (the correspondence is of course different in drainage from in imbibition because the mechanisms are different). We will further assume that the connectedness of the non-wetting fluid (though not necessarily of the wetting fluid) takes place entirely through these occupied pores. We call the fraction p the allowed non-wetting fraction; it is different from the actual fraction of pores occupied by the non-wetting fluid because the connectedness or lack of connectedness of the non-wetting fluid may prevent certain pores from filling or emptying. A simple percolation model of the displacement process will have (in three dimensions) two distinct percolation thresholds. The first (or breakthrough) threshold occurs when the displacing fluid first forms a connected path across the system. The second (or floodout) threshold occurs when the displaced fluid ceases to form a connected path. Here we will assume that only the non-wetting phase exhibits percolation behavior -- thus we will consider the first threshold in drainage and the second threshold in imbibition. If the contact angle is large then both phases may show this percolation behavior, but if we have perfect wetting then the connectedness of the wetting phase occurs through surface films and roughness of the pore walls rather than through bulk occupation of a connected set of pores.

In drainage, the non-wetting fluid forms a single connected cluster (the infinite cluster of percolation). This kind of growth in a single cluster we call invasion percolation. As the non-wetting cluster grows, the wetting fluid may be trapped, or may be able to escape via surface films. However, if we are only interested in the behavior near the percolation threshold for the non-wetting fluid, this trapping will not be important, and we will ignore it (this is in three dimensions; in two dimensions the trapping can become very important). Thus in drainage we will treat the process as invasion percolation without trapping of the other phase.

In imbibition the wetting fluid may grow in a single connected cluster, or it may reach all parts of the pore space by means of surface films. However, again in three dimensions, by the time we reach the disconnection threshold for the non-wetting phase, the wetting phase will in either case be considerably above percolation threshold, and the distinction between invasion percolation and ordinary percolation will not matter (at least for the critical behavior). What does matter however is the incompressibility of the non-wetting phase: regions of the non-wetting phase which are surrounded (trapped) can no longer be invaded by the wetting phase, i.e. the finite clusters of the non-wetting phase remain the same size as when they are first formed. Thus we will treat the imbibition process as ordinary percolation but with trapping of the displaced phase taken into account (trapping percolation).

In the appendix we list the basic concepts of percolation which are used in this paper, together with the values of the critical exponents in three dimensions.

2. PercolationPredictions

(a) Microscopic predictions

In drainage, the percolation threshold for the non-wetting phase occurs at breakthrough. The non-wetting fluid is a single percolation cluster at threshold, and so is a fractal. That is, if we choose an origin in the non-wetting fluid then the volume M(R) of non-wetting fluid within a sphere of radius R grows as

$$M(R) \sim R^D \quad , \qquad (9)$$

where $D \sim 2.5$ is the fractal dimension of ordinary percolation. Alternatively, on a sample of linear extent L, the non-wetting fluid saturation shows a finite size scaling behavior

$$S_{nw} \sim L^{-(3-D)} \quad . \qquad (10)$$

In imbibition the percolation threshold for the non-wetting phase is the final configuration where all the non-wetting fluid is trapped. If we let n(s) denote the number of non-wetting clusters containing s pores (normalized per pore), then n(s) has a power law behavior for large s

$$n(s) \sim s^{-\tau} \quad . \tag{11}$$

Previously it was thought, based on computer simulations of invasion percolation with trapping, that the exponent τ was around 2.07, less than the corresponding exponent

$$\tau = \frac{3+D}{D} \sim 2.20 \tag{12}$$

of ordinary percolation. However more recent analysis of trapping percolation [8] suggests that the value should be the same as in ordinary percolation, i.e. 2.20, despite the fact that the actual size distribution is altered by the trapping rule. If we estimate the residual non-wetting saturation S_{nwr} by counting the number of occupied pores (i.e. we neglect the size variation of the pores) then we have

$$S_{nwr} = \sum_{s'=1}^{\infty} s' n(s') \quad . \tag{13}$$

Since τ is close to 2, the residual saturation receives contributions from clusters over a wide size range. The best way to estimate the exponent τ both in computer simulations and experiments, is to compute that part of the residual saturation which is contained in clusters of size greater than s:

$$M(s) = \sum_{s'=s}^{\infty} s' n(s') \sim s^{-(\tau-2)} \tag{14}$$

(b) Capillary Pressure

In drainage we are concerned with the breakthrough threshold. For an infinite system the non-wetting saturation at breakthrough is zero, but as the non-wetting occupation fraction p increases the saturation grows as

$$S_{nw} \sim \Delta p^{\beta} \quad , \tag{15}$$

where $\beta \sim 0.45$ is the order parameter exponent of ordinary percolation. If we define a dimensionless capillary pressure $\hat{p}_{cap}$ by

$$p_{cap} = \frac{\gamma}{R} \hat{p}_{cap} \tag{16}$$

then we have

$$\Delta \hat{p}_{cap} \sim \Delta p \sim (S_{nw})^{\frac{1}{\beta}} \quad . \tag{17}$$

In imbibition, the critical behavior of the capillary pressure at the residual non-wetting threshold is more complicated because of the trapping rule. In trapping percolation, when we decrease the non-wetting allowed fraction from p to $p-dp$, the wetting phase saturation increases by $P(p)\,dp$, since the wetting fluid can only enter the infinite cluster (i.e. the connected portion) of the non-wetting phase. Thus we have

$$\Delta S_{nw} \equiv S_{nw}(p) - S_{nw}(p_c) = \int_{p_c}^{p} P(p)\,dp \sim \Delta p^{1+\beta} \quad , \tag{18}$$

where $S_{nw}(p_c)$ is the residual non-wetting saturation S_{nwr}. Thus

$$\Delta \hat{p}_{cap} \sim \Delta p \sim (\Delta S_{nw})^{\frac{1}{1+\beta}} \quad . \tag{19}$$

The capillary pressure curves for the drainage and imbibition cases are shown schematically in Fig. 1.

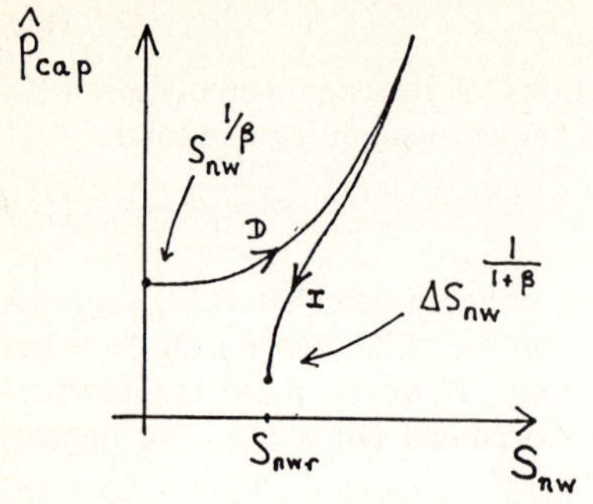

Figure 1. Schematic plot of the dimensionless capillary pressure $\hat{p}_{cap}$ as a function of non-wetting saturation S_{nw}, showing the critical behavior in both drainage and imbibition.

(c) Relative permeabilities

When there are viscous pressure gradients in the system we will assume that locally the system responds to the prevailing capillary pressure in the same way as when the flow rate is infinitesimal. Thus the phase pressures are related by

$$p_{nw} - p_w = p_{cap}(S_{nw}) \quad , \tag{20}$$

where $p_{cap}(S_{nw})$ is the same function of saturation as in Fig. 1. The individual phase pressures satisfy the multiphase Darcy equations

$$\frac{\partial p_i}{\partial x} = - \frac{\mu_i v_i}{k k_{ri}(S_{nw})} \tag{21}$$

where $k_{ri}(S_{nw})$ is the relative permeability to phase i at the local non-wetting saturation S_{nw}, and v_i is the Darcy velocity of phase i. In addition we have the mass balance equation for each phase

$$\phi \frac{\partial S_i}{\partial T} + \frac{\partial v_i}{\partial x} = 0 \quad . \tag{22}$$

where T is the time, and ϕ is the porosity.

In drainage, close to $S_{nw} = 0$ the relative permeability to the non-wetting fluid scales as

$$k_{rnw} \sim \Delta p^t \tag{23}$$

where $t \sim 1.9$ is the conductivity exponent. Using (15) we thus obtain

$$k_{rnw} \sim (S_{nw})^{\frac{t}{\beta}} \tag{24}$$

In imbibition, close to the non-wetting residual saturation k_{rnw} again scales as in (23), since the infinite (connected) non-wetting cluster in trapping percolation is the same as that in ordinary percolation - only the finite clusters are affected by the trapping rule. Using (18) we find

$$k_{rnw} \sim (\Delta S_{nw})^{\frac{t}{1+\beta}} \quad . \tag{25}$$

The relative permeability curves in both drainage and imbibition are sketched in Fig. 2.

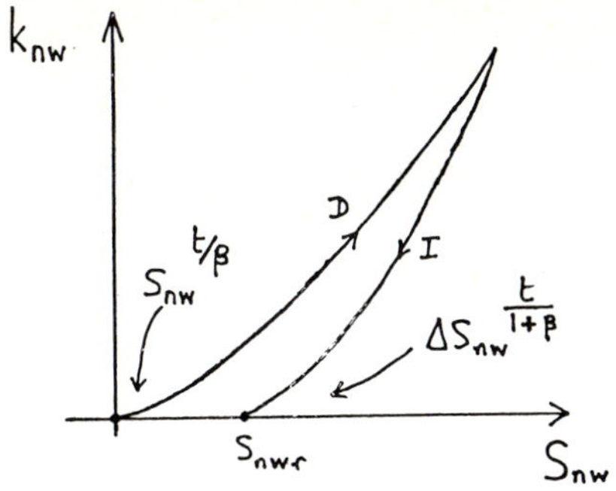

Figure 2. Schematic plot of the non-wetting fluid relative permeability k_{nw} as a function of non-wetting saturation S_{nw}, showing the critical behavior in both drainage and imbibition.

(d) Saturation Profiles due to Buoyancy

Using the above concepts we may determine the critical behaviors of the macroscopic saturation profiles in the presence of pressure gradients due to buoyancy or viscosity. In the buoyancy case the capillary pressure varies linearly with the height

$$p_{cap}(x) = p_{cap}(0) + \Delta\rho\, g\, x \tag{26}$$

where x=0 is some reference height and we have assumed that the wetting fluid is the heavier fluid. Thus the dimensionless capillary pressure satisfies

$$\hat{p}_{cap}(x) = \hat{p}_{cap}(0) + B\frac{x}{R} \tag{27}$$

In drainage we perform the flood in a vertical direction with the lighter (non-wetting) fluid introduced from above. Thus near the leading edge of the front we have from (17)

$$S_{nw} \sim \left(B\frac{\Delta x}{R}\right)^{\beta} \tag{28}$$

where Δx is the height above the leading edge of the front.

In imbibition we perform the flood vertically with the heavier (wetting) fluid introduced from below. From (19) we have

$$\Delta S_{nw} \sim \left(B\frac{\Delta x}{R}\right)^{1+\beta} \tag{29}$$

(e) Saturation Profiles due to Viscosity

The computation of the saturation profiles in the case of viscosity pressure gradients is more complex, since it involves the solution to the multiphase Darcy equations (20-22) with the inlet face boundary conditions

$$v_1 = 0$$

$$v_2 = v \tag{30}$$

where phase 1 is the displaced phase and phase 2 the displacing phase. For a discussion of these equations and the nature of their solutions, see for example MOREL-SEYTOUX [9].

In drainage the small values of non-wetting saturation travel with a common front velocity v_F, which is of the same order as the imposed total flow rate v. In this region the individual flow velocities are expressed in terms of this front velocity by

$$v_{nw} = v_F S_{nw} \quad ,$$

$$v_w = v - v_{nw} \tag{31}$$

It is clear that near the leading edge of the front the pressure gradient in the non-wetting fluid is much greater than that in the wetting fluid. Thus we have from (16) and (21)

$$\frac{\partial \hat{p}_{cap}}{\partial x} = -\frac{R}{\gamma} \frac{\mu_{nw} v_{nw}}{k k_{nw}} \quad . \tag{32}$$

Rearranging and using (31):

$$\frac{k_{nw}}{S_{nw}} \frac{d\hat{p}_{cap}}{dS_{nw}} \frac{\partial S_{nw}}{\partial x} = -\frac{\mu_{nw} v_F R}{\gamma k} \quad . \tag{33}$$

Using the critical behaviors (17) and (24)

$$(S_{nw})^{\frac{t+1-2\beta}{\beta}} \frac{\partial S_{nw}}{\partial x} \sim -\frac{\mu_{nw} v_F R}{\gamma k} \quad . \tag{34}$$

Integrating, we find the critical behavior of the saturation profile near the leading edge of the front as:

$$S_{nw} \sim \left(\frac{Ca_F}{K} \frac{\Delta x}{R} \right)^{\frac{\beta}{t+1-\beta}} \quad , \tag{35}$$

where Ca_F is the capillary number expressed in terms of the front velocity v_F, and Δx is measured backwards from the leading edge of the front.

In imbibition, the values of saturation close to residual non-wetting saturation travel with a velocity which depends on the saturation

$$v(S_{nw}) = v \frac{df_{nw}}{dS_{nw}} \quad , \tag{36}$$

where $f_{nw}(S_{nw})$ is the fractional flow of the non-wetting fluid on the imbibition curve:

$$f_{nw} = \frac{k_{nw}/\mu_{nw}}{k_{nw}/\mu_{nw} + k_w/\mu_w} \quad . \tag{37}$$

In this case the shape of the profile (at long times) is independent of the capillary pressure curve; the pressure gradients in the two fluids are almost equal, and the capillary pressure gradient is very small. From (25) and (36) we have

$$v(S_{nw}) \sim \frac{v}{k_{wr}} (\Delta S_{nw})^{\frac{t}{1+\beta} - 1} \quad , \tag{38}$$

where k_{wr} (which is of order unity) is the wetting fluid relative permeability at residual non-wetting saturation. Thus after a time T, the non-wetting saturation at a distance x from the inlet face is given by

$$S_{nw} - S_{nwr} \sim \left(\frac{k_{wr}}{v} \frac{x}{T} \right)^{\frac{1+\beta}{t-\beta-1}} \qquad (39)$$

3. Discussion

In this talk I have listed the major predictions of percolation models of immiscible displacement. The beauty of these models is that they provide universal predictions with critical exponents which are independent of the detailed nature of the pore space. Some of these predictions are microscopic and require pore-level observation of the system, which is very difficult in a real rock. Others, such as the predictions for the capillary pressure and relative permeability functions, are macroscopic in nature. The main new results of this talk are the predictions for the shape of the saturation profiles in the presence of buoyancy or viscous pressure gradients. When the capillary and Bond numbers are sufficiently small, the length scale over which the critical behavior is seen can become large enough that the behavior may be seen in macroscopic imaging devices such as CAT or NMR scanners.

Appendix

In this appendix we summarize our definitions of the critical exponents of percolation, and their approximate values in three dimensions. We denote the occupation fraction by p and the critical fraction by p_c and write

$$\Delta p = p - p_c \quad .$$

In our applications we are always above threshold, so Δp is positive. The order parameter P(p) is the fraction of occupied sites in the infinite cluster and scales as

$$P(p) \sim (\Delta p)^{\beta} \quad ,$$

where $\beta \sim 0.45$. If the occupied sites are conducting and the empty sites insulating, then the conductivity Σ scales above threshold as

$$\Sigma \sim (\Delta p)^{t} \quad ,$$

where $t \sim 1.9$ is the conductivity exponent. The correlation length L, which may be taken as the typical size of the finite clusters, diverges as

$$L \sim (\Delta p)^{-\nu} \quad ,$$

where $\nu \sim 0.88$. At threshold the infinite cluster is a fractal with fractal dimension

$$D = 3 - \frac{\beta}{\nu} \sim 2.5 \quad .$$

Finally, at threshold the number of finite clusters of size s scales as

$$n(s) \sim s^{-\tau} \quad ,$$

where

$$\tau = \frac{3+D}{D} = \frac{6\nu - \beta}{3\nu - \beta} \sim 2.20 \quad .$$

References

1. J. C. Melrose and C. F. Brandner, Can. J. Petrol. Tech. 13, 54 (1974).
2. R. G. Larson, L. E. Scriven and H. T. Davis, Chem. Eng. Sci. 36, 57 (1981).
3. P. G. de Gennes and E. Guyon, J. de Mechanique 17, 403 (1978).
4. R. Lenormand and S. Bories, C. R. Acad. Sci. Paris 291, 279 (1980).
5. R. Chandler, J. Koplik, K. Lerman and J. Willemsen, J. Fluid Mech. 119, 249 (1982).
6. D. Wilkinson, Phys. Rev. A30, 520 (1984).
7. R. Lenormand and C. Zarcone, SPE paper number 13264, Houston Texas, September 1984.
8. M. Dias and D. Wilkinson, in preparation.
9. H. J. Morel-Seytoux, in *Flow Through Porous Media,* Edited by J. M. de Wiest (Academic, New York 1969).

Capillary and Viscous Fingering in an Etched Network

Roland Lenormand*
Schlumberger-Doll Research, Rigdefield, CT 06877, USA

We present an original set-up for studying the different mechanisms associated with the displacement of immiscible fluids in porous media. A molding technique, using a transparent polyester resin and a photographically etched network has been developed to visualize the front during the displacement. At very low flow-rate we can observe the growth of very thin fingers, even when the displacing fluid is the more viscous. This capillary fingering is very well described by invasion percolation theory and the measured fractal dimension agrees with computer simulations. When the injected fluid is the less viscous fluid, the fingers become more and more dendritic as the flow rate increases; this mechanism seems to be related to a cross-over between invasion percolation and diffusion-limited aggregation (D.L.A.). When the injected fluid is the more viscous fluid, increasing the flow rate decreases the fingering. This mechanism can be described as a cross-over between invasion percolation and a flat interface.

1. Introduction

The displacement of one fluid by another non-miscible fluid in a porous medium is of importance in many processes, especially petroleum recovery. In this domain, an understanding of the relevant mechanisms is very important, because fingering leads to very inefficient recoveries .

The purpose of this paper is to describe experimental displacements in a 2-dimensional permeable medium when both capillary and viscous effects are present. The first part of this study presents the set-up (a 2-dimensional etched network) and the experimental results. In the second part, I show how these results can be related to the statistical theories of diffusion-limited aggregation (D.L.A) and invasion percolation (I.P). In the third part, I present a technique for measuring the fractal dimension of the injected cluster and give the results in the case of capillary displacement.

Generally, two different approaches are used to describe the shape of the interface between two immiscible fluids during a displacement in a porous medium: one emphasizes the role of viscous forces and the other is based on capillary mechanisms. These different approaches have been described in detail in a previous publication [1]:

- "Hele Shaw" displacement.
  In this approach the porous medium is assumed to be equivalent to a "continuum" and the velocity of each fluid to be proportional to the local pressure gradient (Darcy's law). The two-dimensional version of this problem is *mathematically* analogous to the displacement of two immiscible fluids between closely spaced parallel sheets of glass, a set-up known as a "Hele Shaw cell" [2]. Experiments in these cells show instabilities when the less viscous fluid is displacing the more viscous one [3]. In this approach, the fingering is essentially due to the viscous forces.

* present address: Dowell Schlumberger, B.P. 90, 42003 Saint Etienne Cedex, France

- Invasion Percolation.

  This mechanism is related to capillary forces which take place at the ***microscopic*** (pore) scale, when viscous forces are vanishing (quasi-static displacement). The surface tension prevents the non-wetting fluid from entering the smallest throats in the porous medium.and consequently the injected fluid searchs for a continuous path of large throats. It has been shown that this continuous path is analogous to a percolation cluster [4-6].

2. Experiments

We have developed a molding technique [7],[1] using a transparent resin and a photographically etched mold to study two-phase flow in porous media. The cross-section of each duct of the etched network is rectangular with a constant depth $x = 1$mm and a width d which varies from throat to throat (generally $d > 0.1$mm). For this study we used two kinds of networks:

- a 42,000 duct network (150x150 mm.) with seven classes of channels (width from 0.1 to 0.6 mm) distributed with a log-normal law and a random location. The distance between two sites of the network is about 1mm.

- a very large network (300 x 300 mm) containing 250,000 ducts, used for statistical measurement. In previous work [8], we have shown that the structure of the injected cluster is independent of the pore-size distribution, so the size distribution is broadened around 50% to get better accuracy near the bond percolation threshold (0.5 for a square network).

As in a Hele Shaw cell, we are displacing the wetting fluid by the non-wetting fluid (a displacement called "drainage"). The opposite displacement, called "imbibition" is more complex,because the wetting fluid flows along the walls of the ducts at low flow rate [9]. Several different fluids have been used for the experiments:

- Oil (Soltrol 220) colored with Oil Red O; viscosity 4.2 cP or $4.2\text{x}10^{-3}$ N.s/m^2; used as wetting fluid.

- Water-sucrose solution A (WSA); used as wetting fluid (with air as a non-wetting fluid); viscosity 240 cP; colored with Amaranth Red.

- Water-sucrose solution B (WSB); used as non-wetting fluid (with oil as a wetting fluid); viscosity 53 cP; $\theta \sim 50^o$; surface tension $\gamma = 18$ dyne/cm or $18\text{x}10^{-3}$ N/m (non-colored).

- Air ; viscosity 180 μP, used as non-wetting fluid with oil ($\theta = 0°$, $\gamma = 25$ dyne/cm) or with water-sucrose solution A ($\theta = 70°, \gamma = 66$ dyne/cm).

The fluids are pumped through the micromodel using constant flow rate syringe pumps. The micromodel is held horizontal to avoid gravity effects. The experiments are run at different capillary number Ca, which is a dimensionless form of the flow rate q:

$$Ca = \frac{q\,\mu}{\Sigma\,\gamma} \tag{1}$$

In a 3-dimensional medium Σ is the cross-section area of the sample and, for the 2-dimensional network, we will take the product of the total width (150 mm) by the channel depth (1mm), q/Σ being a mean velocity of the fluid in the channels. This capillary number characterizes the ratio between viscous forces and capillary forces. The second parameter of the experiments is the viscosity ratio M, the ratio between the viscosities of the injected fluid and the displaced fluid.

To study the effect of the viscosity ratio and the capillary number, we ran two sets of experiments with the 42,000 duct network, one with a viscosity ratio very low (air displacing WSA), the second

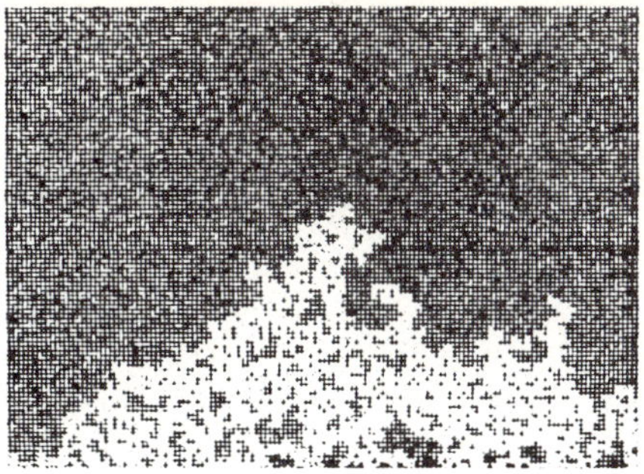

a.1) $M = 7.6 \times 10^{-5}$ $Ca = 23 \times 10^{-11}$

b.1) $M = 13$ $Ca = 3 \times 10^{-6}$

a.2) $M = 7.6 \times 10^{-5}$ $Ca = 11 \times 10^{-9}$

b.2) $M = 13$ $Ca = 1.5 \times 10^{-4}$

a.3) $M = 7.6 \times 10^{-5}$ $Ca = 11 \times 10^{-7}$

b.3) $M = 13$ $Ca = 1.5 \times 10^{-2}$

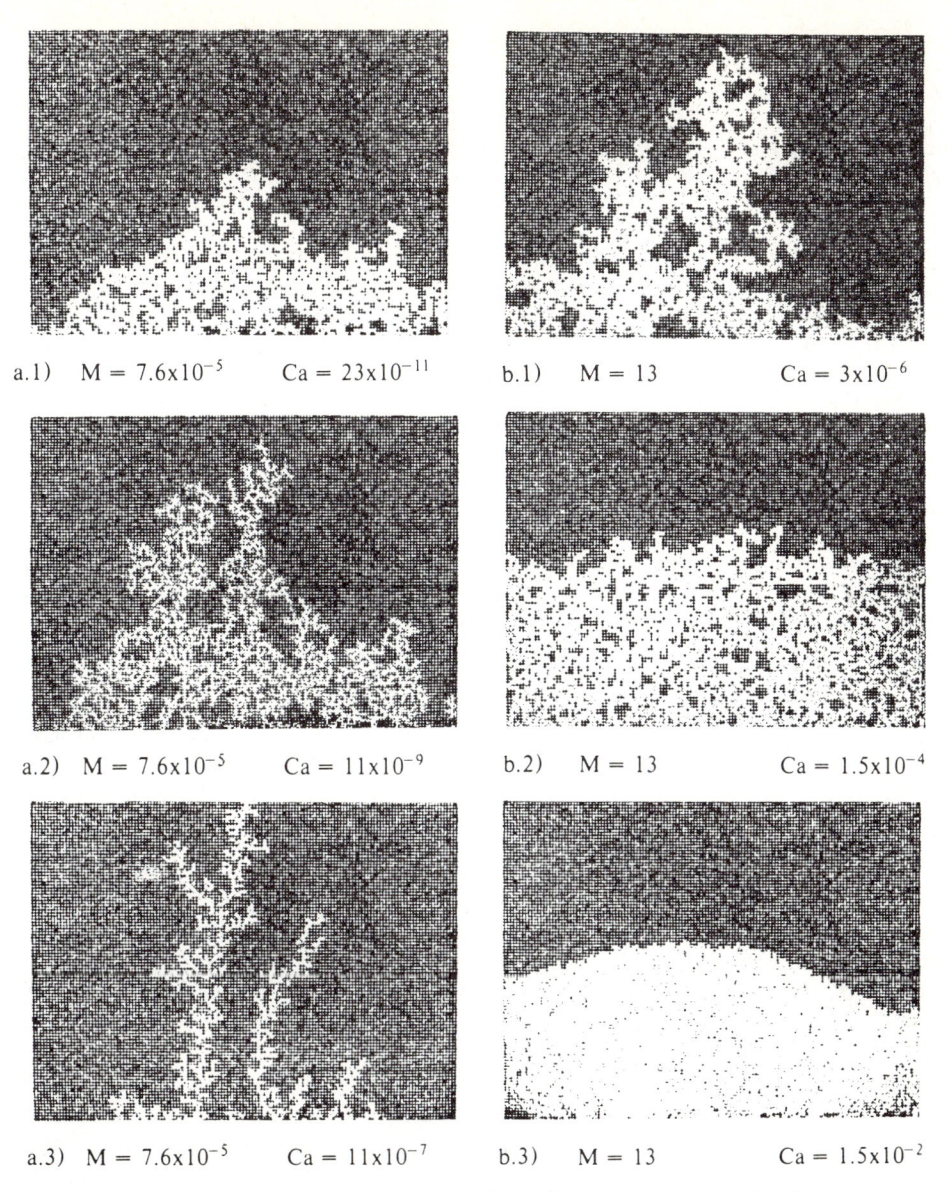

Fig. 1. Displacement of the wetting fluid (black) filling the network by the non-wetting fluid (white) injected at the bottom of each picture for different capillary numbers Ca and viscosity ratio M.

at M=13 (WSB displacing oil). In each set of experiments we varied the flow rate (or the Capillary number). The results are shown in Fig. 1.

- Injecting the LESS viscous fluid (Fig. 1.a)

  In these experiments (air displacing WSA), the viscosity ratio is about $7.6\text{x}10^{-5}$. In the first two experiments ($Ca = 23\text{x}10^{-11}$ and $Ca = 11\text{x}10^{-9}$), shown in Fig. 1a.1 and 1a.2, the patterns of the injected fluid are close to invasion percolation simulations. The fingers spread across the whole network and the size of the trapped clusters ranges from the pore size to a macroscopic scale (of the order of the network size). Furthermore, at the microscopic scale, the fingers grow in all directions. An important feature of these experiments is the stability of the fingers, which remain when the flow is stopped. The last experiment (Fig. 1a.3) is run at $Ca = 11x10^{-7}$. The pattern is different (the fingers are thinner and the fraction of fingers growing backward is very small) and looks like D.L.A simulations [10],[11]. This analogy has already be suggested by PATERSON [12] and some experiments in Hele Shaw cells also show this kind of fingering [13].

- Injecting the MORE viscous fluid (Fig. 1b).
  The first experiment (Fig. 1b.1), run at $Ca = 3x10^{-6}$, also looks like invasion percolation. When the capillary number is increased, the injected cluster becomes more and more compact: for $Ca = 1.5x10^{-4}$ (Fig. 1b.2) the finger length is about 10 mesh sizes, and for $Ca = 1.5x10^{-2}$ (Fig. 1b.3), the finger length is about 1 mesh size, which I may take to characterize a flat interface (the curvature at large scale is due to the non-homogeneity of the injection at the entrance of the network). The size of the trapped clusters decreases when the capillary number increases, and the trapped fraction is almost zero when the front is flat. An important problem is to understand the trapped cluster size distribution at intermediate values of the Capillary number.

However, these results are purely qualitative and I have tried to be more accurate in measuring the fractal dimension of the injected cluster. So far, I have only obtained results for the capillary displacements (invasion percolation).

3. Measurement of a Fractal Dimension

This study has been published with more details in [14]. The experiments are run in a very large network (250,000 ducts) and air is displacing oil. For a given capillary number, the experiments are reproducible, and the structure of the injected cluster qualitatively agrees with computer simulations [15]: during the displacement, the non-wetting fluid presents very thin and dendritic fingers and at the end of the experiment, the cluster size of the trapped phase varies from the pore scale to the network scale.

Digitization of the photographs is quite impossible, because of the black meniscus which surrounds the non-wetting phase in each pore, and we have to use a simple but laborious technique: from an origin O roughly at the center of the network, we count the number N of invaded ducts in an LxL square centered on O. A duct is counted only when the non-wetting fluid has invaded both the duct and also the pore (intersection) next to this duct (ducts where the meniscus remains at one or both ends are not counted).

When the capillary number decreases, the final saturation S increases. This phenomenon is due to the possibility for a fraction of the wetting fluid to "escape" by flowing along the roughness of the pores when trapping occurs [16]. This mechanism seems not to be relevant at large scales and does not change the fractal dimension of the cluster.

At the end of the displacement, the variation of N as a function of L (measured in units of the mesh size) for different capillary numbers is plotted on a log-log scale in Fig. 2. The curves are linear when the size L is greater than about 70 meshes and a least square fit for the slope leads to $D = 1.83 \pm 0.01$ for the three slowest displacements ($Ca = 3.3x10^{-8}, 6.5x10^{-8}, 1.2x10^{-7}$) and $D = 1.80$ for $Ca = 6.2x10^{-7}$. These measurements are in good agreement with the theoretical value $D = 1.82$.

The highest capillary number ($Ca = 1.5x10^{-6}$) leads to a different value ($D > 2$) and this experiment reveals the weakness of the method used to calculate the fractal dimension (in this case, a large cluster of wetting fluid remains in the central part of the network and the origin does not belong to the cluster).

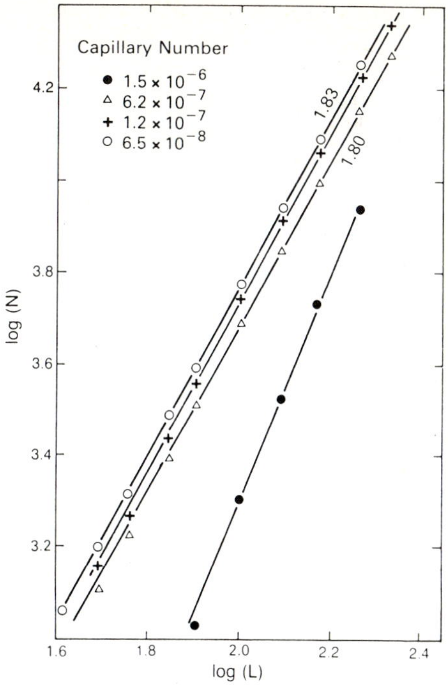

Fig. 2. Number N of filled ducts versus size of the LxL square for different capillary numbers Ca. The data for Ca=$3.3x10^{-8}$ are exactly the same as for Ca=$6.5x10^{-8}$ and are not shown in this figure.

Thus, the main problem is not the finite size of the network (for instance, the number of filled ducts is of the same order as the number of particles used in computer studies of diffusion-limited aggregation [10]) but the difficulty of measuring the fractal dimension of a cluster obtained by injection through a side of the network. Consequently, it seems possible to improve the accuracy of our experiments by injecting the non-wetting fluid through one point in the central part of the network.

Conclusion

We obtained some very interesting results concerning the displacement of a wetting fluid by a non-wetting fluid in a two-dimensional etched network.

- At low flow rate (quasi-static displacement) the non-wetting fluid forms very thin fingers *even if the displacing fluid is more viscous* than the fluid filling the network. This kind of *capillary fingering* is well described by *invasion percolation* theory. The measured fractal dimension is consistent with computer simulations $D = 1.82$.

- When the injected fluid is the *less viscous* fluid, the fingers become more and more dendritic when the flow rate increases. This mechanism seems to be related to a cross-over between invasion percolation and *diffusion-limited aggregation* (D.L.A). However, further theoretical work and measurements on experimental structures have to be done on this problem.

- When the injected fluid is the *more viscous* fluid, increase of the flow rate decreases the fingering. This mechanism can be describe as a cross-over between invasion percolation and a *flat interface.*

Consequently, it seems to us that a statistical approach would be more suitable for describing non-miscible displacement in porous media than the classical equations based on an analogy with the Hele Shaw cell.

Acknowledgements

We would like to thank J. Piguemal for suggesting the micromodel technique and A. Libchaber, Y. Pomeau, L. Kadanoff and L. Schwartz for very useful discussions. This research was supported in part by Centre National de la Recherche Scientifique.

1. R. Lenormand and C. Zarcone, Two-phase flow experiments in a two-dimensional medium, *in proceedings of the Vth Int. Meet. Phys. Chem. Hydro.*, Tel-Aviv, (Dec. 1984).

2. H. S. Hele Shaw, Investigation of the nature of surface resistance of water and of stream-line motion under certain experimental conditions, *Trans. Instn. Nav. Archit., Lond,* **40** , 21, (1898).

3. P. G. Saffman and G. I. Taylor, The penetration of a fluid into a porous medium or Hele Shaw cell containing a more viscous liquid, *Proc. R. Soc. Lond.,* **A 245** , 311-329, (1958).

4. R. Lenormand and S. Bories, Description d'un mecanisme de connexion de liaison destine a l'etude du drainage avec piegeage en milieu poreux, *C. R. Acad. Sci. Paris,* **291 B** , 279-280 (1980).

5. R. Chandler, J. Koplik, K. Lerman and J. F. Willemsen, Capillary displacement and percolation in porous media, *J. Fluid Mech.,* **119** , 249-267, (1982).

6. P. G. De Gennes and E. Guyon, Lois generales pour l'injection d'un fluide dans un milieu poreux aleatoire, *J. Mecanique,* **17** , 403-442, (1977).

7. J. Bonnet and R. Lenormand, Realisation de micromodeles pour l'etude des ecoulements polyphasiques en milieu poreux, *Rev. Inst. Franc. Petr.,* **42** , 447-480, (1977).

8. R. Lenormand, Deplacements polyphasiques en milieu poreux sous l'influence des forces capillaires. Modelisation de type percolation, *Thesis, University of Toulouse,* (1981).

9. R. Lenormand and C. Zarcone, Role of roughness and edges during imbibition in square capillaries, S.P.E paper no 13264, (1984).

10. T. A. Witten and L. M. Sander, Diffusion-limited aggregation, *Phys. Rev. B,* **27** , 9, 5686-5697, (1983).

11. P. Meakin, Diffusion-controlled cluster formation in 2-6 dimensional space, *Phys. Rev. A,* **27** , 3, 1495-1507, (1983).

12. L. Paterson, Diffusion-limited aggregation and two-fluid displacements in porous media, *Phys. Rev. let.,* **52** , 18, 1621-1624, (1984).

13. J. Nittmann, G. Daccord and H. E. Stanley, Fractal growth of viscous fingers: quantitative characterization of a fluid instability phenomenon, *Nature,* **314,** 14, 141-144, (1985).

14. R. Lenormand and C. Zarcone, Invasion Percolation in an etched network: measurement of a fractal dimension, *to be published in Phys. Rev. Let.,* (1985).

15. D. Wilkinson and J. F. Willemsen, Invasion Percolation: a new form of percolation theory, *J. Phys. A,* **16** , 3365-3376, (1983).

16. R. Lenormand, C. Zarcone and A. Sarr, Mechanisms of the displacement of one fluid by another in a network of capillary ducts, *J. Fluid Mech.,* **135** , 337-353, (1983).

Acoustic and Hydrodynamic Flows in Porous Media

Jean-Claude Bacri and Dominique Salin
Laboratoire d'Utrasons, U.A. 789, Université Pierre et Marie Curie, Tour 13,
4, place Jussieu, F-57230 Paris Cedex 05, France

Abstract. — We review the Biot theory of the propagation of acoustic waves in multiconnected porous media filled with fluid. Theoretical predictions are compared to experiments in many different physical systems such as packed and sintered glass beads, fourth sound in ^4He, polymeric gels which allow study of the critical behaviour of gel in the sol - gel transition. Using acoustics as a local probe of fluid concentration variations in porous media, we can determine the concentration profiles during miscible and immiscible fluid flows through the medium, and analyse our results in terms of dispersion and capillary diffusion.

1 INTRODUCTION

As the purpose of these lectures is devoted to show in what way acoustic measurements can be used as a tool for experimental investigations in physical system such as porous media, gels and the sol - gel transition, fourth sound of ^4He and hydrodynamic flows of miscible and immiscible flows in porous media, we will first review the essential features of Biot's theory and its improvements ; we will then compare predictions and experimental results in the physical systems described above. Acoustics allows to derive from velocity measurements the fluid concentration time and space-dependence during miscible and immiscible flows through porous media. An analysis of respectively dispersion and capillary diffusion follows.

2 ACOUSTIC WAVES IN POROUS MEDIA

We will use Biot's notations [1] and emphasize the main physical assumptions and consequences of his approach. A more elegant Lagrangian derivation of Biot's equations has been recently achieved [2] ; we will also refer to it.

2.1 BASIC EQUATIONS

We will assume the existence of volumes large compared to the typical grain size but small compared to the acoustical wave length over which an average fluid $\vec{U}(\vec{r},t)$ displacement and an average solid $\vec{u}(\vec{r},t)$ displacement can be defined. The two coupled motion equations are then :

$$\rho_{11}\frac{\partial^2\vec{u}}{\partial t^2} + \rho_{12}\frac{\partial^2\vec{U}}{\partial t^2} = P\vec{\nabla}(\vec{\nabla}\cdot\vec{u}) + Q\vec{\nabla}(\vec{\nabla}\cdot\vec{U}) - N\vec{\nabla}\wedge\vec{\nabla}\wedge\vec{u} + b\,F(\omega)\left(\frac{\partial\vec{U}}{\partial t} - \frac{\partial\vec{u}}{\partial t}\right) \qquad (1)$$

$$\rho_{22}\frac{\partial^2\vec{U}}{\partial t^2} + \rho_{12}\frac{\partial^2\vec{u}}{\partial t^2} = R\vec{\nabla}(\vec{\nabla}\cdot\vec{U}) + Q\vec{\nabla}(\vec{\nabla}\cdot\vec{u}) - b\,F(\omega)\left(\frac{\partial\vec{U}}{\partial t} - \frac{\partial\vec{u}}{\partial t}\right) \qquad (2)$$

This set of equations implies that the fluid neither creates nor experiences a shear force.and that the fluid is multiconnected throughout the sample. The elastic constants P, Q, R, are related [2] to the bulk modulus of the solid K_s, of the porous frame K_b, of the fluid K_f ; N is the shear modulus of the porous frame. ρ_s and ρ_f

are the solid and fluid densities and

$$\rho_{11} + \rho_{12} = (1 - \phi)\, \rho_s \tag{3}$$

$$\rho_{22} + \rho_{12} = \phi\, \rho_f \tag{4}$$

The term ρ_{12} corresponds to the inertial drag that the fluid exerts on the solid as the solid is accelerated relative to the fluid and vice-versa. The induced mass coefficient is generally written as

$$\rho_{12} = -(\alpha - 1)\, \phi\, \rho_f \tag{5}$$

where α is a purely geometrical factor ($\alpha > 1$) accounting for the tortuousity of the medium [3]. This purely inertial drag, which has nothing to do with viscosity, is well known for a single solid sphere oscillating in a perfect solid [4] ; its extension [5] to an assembly of isolated spherical solid particles in a fluid (suspension) leads to $\alpha = [1 + \phi^{-1}]/2$. Moreover, this tortuousity α is related to the formation factor F of the medium [6], $\alpha = F\phi$, F standing as the ratio of the sample conductivity, saturated with a conducting fluid, to the conductivity of the fluid.

Viscous coupling. The general relation between flux (fluid flow-rate per unit area) and forces ($-\vec{\nabla} p_f$) leads to

$$\phi\, \frac{\partial \vec{U}}{\partial t} = -\frac{k}{\eta} \vec{\nabla} p_f \tag{6}$$

where η is the fluid viscosity and k the permeability. Relation (6) is a generalization of Poiseuille's law for a straight tube of radius a ($k = \phi\, a^2/8$). This relation is equivalent to relation (2) when the pressure p_f is applied while solid is prevented from moving and $b = \eta\, \phi^2/k$, with $F(\omega = 0) = 1$. As the frequency ω increases, the viscous depth [4] $\delta = \sqrt{2\eta/p_f\, \omega}$, decreases and becomes smaller than the typical pore size d when ω reaches the value

$$\omega_c = 2\eta / \rho_f\, d^2 \tag{7}$$

$F(\omega)$ is a complex function of ω/ω_c ; at high frequencies ($\omega \gg \omega_c$) $F(\omega) \sim (1+i)(\omega/\omega_c)^{1/2}$, $i^2 = -1$. We have to comment on the derivation of b which involves steady flow throughout the sample, or at least over a distance large compared to d ; but in acoustic experiments solid displacements are very small ($\sim$ 1 mm/Hz, i.e. $\sim$ 10 Å at 1 MHz) and b seems to be rather related to local friction between solid and fluid rather than to permeability.

2.2 WAVE SOLUTIONS

We have to solve the basic equations (1-2) for displacements ($\vec{U}$ and $\vec{u}$) varying in time and space with the propagating factor $\exp i(\vec{K}\vec{r} - \omega t)$; $\vec{K}$ is the complex wave vector ; its real part K' leads to the phase velocity V through $V = \omega/K'$; its imaginary part K" corresponds to the attenuation of the medium. Inserting this propagatory factor in the set (1-2) and using the decomposition [7] of the displacement vector into longitudinal modes ($\vec{\nabla} \wedge \vec{U} = \vec{\nabla} \wedge \vec{u} = 0$) and transverse modes ($\vec{\nabla} \cdot \vec{U} = \vec{\nabla} \cdot \vec{u} = 0$) one can find one transverse mode corresponding to a shear wave and two longitudinal modes (fast and slow) corresponding to compressional waves. This is the central result of Biot's theory. As experiments concern essentially compressional waves, we will then discuss these two modes relatively to the characteristic frequency ω_c. Their general dispersion relation is

$$K^2_{\substack{Fast\\Slow}} / \omega^2 = \left\{ \Delta \pm [\Delta^2 - 4(\tilde{\rho}_{11}\tilde{\rho}_{22} - \tilde{\rho}_{12}^2)(PR - Q^2)]^{1/2} \right\} / 2(PR - Q^2) \qquad (8)$$

where $\Delta = P\tilde{\rho}_{22} + R\tilde{\rho}_{11} - 2\tilde{\rho}_{12}Q$, $\tilde{\alpha} = \alpha + i\,\dfrac{b\,F(\omega)}{\omega\,\phi\,\rho_f}$, $\rho = (1-\phi)\,\rho_s + \phi\,\rho_f$ and $\tilde{\rho}_{ij}$ corresponds to ρ_{ij} with $\tilde{\alpha}$ instead of α.

2.2.1 *Low-frequency regime* ($\omega \ll \omega_c$)

The fast mode is propagatory,with the whole fluid dragged along with the solid. This mode corresponds to solid and fluid moving in phase ($\vec{U} = \vec{u}$) ; the slow mode corresponds to displacements in opposite phase ($\vec{U} + \vec{u} = 0$) ; this implies a large friction between solid and fluid,

$$K^2 = i\omega / C_D \qquad (9)$$

with $C_D = b/(PR - Q^2)/H$, $H = P + 2Q + R$; K^2 is purely imaginary and then $K' = K''$; the mode is overdamped (the wave does not propagate over more than a few wavelengths). Expression (9) corresponds to a diffusion equation which one can get from the basic ones (1-2).

2.2.2 *High-frequency regime* ($\omega \gg \omega_c$)

In this limit, the viscous coupling between fluid and solid disappears,and the remaining coupling is inertial (α) and elastic. The two longitudinal modes are propagatory,with attenuations proportional to the square root of the frequency. For very stiff skeletal porous frames (K_b, $N \gg K_f$)

$$V^2_{Fast} = V_s^2 / [1 + \phi\,\rho_f(\alpha - 1)/(1-\phi)\,\rho_s\,\alpha] \qquad (10)$$

$$V^2_{Slow} = V_f^2 / \alpha \qquad (11)$$

where V_f and V_s are fluid and porous frame velocities. The fast mode corresponds to oscillations of the solid with a part of the fluid inertially dragged along, the slow mode to oscillations of the fluid renormalized by the tortuousity α of the porous medium.

As its viscosity is zero ($\omega_c \to 0$) the superfluid component of ^4He at low temperature concerns the high-frequency regime. We have observed [8] the fast mode in porous Vycor glass filled with ^4He. The other propagatory mode in superfluid helium confined in porous medium is called fourth sound ; since its first observation [9] it was known that fourth sound measured velocity $C_4^{exp}(T)$ has to be renormalized by an index of refraction, n, in order to follow the theoretical predicted velocity

$$c^{exp}(T) = c^{theo} / n \qquad (12)$$

As it was first shown by JOHNSON [10],this expression is equivalent to (11) with $n = \alpha^2$ and fourth sound velocity measurements allow a determination of α.

3 EXPERIMENTAL CONFRONTATION

3.1 GLASS BEADS

Fused or packed glass beads filled with one fluid are synthetized porous media in which one can control the pore size. Common bead diameters used are in the range of 100 µm which gives a critical frequency ω_c (7) of less than 1 kHz ; this means that in classical experiments (∿ 1 MHz) we are in a high-frequency regime where both longitudinal modes are expected to be propagatory.

3.1.1 *SINTERED GLASS BEADS*

Using a set of sintered glass beads sample (diameter 200 μm) of various porosities, PLONA [11] has observed simultaneously the three predicted modes (shear, fast and slow compressional). From the known values of the system (K_s, K_f, ρ_s, ρ_f, ϕ) and velocity measurements of the dry samples (K_b, N), the measured velocities (at 500 MHz) match the predicted ones with only one adjustable parameter α for each ϕ, the values of which were confirmed later through fourth sound experiments [3] ; as was noticed, the larger the porosity, the better the agreement between measured velocity values and high frequency predicted ones ; we can interpret this remark : as the porosity decreases, the pore dimension decreases and the high-frequency condition ($\omega \gg \omega_c$) is less and less fulfilled, and then dispersion effects might affect velocity measurements, especially for the slow wave, for which the agreement is worse. As we will see in the next paragraph, attenuation measurements might have provided further information.

3.1.2 *PACKED GLASS BEADS*

Fused glass beads correspond to a stiff frame ; the velocity of the fast mode is close to the one in the dry sample, whereas the velocity of the slow mode is close to the one in fluid. On the contrary, for packed beads the frame is weak and the sound velocity in the dry sample is small (∿ 40 m/s) and difficult to observe. The slow mode, the velocity of which is close to the small one, has not yet been observed. We have observed the fast mode [12], the velocity of which is close to the one of the fluid, and studied its attenuation. Velocity measurements are in agreement with the predictions and with a value of the tortuousity $\alpha = \phi^{-1/2}$ [3] as expected for spheres. This value of α remains unchanged whatever the fluid saturating the packing is, as it has been showed with water-ethanol mixtures of different concentrations [13]. Frequency-dependence of the attenuation over a wide range of glass bead diameters (d from 50 to 500 μm) are given in figure 1. The expected K" variation in the high-frequency regime along with ω and d is $K'' \sim \sqrt{\omega}/d$. Obviously this dependence is not followed by the data, especially for the smallest diameters ; this means that even at 200 MHz, ω is not high enough compared to ω_c (ω_c varies from 2 kHz to 20 Hz as d varies from 50 to 500 μm) and Biot's relation (8) has to be solved with the function F(ω) : these solutions correspond to the full lines through the data with only one adjustable parameter.

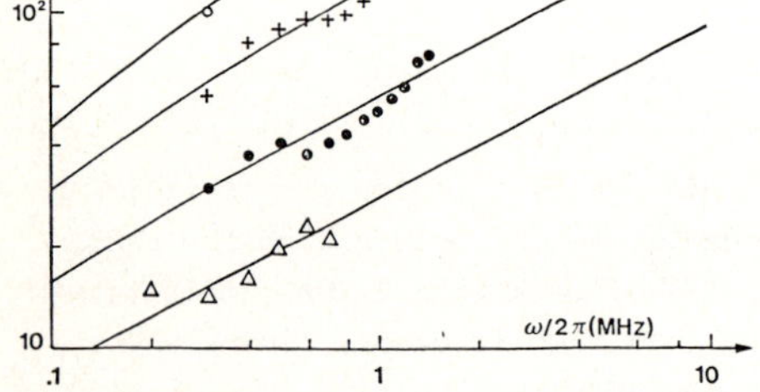

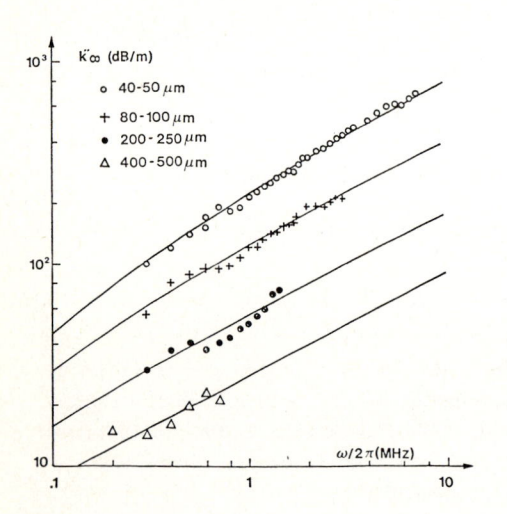

FIGURE 1. Log-log plot of the remaining attenuation of sound $K''_\infty = K''_{exp} - K''_{BC}$ versus frequency.

3.2 GELS AND THE SOL - GEL TRANSITION

A gel is an infinite cross-linked polymer network imbedded in a fluid solvent. Typical distance between two nodes is 100 Å ; the associated critical frequency, $\omega_c \sim 10^{10}$ Hz for $\eta \sim 10^{-2}$ c.g.s. and $\rho_f = 1$ c.g.s., is large compared to laboratory ones : acoustics in gel will provide a test of the theory in the low-frequency regime. The polymer network is a very weak frame (K_b, $N \ll K_f$, K_s) and its volume fraction $c = 1 - \phi$ is only a few per cent.

The overdamped slow mode corresponds to the diffusion equation (9) with a diffusivity

$$C_D = \frac{k}{\eta}\left[K_b + \frac{4}{3}N\right] \tag{13}$$

This expression was first derived by TANAKA et al [14] and by de GENNES [15]. Tanaka et al have directly measured the static permeability k of the gel and its modulus (also in ref. [16]) ($k_b + \frac{4}{3}N \sim 5 \times 10^4$ c.g.s., $\frac{k}{\eta} \sim 10^{-11}$ c.g.s. for $c \sim 5$ %) and using a light scattering technique they observed a mode which was diffusive and with a diffusion coefficient in accordance with (13). The fast mode is propagatory

$$V_{Fast} = H/\rho \quad \text{and} \quad K''_{Fast} = \frac{(\rho_s - \rho_f)^2}{2 V_f \rho} \frac{k}{\eta} \omega^2 c^2 \tag{14}$$

We have previously derived these expressions directly [17]. From the experimental point of view, attenuation provides a good test on both the friction $f \equiv \eta/k$ and the concentration c and their relation to the connectivity of the porous medium. We have performed two experiments which, in our opinion, are able to clarify these relations.

3.2.1 *SOL - GEL TRANSITION*

In this first experiment [18] we look at the gel formation process : from the monomer solutions of concentration c_0 to the gel formed with its reticulating points between different polymer chains. The transition is described through the parameter S_∞ which varies from 0 in the sol phase to 1 in the formed gel ; S_∞ represents the gel fraction and accounts for the connectivity of the medium ; $c = S_\infty c_0$ is the reticulated gel concentration.

Two kinds of gelation have been examined :

- the chemical gelation of a 10 % polyacrylamid gel is observed by recording the ultrasonic attenuation at 110 MHz as a function of time, for two constant temperatures, 20°C and 16°C (Fig. 2). The gelation starts 240 s and 460 s respectively after the mixing. The attenuation shows a sudden increase, and tends to saturate with the slowing down of the reaction. The maximum attenuation attained after a sufficiently long time (≃ 4000 s) is the same for the two different temperatures, when the gel is completely formed. The variation ΔK" as a function of $t - t_g$, t_g being the value where the attenuation starts increasing, is linear from 0 to 150 s and 400 s, at 20°C and 16°C respectively.

- the physical gelation of gelatin occurs when the temperature of the melt is lowered : the ultrasonic attenuation is recorded as a function of temperature. The measurements were made at 790 MHz for four différent concentrations (11 %, 7 %, 5.3 %, 2 %) (Fig. 3). The curves obtained are far from being as demonstrative of the effect as in the case of polyacrylamid gel. However, for the 7 % and 11 % concentrations, the change of slope defining T_g is quite clearly seen. From this point, the variation of ΔK" versus $T_g - T$ is roughly linear. The attenuation plateau obtained at 10°C in

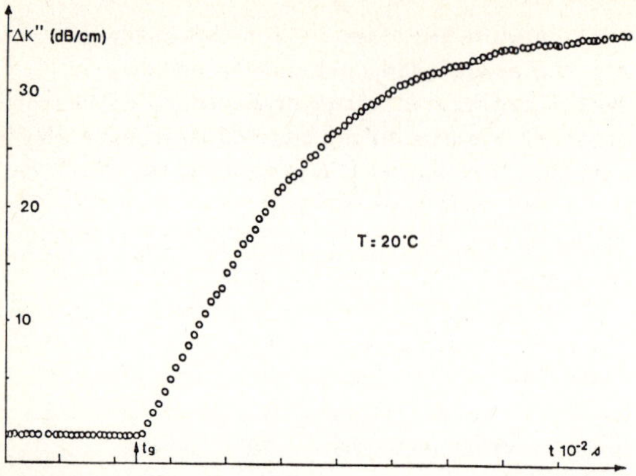

FIGURE 2. Ultrasonic attenuation variation during the gelation process in a polyacrylamid gel. t_g is the gelation threshold. At t = 0 the attenuation is that of the monomer solution

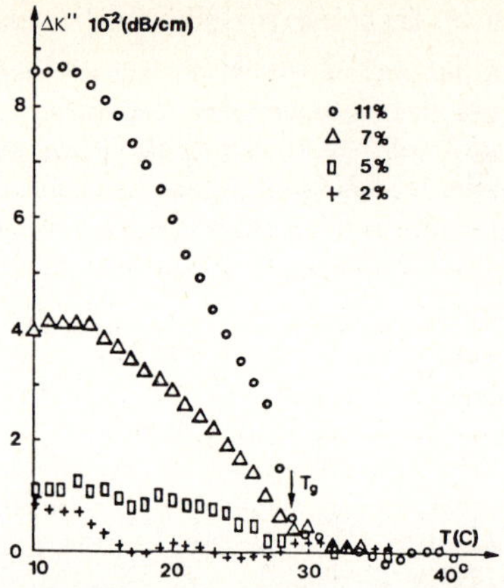

FIGURE 3. Difference of the ultrasonic attenuation between the gelatin and water versus temperature. T_g is the gelation threshold for the 11 % gelatin

the gelatin has been plotted as a function of c_0^2. The attenuation plateau value is linear as a function of c_0^2.

The ratio $K''_{plateau}/\omega^2$ (ω : ultrasonic frequency) has the same order of magnitude in the two gelating systems (near 1 and 4×10^{15} dB s^2/cm).

We can interpret these experiments in the light of the following remarks.

- As the monomer concentration c_0 is constant during the gelation, c will be the parameter involved in the attenuation expression $\left(K'' \sim \omega^2 \; S_\infty^2 \, c_0^2/f\right)$.
- As the attenuation of the plateau occurring at the end of the gelation ($S_\infty = 1$) varies as c_0^2, the friction seems not to depend on concentration ; f would stand as a friction per monomer rather than a permeability ($f = \eta/k$) which varies widely with concentration [14]. This would explain the fact that the measured attenuation of the plateau is three orders of magnitude larger than the expected one.
- The attenuation during gelation is then a measurement of S_∞^2 (or at least S_∞^2/f) and as K'' is linear, in the vicinity of the transition (t_c or T_c) with the difference ($t - t_c$) or ($T_c - T$), we can conclude that the critical exponent of S_∞ is close to 0.5 in reasonable agreement with percolation theory.

3.2.2 *SPINODAL DECOMPOSITION OF GEL*

We consider a gel which is built (S_∞ = 1 definitively) ; when the temperature is lowered under a critical temperature T_s, the gel presents a phase separation between the solvent and the polymeric network. Before reaching the spinodal temperature T_s, pretransitional divergence occurs : the coherence length ξ diverges with a critical exponent $\nu = 0.5$ in a mean field theory. The friction coefficient f is related to ξ through the diffusion coefficient of the concentration fluctuation [17] : $f \sim \xi^{-1} \sim (T - T_s)^{1/2}$. Attenuation measurement [17] versus temperature presents a divergence which can be fitted as $(T - T_s)^{-1/2}$ in accordance with the predicted mean field one through $K'' \sim 1/f$, with the same remark as above concerning the order of magnitude of the attenuation.

4 ACOUSTIC MEASUREMENTS OF FLUID CONCENTRATION VARIATIONS DURING FLOWS

An important experimental problem of biphasic flows through porous media is to determine the time and space dependence of fluid concentrations (concentration profiles) during fluid displacements. Most experiments were carried out by using an effluent analysis technique which gives out only concentrations averaged all over the sample. As the velocity of the fast wave in porous medium varies with the fluid filling, velocity measurements give the concentration in a cross-section of the sample : acoustic waves are generated and analysed thanks to a set of transmitter-receiver pairs of transducers laid along the sample in the flow direction of the sample with a spatial resolution of 3 mm . This gives us a powerful means of profile determination during miscible [13] and immiscible fluid flows [19].

4.1 MISCIBLE FLUID FLOWS

4.1.1 *THEORETICAL SURVEY*

Dispersion in porous media (i.e. the mixing which goes along with the flow of miscible fluid) has been intensively reviewed [20-22]. In the particular case of incompressible fluids, with the same viscosities and densities, flowing at a constant velocity u in the Ox direction, the space and time-dependent concentration c(x,t) obeys a diffusion equation :

$$\frac{\partial c}{\partial t} = - u \frac{\partial c}{\partial x} + K_L \frac{\partial^2 c}{\partial x^2} \qquad (15)$$

where K_L stands as the effective longitudinal diffusion coefficient. For a concentration-independent coefficient K_L, simple initial conditions (at $t = 0$, $c = 1$ for $x < 0$, $c = 0$ for $x > 0$) and an infinite sample, the solution of (15) is :

$$c(x,t) = \frac{1}{2}\left(1 - \mathrm{erf}\left(\frac{x - ut}{2\sqrt{K_L t}}\right)\right) \qquad (16)$$

Basically the most interesting point in miscible flows through porous media is to find out the flow velocity dependence of the effective longitudinal diffusion coefficient K_L. This coefficient accounts for the mixing of liquids through molecular diffusion and convection (i.e. the stirring of the fluid making its way through the complex geometry of the pores). For a cylindrical tube of radius a, the dispersion is essentially due to the velocity profile of Poiseuille 's flow through the tube (convection). The ratio of the longitudinal diffusion coefficient K_L to the molecular diffusion coefficient D is a simple function [23] of the dimensionless Péclet number $Pe = uL/D$. In fact, a bundle of straight tubes is not a good representation of a porous medium. SAFFMAN [24] studied a random network of capillaries with equal diameters, which leads to

$$\frac{K_L}{D} = \frac{1}{F\varphi} + f\left(\frac{a}{\ell}, Pe\right) \qquad (17)$$

where ℓ is the length scale of the capillaries, a the radius of a pore. When $u = 0$, $K_L/D = 1/F\varphi$ (F is the formation factor of the porous medium).

4.1.2 *DISPERSION COEFFICIENT MEASUREMENTS*

In preparation for a run, the column, packed with glass beads is saturated with the fluid to be displaced (a water-ethanol mixture) by a vacuum impregnation technique. The relationship between the sound velocity and the concentration of the mixture saturating the porous medium is determined in a previous experiment, which enables us to determine α acoustically. The displacing fluid (another water-ethanol mixture) is injected at a constant flow-rate varying from 2.5×10^{-5} cm/s to 5×10^{-2} cm/s (laminar flows). At a distance x_0 from the entrance end of the column, we obtain

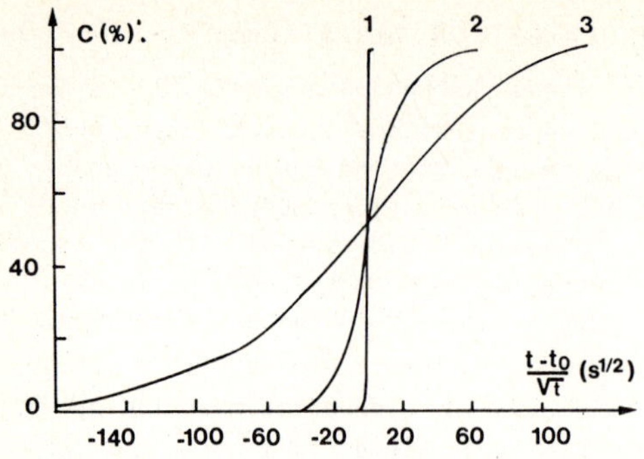

FIGURE 4. Concentration profiles $c(x_0,t)$ versus $(t-t_0)/\sqrt{t}$ in a given cross-section of the column for different flow rates :
(1) $u_1 = 2\times10^{-4}$ m/s ; $x_0 = 4.6$ cm
(2) $u_2 = 10{,}4\times10^{-6}$ m/s ; $x_0 = 9.68$ cm
(3) $u_3 = 4.2\times10^{-7}$ m/s ; $x_0 = 4.6$ cm

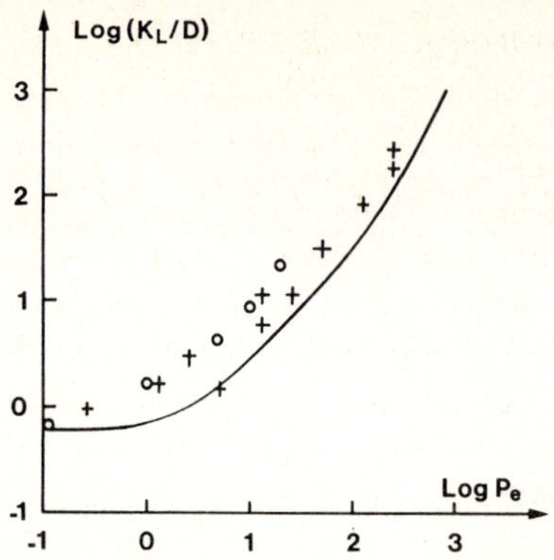

FIGURE 5. Ratio of the longitudinal dispersion coefficient K_L to the molecular diffusion coefficient D versus Péclet number

the time-variation of the concentration c ; the experimental data for different runs corresponding to various flow rates are plotted in Fig. 4 ($t_0 = x_0/u$). The effective diffusion coefficient is derived by matching the theoretical expression (16) of $c(x_0,t)$ to the data. In Fig. 5, we plotted K_L/D versus the Péclet number $Pe = U\ell/D$ (ℓ being here the diameter of the beads, $\ell = 200$ µm), which covers a range of 3 and a half decades.

For small Péclet values, K_L/D falls off to a constant close to 0.6, which is in agreement with the value of α obtained from acoustic measurements ($1/F\varphi = 1/\alpha = 0.57$). Then the ratio K_L/D increases from Pe = 10 onwards. The solid line represents Saffman's model solution with a ratio $\ell/a = 5$ (equ. (17)). It systematically lies below the experimental points, still displaying the same characteristics. But in fact, Saffman's theory corresponds to a specific random network of tubes with a mean length ℓ much larger than a and the effect of pore-size dispersion is not accounted for in this model.

4.2 IMMISCIBLE FLUID FLOWS

Due to the importance of oil recovery, an understanding of immiscible flows through porous media is required. The main basic flows are imbibition and drainage. We give an example of our measurements [19] during second imbibition process, performed on a sandstone initially at irreducible water saturation with a constant water (totally wetting fluid) injection rate. From our saturation measurements,we are able to analyse the basic processes involved in the capillary diffusion occurring during the flow.

4.2.1 *THEORETICAL SURVEY*

Let us first recall some aspects of the theory described by P. G. de GENNES [25]. In the quasi-static limit, the flow equations are derived from water J_w and oil J_o currents. The more general linear relations between the currents and local pressures (p_w and p_o) of the two fluids are

$$\vec{J}_w = \Sigma_w(\Phi)\,(-\vec{\nabla} p_w) - \Lambda(\Phi)\,(\vec{\nabla} p_o) \qquad (18)$$

$$\vec{J}_o = -\Lambda(\Phi)\,(\vec{\nabla} p_w) + \Sigma_o(\Phi)\,(-\vec{\nabla} p_o) \qquad (19)$$

where Σ_w and Σ_o are the permeability to viscosity ratios for each fluid. The cross term coefficient is expected to be zero if each pore is occupied by a single fluid, i.e. in the case of partial wetting ; if one of the fluids completely wets the solid, Λ can be of importance. The last ingredient in the theory is the capillary pressure $\pi(\Phi) = p_w - p_o$ which can be determined experimentally [20-22]. When neglecting gravity but including the cross term (Λ) for a one-dimensional flow and with an imposed injection rate, the following differential equation in $\Phi(x,t)$ can be drawn from the set of equations (18-19) and the incompressible fluid conservation ones :

$$\frac{\partial}{\partial t}\Phi(x,t) = -J\frac{d\rho(\Phi)}{d\Phi}\frac{\partial\Phi}{\partial x} + \frac{\partial}{\partial x}\left(\Sigma(\Phi)\frac{d\Pi(\Phi)}{d\Phi}\frac{\partial\Phi}{\partial x}\right) \tag{20}$$

$$\text{with}\quad \rho(\Phi) = \frac{\Sigma_w + \Lambda}{\Sigma_w + \Sigma_o + 2\Lambda} \quad\text{and}\quad \Sigma(\Phi) = \frac{\Sigma_o\Sigma_w - \Lambda^2}{\Sigma_w + \Sigma_o + 2\Lambda} \tag{21}$$

Ignoring the second term of the RHS in (20) we have a simple convection equation : the profile $\Phi(x,t)$ is then controlled by a local concentration-dependent drift velocity $c(\Phi) = J\frac{d\rho}{d\Phi}$. Generally speaking ($\Lambda = 0$), $c(\Phi)$ vanishes for the two extremal values of Φ which are $\Phi = \Phi_{IR}$ (irreducible water-saturation) and $\Phi = \Phi^*$ (residual oil saturation ; $c(\Phi)$ reaches a maximum in between. As $\Phi(x,t)$ cannot be double-valued for any (x,t), BUCKLEY and LEVERETT [20-22] introduce a discontinuity ("shock wave") corresponding to a vertical profile drifting at a velocity $c(\Phi_s)$.

The second term of the RHS of (20) corresponds to non-linear diffusion [26], with a concentration-dependent diffusion coefficient $D(\Phi) = \Sigma(\Phi)\frac{d\Pi(\Phi)}{d\Phi}$; this term is responsible for the broadening of the shock front, but in very different ways depending on how $D(\Phi)$ varies with Φ. Two cases are to be distinguished :

- If $\Lambda = 0$, $\Sigma(\Phi)$ vanishes at both ends of the saturation range, generally more drastically than $d\Pi/d\Phi$ diverges. Then $D(\Phi)$ falls down to zero for both Φ_{IR} and Φ^* saturation values : this is an "*hypodiffusive*" case for which the profile is sharper than in the classical case.
- If $\Lambda \neq 0$, Σ can be different from zero when Φ gets close to Φ_{IR}, in which case the capillary pressure imposes its divergence on $D(\Phi)$: in this "*hyperdiffusive*" case diffusion will then rapidly increase when getting close to Φ_{IR}.

4.2.2 *EXPERIMENT AND RESULTS*

The totally wetting fluid is a sodium chloride brine, the other fluid is an oil. Flows are driven vertically along the larger dimension of the sandstone core at constant flow rates varying from 0.6 cm^3/h to 24 cm^3/h. Our calibration curve of velocity variations with Φ and the accuracy of relative velocity measurements (better than 5×10^{-4}) provide an overall precision in Φ better than 1 %. The three figures (a, b, c) in figure 6 correspond to imbibition processes at different flow rates. The four curves in each figure correspond to the saturation time-dependence $\Phi(t)$ for four different pairs of transducers. We observe two main features : *i*) when Φ tends to Φ^* every curve presents a cusp. *ii*) On the opposite side ($\Phi \to \Phi_{IR}$), as the flow rate is decreased, no shock profile is observed any more, and for the smallest injection rate, the diffusion process expands all over the sample, as soon as the imbibition is started. We first reason in terms of a shock wave (moving at the velocity $u = c(\Phi_s)$) to which we add diffusion. A natural variable change in (20) is then $\xi = (x - ut)/\sqrt{t}$ where velocities u involved in ξ are deduced from large flow rate values for which the shock wave is dominant, and then reduced in the flow rate proportion. Figure 7 displays a plot of Φ versus ξ for the two smallest flow rates and shows that ξ stands as a good variable which allows [26] the determination of $D(\Phi)$ (Fig. 8).

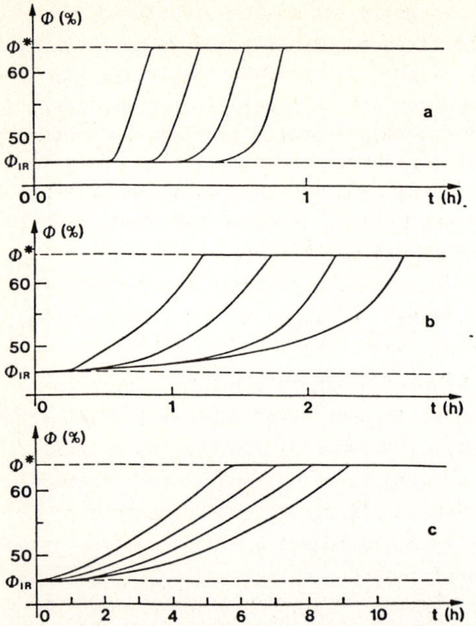

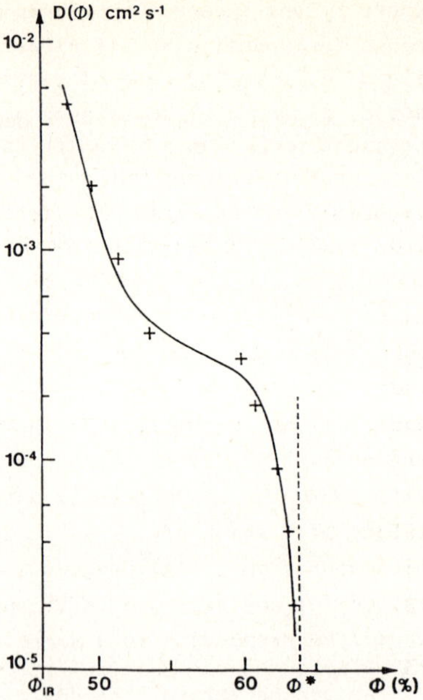

FIGURE 6. Time-dependence of the water saturation $\Phi(t)$ in four cross-sections of the sample (from left to right, x = 3.8 ; 6.8 ; 9.8 ; 12.8 cm, from the inlet onwards) recorded during three imbibition runs performed on a prewet sandstone (Φ_{IR} = 46 %) with three different constant water injection rates (a : 12 cm^3/h ; b : 3 cm^3/h ; c : 1.2 cm^3/h). Φ^* = 64 % corresponds to the residual oil saturation.

FIGURE 8. Variation of the diffusion coefficient $D(\Phi)$, in a semi-logarithmic scale versus the water saturation Φ

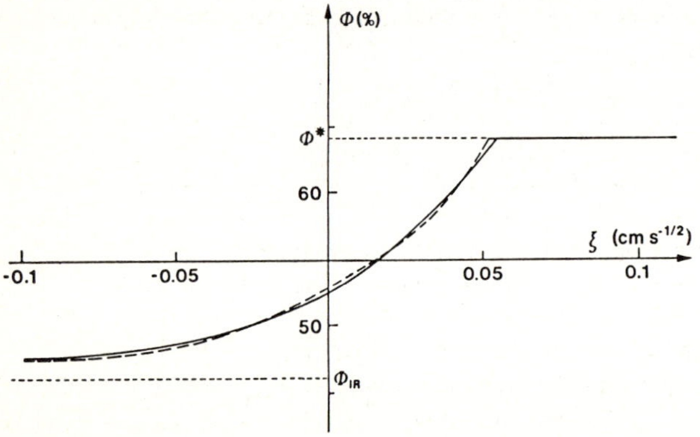

FIGURE 7. Variation of the water saturation Φ with the reduced variable $\xi = (x - ut)/\sqrt{t}$, for our two smallest flow rates (-- : u = 1.25 cm/h , x = 6.8 cm and — : u = 2.5 cm/h , x = 9.8 cm)

4.2.3 *DISCUSSION*

Obviously $D(\Phi)$ diverges close to Φ_{IR} (hyperdiffusion) and it disappears close to Φ^* (hypodiffusion) ; this is our central experimental result.

- When Φ increases to Φ^*, $D(\Phi)$ behaves as $D(\Phi) \sim D^*(\Phi^* - \Phi)^{n^*}$, with $n^* = 1$ (*hypodiffusion*) and $D^* \simeq 10^{-3}$ cm^2/s . Let us recall that a percolation type description [25, 27] concludes in favour of a $D(\Phi) \sim (\Phi^* - \Phi)^{\tilde{m}-1}$ (with $\tilde{m} \gtrsim 4$) behaviour. Indeed, the predicted decrease is more drastic than what is observed here, but there is no way to identify a more or less drastic behaviour under 2 % in our experiment.

- In the neighbourhood of Φ_{IR}, the $D(\Phi)$ divergence can be directly deduced from the time and space saturation dependence : $\Phi - \Phi_{IR} \sim \xi^{-2}$, which leads to $D(\Phi - \Phi_{IR})_{\Phi \to \Phi_{IR}} \sim D_{IR}(\Phi - \Phi_{IR})^{n_{IR}}$ with $n_{IR} = -1$ (*hyperdiffusion*) and $D_{IR} \simeq 10^{-4}$ cm^2/s . The only case yet known of predicted hyperdiffusion in literature [28] concerns water-film growth in a fractal material ; the power law ($n_{IR} = -1$) observed close to Φ_{IR} is indeed in good accordance with this prediction, contrary to the corresponding prefactor (10^{-10} cm^2/s) which is in complete disagreement with our 10^{-4} cm^2/s observed value, but the precited case concerns a single wetting film whereas we deal here with two fluids. An evaluation of the cross-term Λ in the basic equations(18-19) might provide theoretical predictions coherent with the observed effect.

REFERENCES

1 M.A. BIOT : *J. Acoust. Soc. Am.* **28**, 168 (1956) ; **28**, 179 (1956)

2 D.L. JOHNSON : *Proc. of Enrico Fermi Summer School "Frontiers of Physical Acoustics"* (Varenna, 1984)

3 D.L. JOHNSON, T.J. PLONA, C. SCALA, F. PASIERB and H. KOJIMA : *Phys. Rev. Lett.* **49**, 1840 (1982)

4 L. LANDAU and E.M. LIFSHITZ : *Fluid Mechanics* (Pergamon Press, N.Y., 1959)

5 J.G. BERRYMAN : *Appl. Phys. Lett.* **37**, 382 (1980)

6 R.J.S. BROWN : *Geophysics* **45**, 1269 (1980)

7 L.D. LANDAU and E.M. LIFSHITZ : *Theory of Elasticity* (Pergamon Press, 1959)

8 B. LAMBERT and D. SALIN : *J. Physique Lett.* **41**, L - 487 (1980)

9 K.A. SHAPIRO and I. RUDNICK : *Phys. Rev.* **A 137**, 1383 (1965)

10 D.L. JOHNSON : *Appl. Phys. Lett.* **37**, 1065 (1980) ; **38**, 827 (1980)

11 T.J. PLONA : *Appl. Phys. Lett.* **36**, 259 (1980)

12 D. SALIN and W. SCHÖN : *J. Physique Lett.* **42**, L - 477 (1981)

13 J.C. BACRI, C. LEYGNAC and D. SALIN : *J. Physique Lett.* **45**, L - 767 (1984)

14 T. TANAKA, L.O. HOCKER and G.B. BENEDEK : *J. Chem. Phys.* **59**, 5151 (1973)

15 P.G. de GENNES : *Macromolecules* **9**, 587 (1976)

16 J.C. BACRI, J. DUMAS and A. LEVELUT : *J. Physique Lett.* **40**, L - 231 (1979)

17 J.C. BACRI and R. RAJAONARISON : *J. Physique Lett.* **40**, L - 5 (1979)

18 J.C. BACRI, J.M. COURDILLE, J. DUMAS and R. RAJAONARISON : *J. Physique Lett.* **41**, L - 36 (1980)

19 J.C. BACRI, C. LEYGNAC and D. SALIN : submitted to *J. Physique Lett.*

20 A. SCHEIDEGGER : *The Physics of Flow through Porous Media*, 3rd Edition (University of Toronto Press, 1974)

21 C.M. MARLE : *Multiphase flows in Porous Media* (Technip, Paris, 1981)

22 F.A.L. DULLIEN : *Porous Media, Fluid Transport and Pore Structure* (Academic Press, N.Y., 1979)

23 G. TAYLOR : *Proc. Royal Soc. London* **A 219**, 186 (1953)

24 P.G. SAFFMANN : *J. Fluid Mechanics* **6**, 321 (1959)

25 P.G. de GENNES : *Physicochemical Hydrodynamics* **4**, 175 (1983)

26 J. GRANK : *The Mathematics of Diffusion* (Clarendon University Press, Oxford, 1956)

27 E. GUYON and P.G. de GENNES : *J. de Mécanique* **17**, 403 (1978)

28 P.G. de GENNES : to be published

Dynamics of Saturated and Deformable Porous Media

Homogenization Theory and Determination of the Solid-Liquid Coupling Coefficients

Guy Bonnet[1,2] and Jean-Louis Auriault[1]

1 - INTRODUCTION

The main purpose of the paper is the description of the dynamics of an elastic deformable porous medium saturated by a newtonian viscous fluid.

The description at the macroscopic level is obviously phenomenological, taking into account the complexity of the pore geometry in practical applications.

The authors which propose phenomenological equations of the motion may be however separated into two groups :

- A first group gives directly phenomenological equations
  - . either by using the classical theory of mixtures |1|
  - . or by using simplifying, physically based assumptions |2|
- A second group uses the homogenization theory |3| |4| |5|. The main assumption of the theory is that the medium is periodic. The theory uses then directly the equations written at the pore level. The structure of the equations obtained at the macroscopic level are useful for a non-periodic medium, taking into account the statistical periodicity at the macroscopic level.

In fact, for the simplest case of a linearly deformable porous medium the different approaches give approximately same results. The homogenization theory allows however additional informations as it will be shown below.

In the following, the principle of the derivation of the phenomenological equations is given for the simplest case of a permanent motion in a non-deformable porous medium. The main results of the theory are given for the case of the dynamics of a deformable porous medium.

Stress is next laid on the determination of the coupling coefficients appearing in the equations and the description of recent experimental results obtained by one of us is given.

2 - DYNAMICS OF A SATURATED POROUS MEDIUM

2.1 - Description of the homogenization process : steady flow through a rigid porous medium |7|

Some of the results presented here are, in a different form, given in |6|.

[1]Institut de Mecanique de Grenoble, Domaine Universitaire, BP 68, C.N.R.S. L.A. N° 6, F-38402 St. Martin d'Heres Cedex, France
[2]Laboratoire de Genie Civil U S T L, C.N.R.S. U A N° 0710127, F-34060 Montpellier Cedex, France

A) Notations and geometry of the problem

A rigid porous medium with a periodic porous structure is sketched in figure 1. The period Ω is composed by

. an interconnected fluid volume Ω_ℓ
. a skeleton part Ω_s .

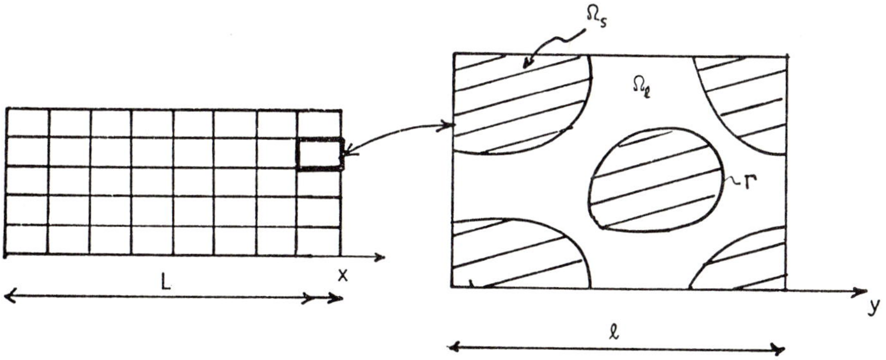

A periodic porous medium and two different scales to represent it.
Figure 1

ℓ is a length which is characteristic of the period
L is a length which is characteristic of the macroscopic medium with $\varepsilon = \ell/L << 1$

The medium is described by a position vector $\underset{\sim}{x}$ at the macroscopic level ("slow" variable) and by a position vector $\underset{\sim}{y}$ at the microscopic level where $\underset{\sim}{y} = \underset{\sim}{x} / \varepsilon$ ("quick" variable).

Any variable is the sum of :

- a function of $\underset{\sim}{x}$, slowly varying
- a function of $\underset{\sim}{y}$, considered as periodic

B) Equations of the motion at the microscopic level

The flow of a newtonian incompressible fluid within the porous medium is studied here.

The equations of the problem are :

B.1) Balance of momentum

$$\operatorname{div} \underset{\sim}{\sigma_\ell} = \rho \left(v_j \frac{\partial}{\partial x_j}\right) v_i$$

where $\underset{\sim}{\sigma_\ell}$ is the stress tensor

ρ the specific mass
$\underset{\sim}{v} = v_i$ the fluid velocity $(i=1,3)$

B.2) Constitutive equations for a newtonian incompressible fluid

$\sigma_{ij} = P \delta_{ij} + 2 \mu D_{ij}$ $\qquad v_{i,i} = 0$ (non-compressible flow) with $D_{ij} = \frac{1}{2}\left(\frac{\partial v_i}{\partial x_j} + \frac{\partial v_j}{\partial x_i}\right)$

where ρ is the specific mass and P is the pressure.
μ is the dynamic viscosity

C) **Boundary problem**

The problem may be described by using only $\underset{\sim}{v}$ and p by :

(1) $\mu \Delta \underset{\sim}{v} = -\,\text{grad}\, P + \underset{\sim}{v}\ \text{grad}\,(\underset{\sim}{v})$

(2) $\text{div}\ \underset{\sim}{v} = 0$

(3) $\underset{\sim}{v}/_\Gamma = 0$ (Boundary condition for the viscous fluid)

D) **Method of resolution**

The solution of (1) to (3) is assumed to have the form

$$\underset{\sim}{v} = \underset{\sim}{v}^{(0)}(\underset{\sim}{x},\underset{\sim}{y}) + \underset{\sim}{v}^{(1)}(\underset{\sim}{x},\underset{\sim}{y}) + \ldots$$

$$p = p^{(0)}(\underset{\sim}{x},\underset{\sim}{y}) + p^{(1)}(\underset{\sim}{x},\underset{\sim}{y}) + \ldots$$

where the $v^{(i)}$ and $p^{(i)}$ are assumed periodic as functions of y

The expansions of $\underset{\sim}{v}$ and p above are incorporated into eqns (1) to (3), taking into account that

$\frac{\partial}{\partial x_i}$ is replaced by $\frac{\partial}{\partial x_i} + \frac{1}{\varepsilon}\frac{\partial}{\partial y_i}$

By identifying each power of ε in the expansions of (1) to (3), it is now possible to obtain each function of the expansion by solving a series of boundary problems with the y variable

For example, equation (1) at ε^{-2} gives

$\mu\, \Delta_y \underset{\sim}{v}^{(0)} = 0$ (the y as a subscript means that the Laplacian operator is derived with respect to $\underset{\sim}{y}$)

E) **Sequel of boundary problems and solutions**

- 1st order

Equations (1) at ε^{-2} order, (2) at ε^{-1} and (3) at ε^0 give :

$\mu\, \Delta_y \underset{\sim}{v}^{(0)} = 0\ ,\quad \underset{\sim}{v}^{(0)}/\,\Gamma = 0\ ,\quad \text{div}_y\, \underset{\sim}{v}^{(0)} = 0$

whose the solution is $\underset{\sim}{v}^{(0)} = 0$

- 2nd order

$\mu \Delta_y \underset{\sim}{v} = -\,\text{grad}_y\, p^{(0)}\ ,\quad \text{div}\ \underset{\sim}{v}^{(1)} = 0\ ,\quad \underset{\sim}{v}^{(1)}/ \quad = 0$

where $\underset{\sim}{v}^{(1)}$ and $p^{(1)}$ are Ω - periodic.

The solution of this boundary problem is
$\underset{\sim}{v}^{(1)} = 0 \qquad p^{(0)} = p^{(0)}(\underset{\sim}{x})$ does not depend on $\underset{\sim}{y}$

- 3rd order

$$\mu \Delta_y \underset{\sim}{v}^{(2)} = - \text{grad}_x\, p^{(0)} - \text{grad}_y\, p^{(1)} \quad , \text{div}\ \underset{\sim}{v}^{(2)} = 0 \quad , \underset{\sim}{v}^{(2)}/\Gamma = 0$$

where $\underset{\sim}{v}^{(2)}$ and $p^{(2)}$ are Ω periodic

$\underset{\sim}{v}^{(2)}$ and $p^{(1)}$ appear to be linear functions of $\text{grad}_x\, p^{(0)}$

The velocity vector which is the particular solution of ($\text{grad}_x\, p^{(0)} = \delta_{ij}$ is denoted k^j . (δ_{ij} is the Kronecker symbol and $\text{grad}_{x_i}\, p^{(0)}$ is the component of $\text{grad}_x\, p^{(0)}$ on x_i).

The general solution of this boundary problem is then :

$$v_i^{(2)} = k_i^j \ \text{grad}_{x_j} (p^{(0)}) \quad \text{or} \quad \underset{\sim}{v}^{(2)} = \underset{\sim}{k}\ \text{grad}_x\, p^{(0)}$$

where $\underset{\sim}{k}$ depends on $\underset{\sim}{y}$.

At this step it could be thought that the problem is solved at the lowest order for $\underset{\sim}{v}$ and p . In fact, the following order is needed, giving a compatibility condition to $\underset{\sim}{v}^{(2)}$.

- 4th order

The Equations (2) and (3) give :

$$\text{div}_x\, \underset{\sim}{v}^{(2)} + \text{div}_y\, \underset{\sim}{v}^{(2)} = 0 \qquad \underset{\sim}{v}^{(3)}/\Gamma = 0$$

by using the volumic mean : $< . > = |\Omega|^{-1} \int . \, d\Omega$ (computed with $\underset{\sim}{y}$)

$$\text{div}_x <\underset{\sim}{v}^{(2)}> = -\int_{\Omega_p} \text{div}_y\, \underset{\sim}{v}^{(3)} d\Omega = - \int_{\delta\Omega_p} \underset{\sim}{v}^{(3)} \underset{\sim}{N}\, ds$$

taking into account that $\underset{\sim}{v}^{(3)}/\Gamma = 0$ and that $\underset{\sim}{v}^{(3)}$ is Ω-periodic, the last integral is zero and therefore :

$$\text{div} < \underset{\sim}{v}^{(2)} > = 0$$

Remark

By writing the boundary problems with an equivalent variational formulation in Sobolev spaces and by using the Lax-Milgram lemma | 8 |, it is possible to show the existence and uniqueness of each boundary problem | 7 |.

F) Macroscopic constitutive equations

The macroscopic constitutive equations are finally :

$$\text{div}_x <\underset{\sim}{v}> = 0 \qquad <\underset{\sim}{v}> = \underset{\sim}{K}\ \text{grad}_x p$$

with
$$\underset{\sim}{K} = \varepsilon^2 |\Omega|^{-1} \int_{\Omega_e} \underset{\sim}{k}(\underset{\sim}{y})\, d\Omega$$

The final result is that the macroscopic flow is governed by the usual Darcy's law with an incompressible macroscopic flow.

G) **Complementary information obtained by the homogenization process**

1 - The viscous stress tensor is $O(\varepsilon p^{(0)})$: Otherly said : the local gradient of the viscous stress is of the same order as the macroscopic pressure gradient. It is a necessary condition to obtain the flux through the porous medium.

2 - $\underset{\sim}{K}$ is a symmetric positive second order tensor. It is important to notice that the result does not call the irreversible process thermodynamics as often used to obtain this.

3 - $\langle \underset{\sim}{v} \rangle$ defined here as a volumic mean is also a surface mean as is the usual definition of the filtration velocity.

4 - For a weakly compressible liquid, the Darcy's law is still obtained. Only the balance of mass is modified.

2.2 - Dynamic flow of an incompressible fluid through a porous medium

A) Equations at the microscopic level

The motion of the fluid is assumed to have a harmonic motion with a small amplitude $\lambda \ll \ell$. The flow is therefore described by linear Navier-Stokes equations. The velocity and the pressure are searched with the form $\underset{\sim}{v}\, e^{i\omega t}$ and $p\, e^{i\omega t}$

The equations at the microscopic level are :

$$\mu \Delta \underset{\sim}{v} = -\operatorname{grad} p + i\omega\rho_\ell \underset{\sim}{v} \qquad (5)$$

$$\operatorname{div} \underset{\sim}{v} = 0 \qquad (6)$$

$$\underset{\sim}{v}/\Gamma = 0 \qquad (7)$$

In addition, the transient Reynolds number R_T is assumed O(1), in order take into account inertial effects at the microscopic level.

thus $R_T = \ell^2 \omega \rho \mu^{-1} = O(1)$

This condition is checked by writing

$\varepsilon^2 \mu' \Delta \underset{\sim}{v} = -\operatorname{grad} p + i\omega \rho_\ell \underset{\sim}{v}$ (8) where $\mu'\Delta \underset{\sim}{v}$, grad p and $\omega\rho_\ell \underset{\sim}{v}$ are of the same order compared to ε .

B) Result of the homogenization process

The equations of the flow at the macroscopic level are the following :

$$\langle \underset{\sim}{v} \rangle = \underset{\sim}{K}(\omega) \operatorname{grad}_x p \quad (9) \qquad \operatorname{div}(\langle v \rangle) = 0 \quad (10)$$

where $\underset{\sim}{K}(\omega)$ is a complex tensor function of the pulsation ω .

Remarks

1 - if $\Phi = |\Omega_\ell| \cdot |\Omega|^{-1}$ is the porosity, and Φp the partial pressure, the equation (9) may be written.

grad $(\Phi p) = \Phi H_1(\omega) \langle \underset{\sim}{v} \rangle + \Phi H_2(\omega)/\omega \langle \dot{\underset{\sim}{v}} \rangle$ where

The $\frac{\Phi H_2}{\omega}$ term appears to be an apparent specific mass.

The $\Phi H_1 \langle \underset{\sim}{v} \rangle$ term is a dissipative one.

2 - For a transient motion, Darcy's law is then :

$\operatorname{grad}(\Phi p) = F^{-1}(\Phi H) * \langle \underset{\sim}{v} \rangle$, where the star denotes a convolution product and F the Fourier transform.

As a consequence, the constitutive law at the macroscopic level is characterized by memory effects, despite the fact that the motion at the microscopic level is described by instantaneous constitutive equations.

2.3 - Dynamics of a saturated and deformable porous medium

The porous matrix is assumed in the following to be deformable linear elastic with small deformations. The fluid is assumed to be deformable and newtonian.

2.3.1 - Two phase behaviour

A) Basic assumptions

- It was shown previously (§ 2.1 G) that the viscous stress tensor is $O(\varepsilon p)$ where p is the pressure. This condition, needed to obtain the relative displacement between solid and liquid, is checked by writing the stress in the liquid $\underset{\sim}{\sigma}$ with the form :

  $$\underset{\sim}{\sigma}_{\ell} = p\delta_{ij} + 2\mu'\varepsilon^2 D_{ij}(\underset{\sim}{v}) + \lambda'\varepsilon^2 \operatorname{div}(\underset{\sim}{v})$$

  where $\underset{\sim}{\sigma}_{\ell}, p$ and $\mu' D_{ij}$ are of the same order.

- As previously, it is assumed that $R_T = O(1)$.

- The wave length is assumed to be much greater than the length of the period, thus $\lambda \gg \ell$.

B) Equations at the microscopic level

- Balance of momentum and constitutive equations for the elastic solid

$$\operatorname{div}(\underset{\sim}{\sigma}_s) = \operatorname{div}(\underset{\sim}{a}\, \underset{\sim}{e}_x(\underset{\sim}{u}_s)) = -\rho_s \omega^2 \underset{\sim}{u}_s \quad \text{in } \Omega_s \tag{11}$$

where $\underset{\sim}{\sigma}_s$, $\underset{\sim}{e}_x$ ρ_s , $\underset{\sim}{u}_s$ are the stress tensor field, deformation, specific mass, and displacement of the solid.

- Balance of momentum and constitutive equations for the fluid

$$\lambda'\varepsilon^2 \operatorname{grad}(\operatorname{div} \underset{\sim}{v}) + \mu'\varepsilon^2 \Delta \underset{\sim}{v} = -\operatorname{grad} p + i\omega \rho_{\ell} \underset{\sim}{v} \quad \text{in } \Omega_{\ell} \tag{12}$$

$$p = K_w \operatorname{div}(\underset{\sim}{v}) / i\omega \quad \text{in } \Omega_{\ell} \tag{13}$$

where Kw is the bulk modulus of the water.

- Interface conditions

$$\left| \underset{\sim}{\sigma}\, \underset{\sim}{N} \right|_{\Gamma} = 0 \quad (14) \qquad \left| \underset{\sim}{u} \right|_{\Gamma} = 0 \quad (15)$$

where $\underset{\sim}{\sigma}$ is $\underset{\sim}{\sigma}$ or $\underset{\sim}{\sigma}_s$; $\underset{\sim}{u}$ is $\underset{\sim}{u}_s$ or $\underset{\sim}{u}_{\ell}$

$\underset{\sim}{N}$ is a unit vector normal to Γ.

C) <u>Results of the homogenization process</u> | 7 |

to first order, the equations for the motion are :

$$\mathrm{div}_x \langle \underset{\sim}{\sigma}^t \rangle = -\omega^2 \langle \rho_s \rangle \underset{\sim}{u}_s + \rho_\ell \, i\omega \langle \underset{\sim}{v} \rangle \quad (16)$$

$$\langle \underset{\sim}{\sigma}^t \rangle = \underset{\sim}{c} \, \underset{\sim}{e}_x (\underset{\sim}{u}_s) + \underset{\sim}{\alpha} \, p \quad (17)$$

$$\mathrm{div}_x (\langle \underset{\sim}{v} \rangle - \Phi . i\omega \, \underset{\sim}{u}_s) = - \underset{\sim}{\alpha} i\omega \, \underset{\sim}{e}_x + \underset{\sim}{\beta} i\omega p \quad (18)$$

$$\langle \underset{\sim}{v} \rangle - n \, i\omega \, \underset{\sim}{u}_s = \underset{\sim}{K}(\omega) (\mathrm{grad}_x p + \rho_\ell \, \omega^2 \underset{\sim}{u}_s) \quad (19)$$

where $\underset{\sim}{\sigma}^t$ is defined as $= \underset{\sim}{\sigma}_s$ in Ω_s

$= \underset{\sim}{\sigma}_\ell$ in Ω_ℓ

$$\langle \underset{\sim}{\sigma}^t \rangle = \frac{1}{|\Omega|} \int_\Omega \underset{\sim}{\sigma}^t \, d\Omega$$

$\underset{\sim}{C}$, $\underset{\sim}{\alpha}$ and $\underset{\sim}{\beta}$ are elastic constants.

<u>The physical signification</u> of the equations (16)-(18) is the following :

(16) is the macroscopic balance of momentum.
(17) is a macroscopic constitutive elastic equation for the mixture.
(18) is an expression of the balance of mass for the fluid.
(19) is a filtration law at the macroscopic level.

It is possible to show, by writing the equations (16)-(19) as functions of the partial stress-tensors that these equations are strictly equivalent to those obtained by M.A. BIOT | 2 |, written below :

$$N \, u^i_{s,kk} + ((N+A)e+Q\varepsilon)_{,i} - bi\omega(u^i_s - u^i_\ell) = -\rho_{11}\omega^2 u^i_s - \rho_{12}\omega^2 u^i_\ell$$

$$(Qe+R\varepsilon)_{,i} + bi\omega(u^i_s - u^i_\ell) = -\rho_{12}\omega^2 u^i_s - \omega^2 \rho_{22} u^i_\ell$$

where A, Q, R, N are elastic constants and $e = \mathrm{div}\,(u_s)$ $\varepsilon = \mathrm{div}(u_\ell)$

b and ρ_{ij} are obtained from K(or $H = K^{-1} = H_1 + iH_2$) by

$$\rho_{22} = \Phi^2 H_2/\omega \qquad b = \Phi^2 H_1$$

$$\rho_{12} = \Phi \rho_\ell - \Phi^2 H_2/\omega \qquad \rho_{11} = (1-\phi)\rho_s - \rho_{12}$$

The only slight difference obtained by the previous theory (compared to the BIOT theory) is that the ρ_{ij} terms are frequency dependent.

2.3.2 - <u>One-phase behaviours</u>

It was assumed previously (§ 2.3.1) that the viscous stress is $O(\varepsilon p)$ where p is the pressure.

It is interesting to see what happens when the viscous stress is assumed to be more important.

a) if $\underset{\sim}{\sigma}_\ell = 0(p)$

The macroscopic behaviour is not a two-phase behaviour as given above, but an elastic one. It confirms the result obtained in the § 2.1 : the relative displacement between solid and liquid is possible only if $\underset{\sim}{\sigma} = 0(\varepsilon p)$.

b) if $\underset{\sim}{\sigma}_\ell = 0(\varepsilon^{-1} p)$, the macroscopic behaviour is a viscoelastic one.

It has to be noticed that the macroscopic behaviour never contains any term relative to a macroscopically viscous fluid, as it sometimes appears in the literature (for example | 9 |).

3 - EVALUATION AND MEASUREMENT OF THE SOLID-LIQUID COUPLING COEFFICIENTS

a) State of the art

Figure 2 shows the theoretical estimation for 3 different simplified geometries of the quantities.

$$b^* = \frac{b\,k}{\mu\,\Phi^2} \qquad \rho^*_{22} = \rho_{22}/\Phi\rho_\ell$$

functions of the nondimensional frequency $f^* = \frac{f}{f_c} = 2\pi f\,\frac{k\rho_\ell}{\mu\Phi}$

f_c being the characteristic frequency of BIOT.

The b^* term is well approximated by :

$$b^* = \mathrm{Re}\left(1/4\ \frac{K\ T(K)}{1-2\ T(K)/iK}\right) \quad \text{with} \quad T(K) = \frac{\mathrm{ber}'(K)+i\ \mathrm{bei}'(K)}{\mathrm{ber}(K)\ +i\ \mathrm{bei}(K)}$$

$K=\delta(f^*)^{1/2}$ ber and bei being Kelvin functions

where δ is a structure factor, depending on the pore geometry.

The ρ_{22}^* term is well approximated by

$$\rho^*_{22} = (3(\beta-\alpha)(G(K)-1)+1)$$

$$G(K) = -\ \mathrm{Im}\left(\frac{iJ_o(\sqrt{-iK^2})}{J_2(\sqrt{-iK^2})}\right)$$ J_o and J_2 being Bessel functions

where $\alpha = \lim \rho_{22}^* \quad (\omega \to \infty)$

$\beta = \lim \rho_{22}^* \quad (\omega \to 0)$ α and β are constants depending on the pore geometry

The state of the art concerning the estimation and the measurement of the coefficients α , β and δ is given in table 1.

It has to be noticed that the only α term is known for geometries similar to that of natural porous media, often approximated by an assembly of spheres. The lack of information concerning the β term is inconvenient, because most applications are concerned by the frequencies $f < f_c$.

b) Measurement of the coupling coefficents on glass spheres

The purpose of the following is to describe experiments performed to measure ρ^*_{22} and b^* for an assembly of glass spheres at frequencies lower than the critical frequency.

TABLE 1	GEOMETRY	Ref	α	β	δ
Theoretical Evaluation	Plane slits	(2) (7)	1	1.2	2.3
	cylindrical channels	(2) (7)	1	1.33	2.8
	"crenel" slits	(11)	2,7	3.7	6,3
	spheres assembly	(12)	$1/\sqrt{\Phi}$		
Measurement	"crenel" slits	(11)		3.7	6.3
	spheres assembly	(12)	$1/\sqrt{\Phi}$	present work 1.66 ±0.13	 3.2 ±1.2

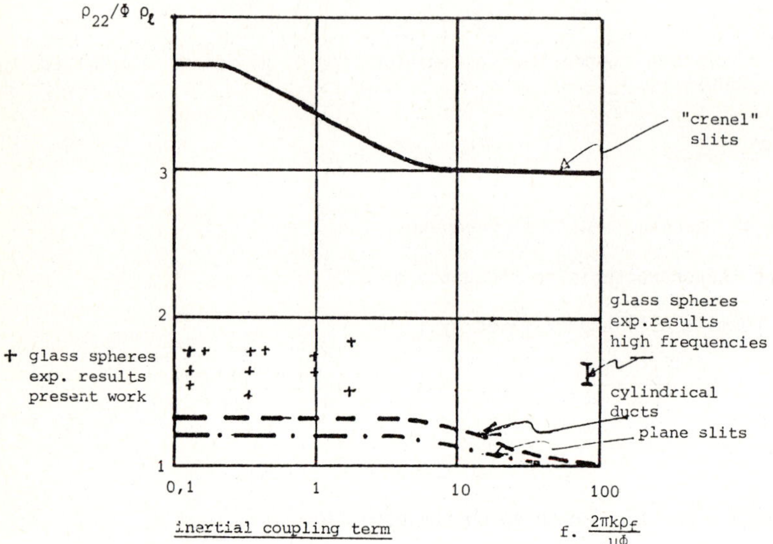

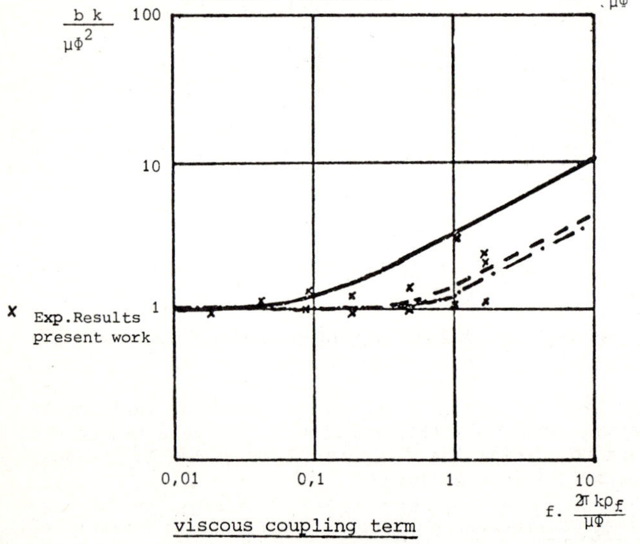

Figure 2 Coupling coefficients

The 6 mm diameter spheres are placed in a 6 cm inner diameter, 1 m length, tube. The spheres are fixed by two grids after compaction by vibration (fig. 3). The spheres are saturated by different liquids (oil, water, water and glucose).

The top of the tube is at atmospheric pressure and the bottom is communicating with an excitation chamber. The excitation is given by a piston moved by a vibrating exciter at frequencies between 10^{-2} et 10 Hz.

The pressure in the chamber is measured by a piezoelectric pressure gauge (resolution 1 Pa) and the displacement of the piston by a capacitive displacement gauge (resolution 1 μm). The complex quotient $p(\omega)/x(\omega)$ is determined by a frequency response analyser ENERTEC SOLARTRON 1250. The coupling coefficients $b(\omega)$ and $\rho_{22}(\omega)$ are proportional to the real and imaginary part of the complex quotient $P(\omega) / Q(\omega)$ where Q is the quantity of flow injected.

The results of the measurements are reported in figure 2 and the interpretation in terms of β and δ is reported in table 1.

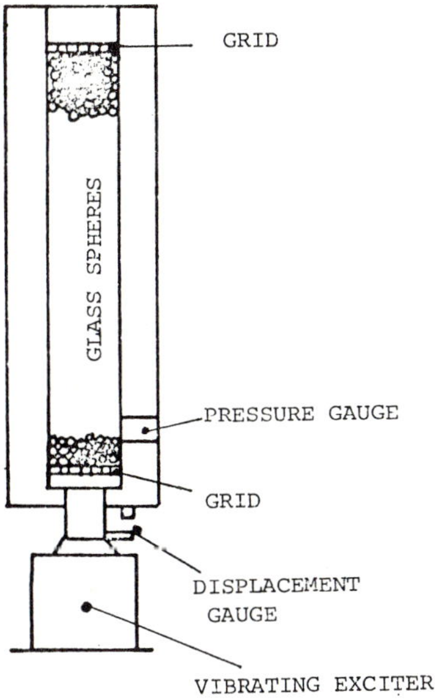

Figure 3 Measurement of the coupling coefficients on glass spheres

As a conclusion : the β term has practically the same value as α found by other authors

- The plage of variation of the δ term includes the approximation of cylindrical ducts.

4 - CONCLUSION

The purpose of the paper was to present results concerning the dynamics of saturated porous media. The phenomenological equations of the dynamics of porous media are obtained by using the homogenization theory. The theory leads to the equations of the classical model of BIOT, except the frequency dependence of the inertial coupling terms.

The first measurement of the coupling terms at low frequencies ($f \leqslant f_c$) on glass spheres is reported here. The application to the numerous fields concerned by the theory (recognition of petroleum reservoirs, aquifers and marine sediments, soil dynamics) will be greatly helped by the extension of the previous measurements to the natural porous media.

REFERENCES

1 - **ATKIN R.J., CRAINE R.E.**, J. Inst. Maths. Appl., 17, 153 (1976).

2 - **BIOT M.A.**, J. Acoust. Soc. Am., 28, 168 (1956).

3 - **SANCHEZ PALENCIA E.**, Non homogeneous media and vibration theory (Springer-Verlag, Berlin 1980).

4 - **LEVY T.**, Int. J. Eng. Sci., 17, 1005 (1979).

5 - **AURIAULT J.L.**, Int. J. Eng. Sci., 18, 775 (1980).

6 - **ENE H.J., SANCHEZ-PALENCIA E.**, J. de Mécanique, 14, 73 (1975).

7 - **AURIAULT J.L.**, Homogenization - Application to porous saturated media. Summer School GDANSK (1983).

8 - **NECAS J.**, Les Méthodes directes en théorie des équations elliptiques (Masson, Paris 1967).

9 - **AYALA MILLIAN G., BREBBIA C.A.**, "Solution of wave propagation problems in a saturated medium" in variational methods in engineering" (BREBBIA & TOTTENHAM ed., London 1975).

10 - **BONNET G.**, Contribution à l'étude de milieux poreux saturés en régime dynamique : application à la reconnaissance par pompage harmonique et à la reconnaissance sismique. (thèse D. es. Sc., Montpellier 1985).

11 - **AURIAULT J.L., BORNE L., CHAMBON R.**, to be published in J. Acoust. Soc. Am. (1985).

12 - **JOHNSON D.L., PLONA J.J., SCALA C., PASIERB F., KOJIMA H.**, Phys. Rev. Lett., 49, 1840 (1982).

The Art of Walking on Fractal Spaces and Random Media

J. Vannimenus
Groupe de Physique des Solides de l'Ecole Normale Supérieure, 24, rue Lhomond, F-75231 Paris Cédex 05, France

1 INTRODUCTION

Since we know that mountain shapes are natural fractals, the title would be appropriate for a lecture on climbing the Mont Blanc, the Aiguille Verte or any of the peaks we can contemplate from Les Houches. My subject is more down to earth and only consists in a short review of recent work on different types of walks on fractal lattices and on random systems such as percolation clusters. The first part is a reminder of geometrical and physical aspects of dimensionality, and discusses why several dimensions are needed to characterize fractal objects. The second part is devoted to "simple" walks, mainly random walks, and gives a short account of classical diffusion and conduction in random media. Several reviews of these aspects have appeared recently, by ALEXANDER |1|, MITESCU and ROUSSENQ |2|, RAMMAL |3| and AHARONY |4|, and the reader is referred to these papers for deeper discussions and for technical points. The following parts deal with "advanced" walks, where additional rules are specified : this leads to a large variety of problems and to a wealth of new physical effects. The examples include random walks in the presence of traps, biased walks (e.g., dispersion of particles in a flow through a porous medium), selfavoiding walks. For all these problems, the lack of translational invariance of the lattice (fractal or random) introduces new features, and even the simplest situations may provide surprises.

2 GEOMETRICAL and PHYSICAL ASPECTS of DIMENSIONALITY

2.1 Fractal Dimension

The notion of fractal dimension that is maybe the most intuitive for physicists dates back to Bouligand and Minkowski : the idea consists in covering an object with the minimum number of "tiles" of linear size ℓ. If the number $N(\ell)$ of tiles needed behaves as a power law for small ℓ, this defines a fractal dimension D which is in general non-integer :

$$N(\ell) \sim \ell^{-D} \qquad (\ell \to 0) \quad . \tag{1}$$

Mathematicians have, however, found out that a more precise formulation is necessary to obtain "good" theorems. Detailed comparisons of various dimensions may be found in the book by MANDELBROT |5| and in the review by FARMER et al. |6|, but it is useful to recall the definition of the Hausdorff dimension D_H :

- Cover the set by balls of radius $\rho_m \leq \rho$, and define its δ-content :

$$M_\delta(\rho) = \operatorname*{Inf}_{(\rho_m \leq \rho)} \left|\sum_m \rho_m^\delta\right| \xrightarrow[\rho \to 0]{} M_\delta \tag{2}$$

- There exists D_H such that

$$M = \begin{cases} 0 & \text{if} \quad \delta < D_H \\ \infty & \text{if} \quad \delta > D_H \end{cases} \tag{3}$$

The important point is that the Hausdorff dimension is always defined, but, on the other hand, it is uneasy to use and to calculate in practice. Also, it is a weak property since very little can be said in general if the value of D_H is known. Stronger properties are required for fractals of physical interest. One usually requires "self-similarity", or rather **dilation invariance** in a range of length scales between a microscopic distance a (particle size, lattice spacing, ...) and a macroscopic cutoff L (sample size, correlation length, ...). This invariance is statistical in most cases, in the sense that it holds only for average properties - crudely speaking, photographs at different magnifications look qualitatively similar.

The other strong property generally assumed in fractal models of random media is **homogeneity**, i.e., that all parts of the fractal are equivalent (statistically). The fractal dimension may then be obtained by measuring the mass M in a sphere of radius R centered on any point O belonging to the fractal :

$$M(R) \sim A\, R^D \quad , \qquad a \ll R \ll L \qquad (4)$$

The prefactor A depends on the origin chosen, but D does not. Examples of such fractals are provided by various models for the backbone of percolation clusters, such as the random model of KIRKPATRICK |7| or the Sierpinski gasket studied by GEFEN et al. |8|. It is important to realize that many strange attractors studied in relation to chaotic systems appear to be nonhomogeneous in the above sense : plots of $\ln\langle M(R)\rangle$ and of $\langle \ln M(R)\rangle$ versus $\ln R$ give different slopes when averages are taken over all origins. A model for the spatial distribution of turbulence using non-homogeneous "multifractals" has been proposed recently by BENZI et al. |9|, and is illustrated in figure (1). The possibility that multi-fractals also occur in random systems should be kept in mind.

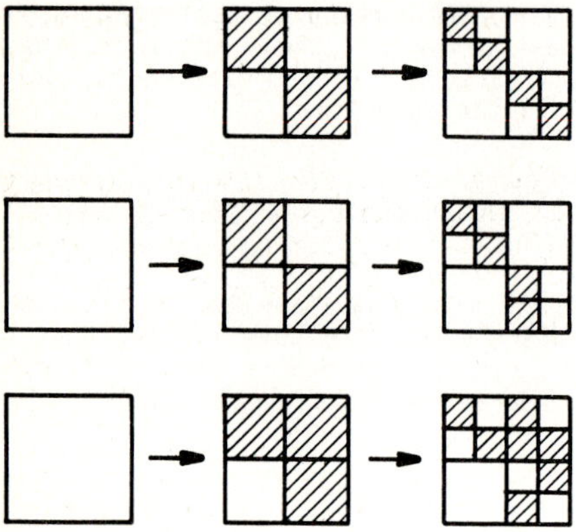

Fig.1. Fractal models for the distribution of active zones (hatched squares) in turbulence : deterministic (top), homogeneous random fractal (middle), multi-fractal with two values, 1/2 and 3/4, for the active volume fraction (bottom)

2.2 The need for other dimensions

Even if scale invariance and homegeneity are assumed, very different fractal objects may have the same D. For instance, the simple value D = 2 corresponds to several standard physical systems in addition to the Euclidean plane : planefilling Peano curves ; random walks (Brownian motion) in any space dimension $d \geqq 2$; Dhar's modified rectangular lattice (which has D = d, see |10| and fig.2) ; lattice animals for d = 3 ; the three-dimensional Sierpinski web (see the picture on page 143 of ref.|5|)

Such lack of uniqueness could cast serious doubts on the value of numerical simulations. Agreement between the value of D obtained numerically for a given model of cluster aggregation and the value measured on colloidal systems, say, could be fortuitous. In fact, the simulations also provide **pictures** of the

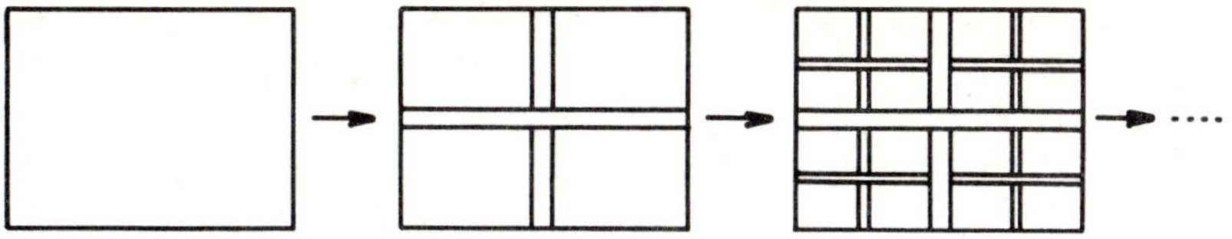

Fig.2 First stages in the construction of Dhar's lattice. Its fractal dimension $D = 2$ is equal to the space dimension, but its spectral dimension is different ($\tilde{d} = 3/2$).

clusters, and this visual check is at least as important in supporting a model as the value of D ! Much of the recent progress on aggregation is due in my opinion to the widespread availability of computer graphics, which reveal a lot of useful detail. Still, one would like to have more quantitative ways to characterize and classify fractals. The geometrical approach to this problem consists in defining other properties that can be directly measured, such as the ramification, the lacunarity... |5|. A more physical approach has been found quite useful for fractal **lattices** : for usual, periodic lattices, the space dimension d appears in many expressions, and it is natural to ask whether it has a simple counterpart in the corresponding expressions for fractals. The first discussion of that point seems to have been given by DHAR |10|, with the conclusion that several "dimensions" may be defined for different physical properties, and that they are different from D in general. This approach is illustrated in the following sections.

3. SIMPLE WALKS

We discuss now two types of walks, which define respectively the spreading dimension $\hat{d}$ and the spectral dimension $\tilde{d}$.

3.1 Hurried walkers

Let us imagine walkers that try to go as far as possible from their starting point, with the constraint to take steps only on neighbouring lattice sites (fig.3). On an Euclidean lattice, the number of sites that can be reached in N steps grows as N^d, and on a fractal lattice the corresponding average number of accessible sites A_N will grow as a power of N :

$$A_N \sim N^{\hat{d}} \qquad , \qquad N \to \infty \tag{5}$$

The dimension $\hat{d}$ has received various names (spreading, chemical, connectivity, topological...) and has been much studied in relation with percolation clusters |11, 12, 13|, but also for models of epidemy propagation |14| and forest fires |15|. An important distinction between $\hat{d}$ and D is that the spreading dimension

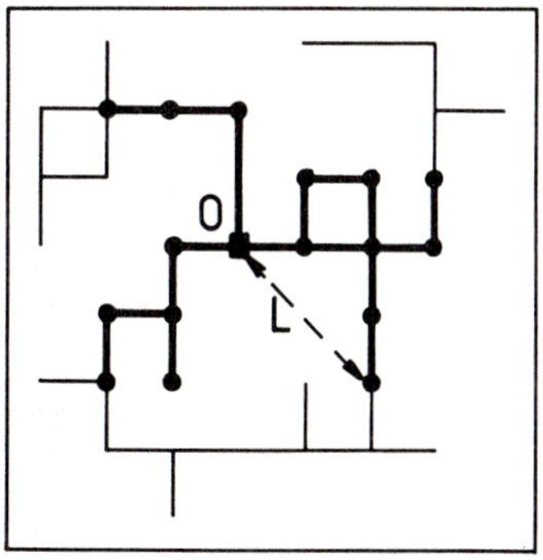

Fig.3. The sites accessible from the origin O in N = 4 steps are indicated by heavy dots.

depends only on the graph structure of the lattice, not on the Euclidean distance : it is an **intrinsic** property.

For example, $\hat{d} = 1$ for a chain, independently of its shape, which may be a Peano curve ($D = 2$), a self-avoiding walk ($D = 4/3$ in two dimensions)... In general, there is no direct relation between $\hat{d}$ and D, just an inequality : $\hat{d} \leq D$. The equality $\hat{d} = D$ holds only when the Euclidean metrics is equivalent to the natural matrics of the graph |12|, e.g., for the Sierpinski gasket or for Witten-Sander aggregates |16|. On the contrary, the strict inequality holds for percolation clusters : for $d = 2$, $\hat{d} = 1.675 \pm 0.005$ |17| and $D = 1.8958...$; for $d = 3$, $\hat{d} = 1.83 \pm 0.03$, $D \simeq 2.5$; for $d \geq 6$, $\hat{d} = 2$ and $D = 4$.

The spreading dimension is related to the **tortuosity** of the lattice. Let us define the tortuosity exponent δ through the length L_{min} of the shortest path through a lattice of linear extent L :

$$L_{min} \sim L^{\delta} \tag{6}$$

For regular enough objects, the number of sites accessible in L_{min} steps is a finite fraction of the total mass, so one has

$$(L_{min})^{\hat{d}} \sim L^{D} \quad \Longrightarrow \quad \delta = D/\hat{d} \tag{7}$$

For percolation clusters, the exponent δ has been determined independently |18| and excellent agreement is found with relation (7).

3.2 Random walks and diffusion

The study of random walks on fractal lattices has been extremely active in the last few years and sriking results have been obtained. Recent discussions can be found in references |1| to |4|, and we just recall some fundamental results :

- "Anomalous diffusion" occurs, i.e., the average square distance from the origin of the walk grows more slowly than linearly with time (= number of elementary steps), for large times :

$$\langle R^2(\tau) \rangle \sim \tau^{2\nu} \quad , \quad \nu = \tilde{d}/2D \tag{8}$$

- In (8), $\tilde{d}$ is a new intrinsic dimension, smaller than D in general, and is called the spectral (or "fracton") dimension, because it also governs the density of states for low-energy scalar harmonic excitations :

$$\rho(\omega) \sim \omega^{\tilde{d}-1} \qquad (\omega \to 0) \tag{9}$$

Note that the problem of elastic vibrations has **vector** character, and is governed by a different dimension |19|.

- The probability of return to the origin after time τ, averaged over all starting points, is given by :

$$P_0(\tau) \sim \tau^{-\tilde{d}/2} \qquad (\tau \to \infty) \quad . \tag{10}$$

An important application of these results is to diffusion on percolation clusters (the problem of the "ant in the labyrinth"). Here anomalous diffusion (8) is observed for times, such that the diffusion length remains smaller than the correlation length ξ corresponding to the given concentration p of present bonds. A crossover occurs when $\langle R^2 \rangle \sim \xi^2$ and for longer times the asymptotic behaviour depends on the position with respect to the percolation threshold p_c (fig.4) :

$$\begin{aligned} &- \text{for } p < p_c \qquad \langle R^2 \rangle \sim \xi^2(p) \\ &- \text{for } p > p_c \qquad \langle R^2 \rangle \sim \tau D_\infty(p) \end{aligned} \tag{11}$$

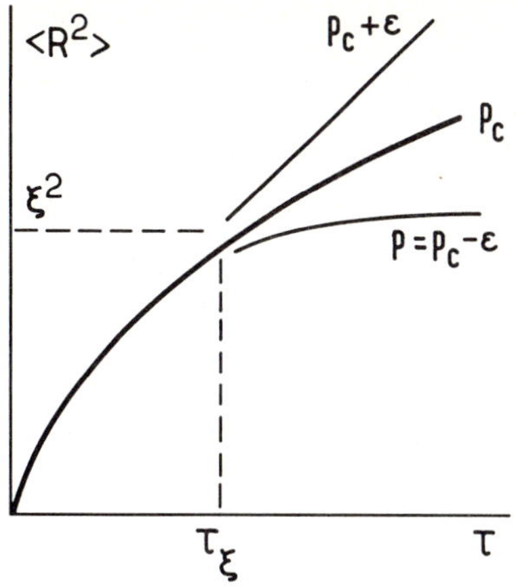

Fig.4 Mean square displacement of a particle diffusing on a percolation cluster, as a function of time, near the threshold p_c. The crossover occurs when the displacement is comparable to the correlation length.

Above p_c, the diffusion constant for particles placed on the infinite cluster (e.g., tracer particles in a porous medium above an injection threshold) has a singularity of the form

$$D_\infty(p) \sim (p - p_c)^\rho \quad , \qquad (p - p_c)/p_c \ll 1 \tag{12}$$

The crossover time is such that

$$\tau^{\tilde{d}/D} \sim \tau(p - p_c)^\rho \sim \xi^2 \sim (p - p_c)^{-2\nu_p} \tag{13}$$

Hence

$$\rho = 2\,\nu_p\,(\frac{D}{\tilde{d}} - 1) \tag{14}$$

(ν_p is the correlation length exponent and should not be confused with ν defined in (8)).

3.3 Conduction in random media

The percolation model has proved very useful to study the transport properties of disordered systems, such as random metal-insulator mixtures with a volume fraction p of metal. The essential fact is that the conductivity of such materials vanishes below a percolation threshold (fig.5) : much above this threshold simple effective medium theories are adequate, but in the critical region near the threshold the conductivity has a non-analytic behaviour of the form

$$\Sigma(p) \sim (p - p_c)^t \tag{15}$$

and more refined theoretical methods are needed.

The critical exponent t can be related to the spectral dimension $\tilde{d}$ of percolation clusters. The usual argument starts from the Einstein relation between conductivity and diffusivity :

$$\Sigma(p) = D_0(p) \tag{16}$$

where $D_0(p)$ is the diffusion constant averaged over **all** starting points for random walks. The validity of this relation for random systems is not obvious, and it has

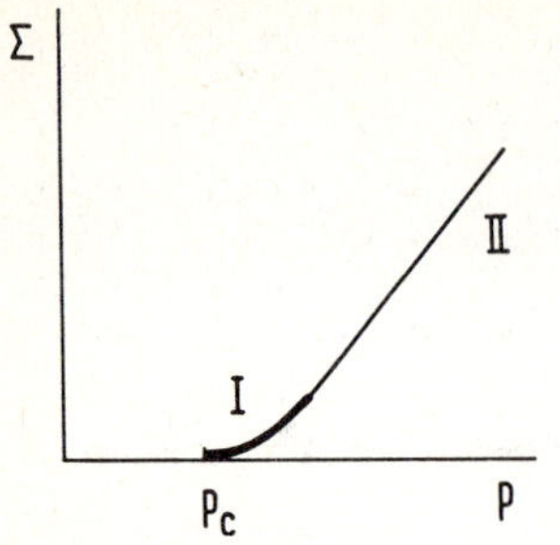

Fig.5. Conductivity of a random metal-insulator mixture. Effective-medium theory is valid far from the percolation threshold (region II), but not in the critical region (I, heavy line).

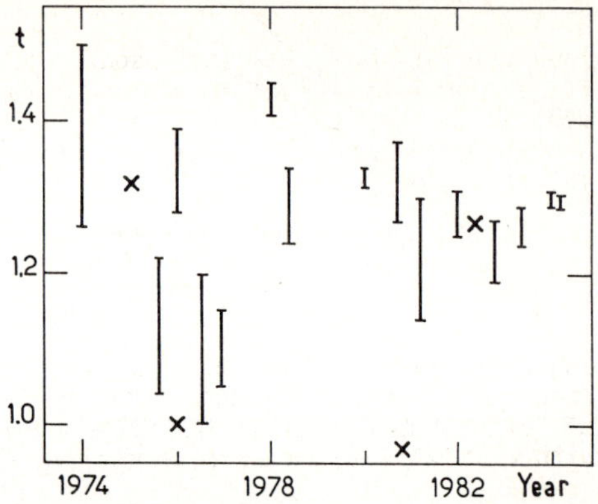

Fig.6. History of the conductivity exponent t, in two dimensions. The bars correspond to numerical determinations, the crosses to analytical predictions

been justified by Derrida (private communication) and by Gefen (to be published). Equation (16) shows that $t = \rho + \beta$, where β is the exponent for the probability that a site belongs to the infinite cluster, and one obtains :

$$\tilde{d} = 2\,\frac{d\nu_p - \beta}{t - \beta + 2\nu_p} \tag{17}$$

The exponents β and ν_p have a purely geometrical origin, while t and $\tilde{d}$ describe dynamic properties of percolation, and many people have tried to reduce the number of independent exponents. Among these proposals, the ALEXANDER-ORBACH conjecture |20| is strikingly simple :

$$\tilde{d} = 4/3 \quad , \qquad \text{for all} \quad d \geq 2 \tag{18}$$

It implies that $t = 91/72 = 1.2639...$ for $d = 2$, using known exact values of β and ν_p. Some other conjectures are close to this value, for instance $t = \nu_p = 4/3$ for $d = 2$, and the scatter of numerical data was rather large in 1982 (fig.6), so this question stood as a challenge for specialists of numerical simulations. Accuracy of better then 1 percent on t is very difficult to obtain, much as in real experiments on critical phenomena, and very large calculations are needed. The results obtained nearly simultaneously by four different groups |21, 22, 23, 24| agree on the value $t = 1.30 \pm 0.01$, which appears to rule out the various conjectures with reasonable confidence. This suggests that $\tilde{d}$ is **not** related to simple geometrical features of percolation. Another strong argument against such a relation comes from the existence of a whole family of exponents, of which $\tilde{d}$ is just one of the first members (see the contributions by Rammal and by Redner in this volume).

The examination of fig.(6) suggests two remarks :

- Error bars are often rather optimistic (even when large),
- There is a clear tendency for the numerical results to be "attracted" by the prevailing theory.

Nevertheless, the most recent results seem quite reliable. A specialized computer is in project at Saclay for this problem, it should give a still better accuracy and provide very strong bounds on any new proposals for exact relations involving t. In three dimensions, the best available result |25| is $t = 1.95 \pm 0.10$, much higher than typical values quoted a few years ago.

4 WALKS In the PRESENCE of TRAPS

4.1 Number of distinct visited sites and compact exploration

In a random walk on a lattice, some sites are visited several times by the walk and it is useful to define the average number of distinct sites visited in N steps, denoted S_N. For Euclidean lattices it is well known that S_N is of order $N^{\frac{1}{2}}$ for d = 1, N/lnN for d = 2, and N for d 3. The equivalent results for fractal lattices |26, 3| are :

$$S_N \sim \begin{cases} N^{\tilde{d}/2} & \text{if} \quad \tilde{d} \leq 2 \\ N & \text{if} \quad \tilde{d} > 2 \end{cases} \qquad (19)$$

and $\tilde{d}$ = 2 may be viewed as a critical dimension for random walks on fractals. The deep qualitative difference between the two cases can be expressed in the notion of "compact exploration" |27|. The average radius of a walk of N steps is $R_0 \sim N^{\nu}$, with $\nu = \tilde{d}/2D$ (8), and the number Σ_N of sites that are effectively accessible is $\Sigma_N \sim R^D$:

- if $\tilde{d} \leq 2$, $\Sigma_N \sim S_N$: a finite fraction of Σ_N is visited, and the walk explores essentially sites that have already been visited - exploration is compact.

- if $\tilde{d} > 2$, $\Sigma_N \gg S_N$: many sites are never visited, exploration is non-compact.

4.2 Capture by traps

The experimental situation considered is typically the hopping diffusion of an exciton in a binary alloy, with a small concentration x of traps where it recombines |28|. The average survival probability $\Phi(N)$ after N steps is directly measurable through the luminescence signal :

$$\Phi(N) = \langle (1 - x)^{Q_N} \rangle \qquad (20)$$

where Q_N is the number of distinct sites visited by a given walk, and $\langle Q_N \rangle = S_N$. Note that there is a double averaging, over different random walks and over the trap distribution. Two regimes have to be considered :

- For relatively short times

$$\Phi(N) \sim \exp(-x\, S_N) \sim \exp(-C\, x\, N^{\tilde{d}/2}) \qquad (21)$$

if $\tilde{d} \leq 2$. In practice, this situation arises for percolation clusters near p_c and is well observed experimentally |28|.

- For **very** long times, the tail is dominated by particles surviving in very rare regions without traps, and decreases more slowly than predicted by (21). On Euclidean lattices one has the rigorous result |29| :

$$\mathrm{Log}\, \Phi(N) \sim -C\, x^{2(d+2)}\, N^{d/(d+2)} \qquad (22)$$

and its extension to fractals is otained simply by replacing d by $\tilde{d}$ in (22) |30,31|. This regime seems inaccessible experimentally, but can be studied numerically, and the crossover between Euclidean and fractal behaviour has been observed on percolation clusters |32|. Physically, it is satisfactory that $\tilde{d}$, rather than D, replaces the space dimension in (22), since $\Phi(N)$ depends only on the structure of the dilute network and is an intrinsic quantity.

5. BIASED WALKS

The introduction of a preferred walking direction has deep consequences, which are sometimes rather surprising, and we present several types of biased walks recently studied.

5.1 Fully directed walks on a dilute lattice

The rule of the game is very simple and is illustrated in fig.7 : the walker can only step along the arrow direction, using allowed bonds (probability p).

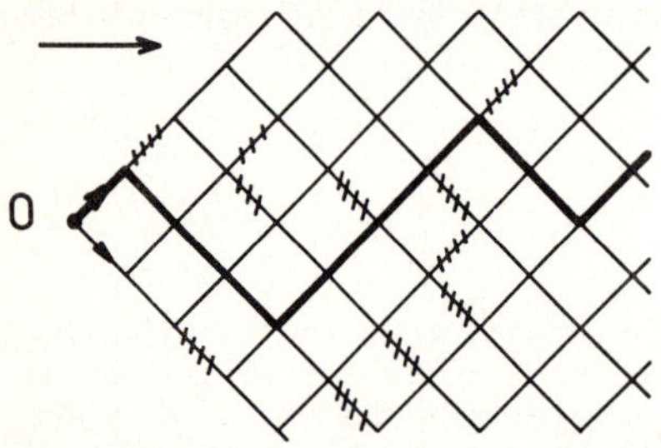

Fig.7 A directed walk (heavy line) on a square lattice. The hatched bonds are forbidden

The statistical problem is to calculate the number Ω_N of N-step walks on the dilute lattice. For the pure system (p = 1), $\Omega_N(1) = 2^N$ and since all of the N steps must be allowed for the walk to survive in the dilute system, one has simply for the average number of walks :

$$\langle \Omega(N) \rangle = p^N \Omega(1) = (2p)^N \tag{23}$$

But wait ! Eq.(23) says that for $p > 1/2$ there is an exponential number of walks of length N, whereas we know that the percolation threshold is $p_c = 0.6445$ on the directed square lattice. For $1/2 < p < p_c$, most samples are **not** connected and $\Omega_N = 0$, in contradiction with (23) ! Where is the catch ? The answer is that the very few connected samples have a **huge** number of walks, and that $\langle\Omega_N\rangle$ is not physically significant : one needs the distribution $P(\Omega_N)$ and in particular the most probable value Ω^*_N. A detailed numerical study can be made |33|, using a recursive approach to calculate Ω_N exactly for many samples, and the results confirm that Ω^*_N can be very different from $\langle\Omega_N\rangle$ even above p_c. This provides a pedagogical example for a subtle problem raised by DERRIDA |34| for self-avoiding walks on dilute lattices.

5.2 Non-symmetric walks

These are walks for which the probability W_{ji} to jump from i to an adjacent site j is different from W_{ij}. In the one-dimensional case, this may lead to a very peculiar situation, where the particle drifts to infinity in a well-defined direction, but its velocity vanishes |34, 35|. The mean position of the particle varies for large times as :

$$\langle x\rangle \sim t^{\alpha}$$
$$\text{with} \quad \alpha < 1 \quad \text{fixed by} \quad \overline{(W_{i,\, i+1}/W_{i+1,\, i})^{\alpha}} = 1 \tag{24}$$

The generalization of this system to higher dimensions is very interesting because it provides a model for diffusion in a flow in a porous medium, when $\vec{v}_0 = \langle\vec{v}\rangle$ is **very small** (limit of small Peclet number) and locally $v \gg |v_0|$. The problem has been studied by sophisticated field-theory methods |36, 37, 38| and the main results are :

- **No** singularity appears for $d > 2$, the diffusion coefficient is well behaved.
- $d = 2$ is "a critical dimension", where logarithmic corrections appear. Their precise form depends whether the fluid is compressible or slightly compressible.

This suggests that the situation is totally different on a fractal when $\tilde{d} < 2$, but the field-theory approach is difficult to generalize to such situations, and does not apply to percolation clusters. The above predictions are therefore restricted to well-connected porous media.

5.3 Dispersion in ill-connected porous media

This situation corresponds to diffusion in a steady flow near an injection threshold and was considered by de Gennes |39|. The difficulty comes from the interplay between diffusion on the dead ends where no flow occurs and convection on the backbone (fig.8). Note that the problem is different from diffusion of a charged particle in a strong electric field, where the field is felt even in the dead ends |40|.

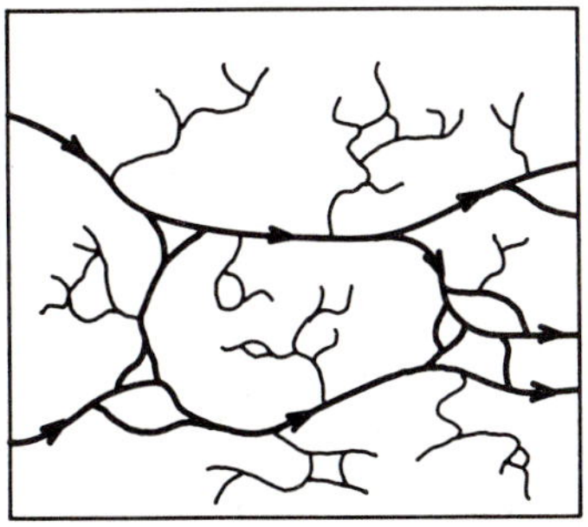

Fig.8 Idealized structure of a porous medium near an injection threshold. The flow is limited to the backbone (heavy lines), but diffusion also occurs on the stagnant dead ends

The mass of fluid contained in the dead ends is much larger than the mass in the backbone, so **stagnation** effects are dominant. Near the threshold, the average velocity $\bar{U}$ (= Flow/Volume occupied by the fluid) has a singularity :

$$\bar{U} = k.|f|/S_\infty \sim (p-p_c)^{t-\beta}\ |f| \tag{25}$$

where k is the permeability which vanishes with the same exponent t as the conductivity in (15), and S_∞ is the analogue of the mass of the infinite cluster in standard percolation and has the same exponent β. The dead ends have typical size ξ and act as traps with a long release time :

$$\tau \sim \xi^2/D_\infty \sim (p - p_c)^{-2\nu_p}/(p - p_c)^\rho \tag{26}$$

where D_∞ is the diffusion constant introduced in (12). Now, the local flow velocity on the backbone fluctuates enormously, and it is not clear which velocity one should use to compute $D_{//}$, the diffusion coefficient along the flow. The discussion given by de GENNES |39| leads to the choice of the global average velocity $\bar{U}$, rather than the average velocity on the backbone $\bar{U}$ for instance, and to an expression :

$$D_{//} \sim \bar{U}^2\xi^2/D_\infty \sim (p - p_c)^\omega \tag{27}$$

$$\omega \sim t - \beta - 2\nu_p \tag{28}$$

The best presently available values for the exponents gives $\omega \simeq -1.5$ for d = 2 and $\omega \simeq -0.25$ for d = 3. However, de Gennes'result depends on several approximations and relies on some subtle compensations. I have argued |41| that the tortuosity of the backbone is important, and that one should replace $\bar{U}$ in (27) by $\bar{U}$ (L_{min}/L), where L_{min} is the shortest path length defined in section 3.1. This leads to a modified relation

$$\omega' = t - \beta - 2\nu_p\, D / \bar{d} \tag{29}$$

and $\omega' \simeq -0.9$ for d = 3. The difference is large enough to be observable, but numerical simulations are an order of magnitude more difficult than for the conductivity problem. A test would be to carry out the calculation in mean-field theory (very large d), where $\omega = 1$ and $\omega' = 0$ (i.e., logarithmic divergence), but even in that case the problem remains intricate and the correct result could well be different from both theories.

6 SELF-AVOIDING WALKS

The principal motivation for studying SAW on fractals or percolation clusters comes from the problem of polymers in a random environment. This is a very subtle problem, and many conflicting theories have been proposed : some argue that disorder has no effect on asymptotic properties, others think that these change only at the percolation threshold, or even that any amount of disorder is relevant. The difficulty has the same origin as for the fully directed walk mentioned above, and comes from the different behaviour of the average and most probable values of the SAW number |33|.

6.1 Self-avoiding walks on fractal lattices

The SAW problem can be solved exactly on some fractal lattices, and this is very useful to understand the effect of the lack of translational invariance |42, 43|. The main result is that the large-scale behaviour of the mean square radius of a SAW :

$$\langle R_N^2 \rangle \sim N^{2\nu_s} \tag{30}$$

is different from its behaviour in a Euclidean system of the same space dimension. For the 2-d Sierpinski gasket, $\nu_s = 0.798...$, while $\nu_s = 3/4$ for the pure square lattice. This strongly suggests that ν_s is also different on percolation clusters.

Morover, no simple connection seems to exist between the exponent ν_s and the fractal and spectral dimensions. Simple Flory-type formulas using D and $\tilde{d}$ can account for the trend in the variation of ν_s among various fractals, but other characteristic properties are clearly involved, and there is presently no good expression to predict ν_s for a general fractal.

6.2 Gelation and self-interacting SAW on fractals

This is related to the problem of physical gelation, where an attractive interaction between monomers of the SAW favors nearest-neighbour positions. The first study on fractals was made by Klein and Seitz |44|, who found no collapse transition at finite temperature on the Sierpinski gasket. This result seems to depend strongly on the peculiar topology of the gasket, and a collapse transition appears to occur on other fractals, such as Dhar's lattice |45|.

A similar transition occurs for lattice animals (which provide a simple model for branched polymers) on various fractals |46|, and here again it is useful to have models on which an exact solution can be exhibited.

6.3 "True" self-avoiding walks (the rat in a labyrinth)

This is a type of walk that never stops, due to a modification of the rules : when a walker is trapped (all neighbouring sites occupied), he chooses the site with smallest occupation number, or picks randomly one of the sites which he has visited the fewest times (fig.9).

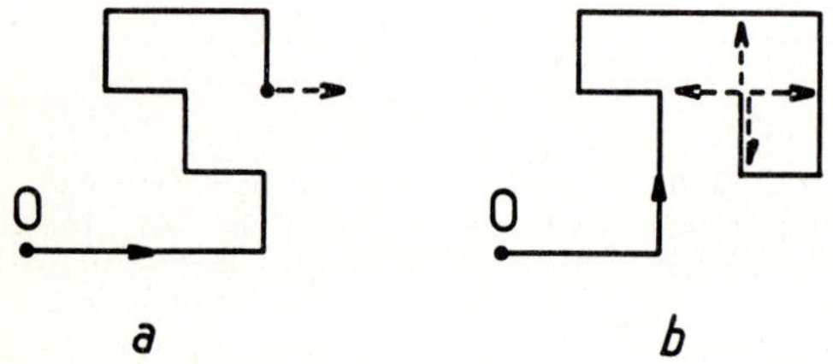

Fig.9 When possible, the walker steps to an empty site (a) ; if not, he chooses a site at random (b)

The "true" SAW was introduced by AMIT et al. |47|, and it behaves very differently from ordinary SAW. On a regular 2-d lattice the exponent ν_S is 1/2, instead of 3/4, and one can show that d = 2 is an upper critical dimension. This suggests that the situation on fractals is non-trivial, and indeed it has been found numerically that $\nu_S \simeq 0.51$ on the Sierpinski gasket |48|, whereas for a random walk $\nu \simeq 0.43$. In other words, a rat will escape faster from a labyrinth than an ant !

In conclusion, let us mention that other types of walks have been recently introduced on standard lattices, and there is no risk in predicting that their study on fractals will bring out new fascinating aspects of a rich domain.

REFERENCES

1 S. Alexander : "Percolation Structures and Processes", p. 149, G. Deutscher, R. Zallen and J. Adler eds. (Higler, Bristol, 1983).
2 C. Mitescu, J. Roussenq : "Percolation Structures and Processes", p. 81.
3 R. Rammal : J. Stat. Phys. 36, 547 (1984).
4 A. Aharony : "Fractals in Statistical Physics", Proceedings of the Int. Conf. on Collective Phenomena, Tel Aviv (1984).
5 B. Mandelbrot : "The fractal Geometry of Nature", Freeman, San Francisco (1982).
6 J.D. Farmer, E. Ott and J.A. Vorke : Physica 7D, 153 (1983).
7 S. Kirkpatrick : "Ill-Condensed Matter", Les Houches 1978, P. 374 (North Holland, 1979).
8 Y. Gefen, A. Aharony, B. Mandelbrot, S. Kirkpatrick : Phys. Rev. Lett. 47, 1771 (1981).
9 R. Benzi, G. Paladin, G. Parisi, A. Vulpiani : J. Phys. A 17, 3521 (1984).
10 D. Dhar : J. Math. Phys. 18, 577 (1977).
11 S. Havlin, R. Nossal : J. Phys. A 17, L427 (1984).
12 R. Rammal, J.C. Angles d'Auriac, A. Benoit : J. Phys. A 17, L491 (1984).
13 J. Vannimenus, J.P. Nadal, H. Martin : J. Phys. A 17, L351 (1984).
14 P. Grassberger : Math. Biosci. 62, 157 (1983).
15 G. McKay, N. Jan : J. Phys. A 17, L757 (1984).
16 P. Meakin, I. Majid, S. Havlin, H.E. Stanley : J. Phys. A 17, L975 (1984).
17 P. Grassberger : J. Phys. A 18, L215 (1985).
18 R. Pike, H. Stanley : J. Phys. A 14, L169 (1981).
19 S. Alexander : in this volume ; P. Sen : ibid ; I. Webman : ibid.
20 S. Alexander, R. Orbach : J. Physique Lettres 44, L-13 (1983).
21 H. Herrmann, B. Derrida, J. Vannimenus : Phys. Rev. B 30, 4080 (1984).
22 J.G. Zabolitzby : Phys. Rev. B 30, 4077 (1984).
23 D. Hong, S. Havlin, H. Herrmann, H. Stanley : Phys. Rev. B 30, 4083 (1984).
24 C.J. Lobb, D.J. Frank : Phys. Rev. B 30, 4090 (1984).
25 B. Derrida, D. Stauffer, H.J. Herrmann, J. Vannimenus : J. Physique Lett. 44, L701 (1983).
26 R. Rammal, G. Toulouse : J. Physique Lett. 44, L13 (1983).
27 P.G. de Gennes : C.R.A.S. 296, 881 (1983).
28 P. Evesque : J. Physique 44, 1217 (1983).
29 M. Donsker, S. Varadhan : Commun. Pure Appl. Math. 32, 721 (1975) ; F Delyon, B. Souillard : Comment Phys. Rev. Lett. 51, 1720 (1983).
30 I. Webman : Phys. Rev. Lett. 52, 220 (1984).
31 G. Zumofen, A. BLumen, J. Klafter : J. Phys. A 17, L479 (1984).
32 I. Webman : J. Stat. Phys. 36, 603 (1984).
33 J.P. Nadal, J. Vannimenus : J. Physique 46, 17 (1985).
34 B. Derrida : Phys. Reports 103, 29 (1984).
35 B. Derrida, Y. Pomeau : Phys. Rev. Lett. 48, 627 (1982).
36 J.M. Luck : J. Phys. A 17, 2069 (1984).
37 J.A. Aronovitz, D.R. Nelson : Phys. Rev. A 30, 2948 (1984).
38 D.S. Fisher : Phys. Rev. A 30, 960 (1984).
39 P.G. de Gennes : J. Fluid Mech. 136, 189 (1983).
40 D. Dhar : J. Phys. A 17, L257 (1978).

41 J. Vannimenus : J. Physique Lett. 45, L-1071 (1984).
42 D. Dhar : J. Math. Phys. 19, 5 (1984).
43 R. Rammal, G. Toulouse, J. Vannimenus : J. Physique 45, 389 (1984).
44 D.J. Klein, W.A. Seitz : J. Physique Lett. 45, L241 (1984).
45 J. Vannimenus, D. Dhar, M. Knezevic, D. d'Humières, to be published.
46 M. Knezevic, J. Vannimenus, D. Dhar, to be published.
47 D. Amit, G. Parisi, L. Peliti : Phys. Rev. B 27, 1635 (1983).
48 J.C. Angles d'Auriac, R. Rammal : J. Phys. A 17, L15 (1984).

Flow in Porous Media and Residence Time Distribution

D. Schweich
LSGC - CNRS, ENSIC, 1 rue de Granville, F-54042 Nancy Cedex, France

Although a fluid obeys locally the Navier-Stokes equations, flow in a porous medium is a very complex problem to solve. Residence time distributions (R.T.D.) E(t) provide an alternative method. We assume the following simplifying assumptions:

- The porous medium has only one entrance and one exit, without any leak.
- The flow is steady, isothermal, without any dilation, it is unidirectional in a constant cross-section.

It is straightforward to calculate E(t) for perfect unidirectional piston flow with constant velocity, or in a perfect mixer where composition is uniform. These models are generalized for flows in porous media.

(i) - For regular media totally saturated by the flowing fluid, a simple piston dispersion model, including Fickian dispersion, is usually sufficient. It is however practically unconvenient.

- A similar model consists in a cascade of perfect mixers. Both models however lead to symmetrical R.T.D.

(ii) For porous media with several structural scales, however RTD_s are asymmetric with a slow decay. The simplest model separates the fluid in two parts:

- a fraction ε of the accessible volume obeys the cascade of mixers.
- a fraction $(1-\varepsilon)$ of the volume is motionless, exchanging with the first phase by molecular diffusion only, with a characteristic transfer time t_m.

For real systems however, one has to use a distribution $f(t_m)$ of transfer times. two cases are reviewed:

1) $f(t_m) = X \quad (t_m - t_{m_1}) + (1-x) \quad (t-t_{m_2})$

where it is possible to define an average tranfer time t_m. Then for large times, the R.T.D. is symmetric.

2) $f(t_m) \quad t_m^{1-\alpha} \qquad (0 < \alpha < 1)$,

where it is not possible to define t_m. Here the RTD is always asymmetric.
We show some practical uses for these models and discuss their weaknesses.

1. Introduction

L'écoulement d'un ou plusieurs fluides dans un milieu poreux, qu'il soit artificiel (lit catalytique de l'industrie chimique) ou qu'il soit naturel (réservoir pétrolier), constitue un problème excessivement complexe si on veut employer les lois fondamentales de la mécanique des fluides. Si le fluide obéit localement à l'équation de Navier-Stokes, la complexité des conditions limites rend toute solution impossible. Nous présentons ici l'alternative offerte par la méthode de la distribution des temps de séjour qui fournit simultanément les principes d'expérimentation et de modélisation d'un écoulement quelconque. Nous nous placerons toutefois dans le cadre des hypothèses simplificatrices suivantes :

- l'écoulement est permanent, isotherme et sans dilatation,
- le milieu poreux ne possède qu'une entrée et une sortie, aucune fuite n'est possible,
- l'écoulement est unidirectionnel et dans une section constante (à partir du paragraphe 3).

Nous allons examiner successivement divers modèles d'écoulement du plus simple à de plus raffinés en indiquant leurs propriétés et leurs limites qui mériteraient d'être repoussées.

2. La distribution des temps de séjour

2.1. Principes et propriétés élémentaires

On se propose de caractériser l'écoulement dans un système tel que celui illustré dans la figure 1. Pour cela on effectue dans la section d'entrée une perturbation de concentration x(t) au moyen d'un traceur parfait. Ce dernier terme désigne une substance détectable mais qui a un comportement aussi proche que possible de celui du fluide en écoulement permanent. Dans la section de sortie, un détecteur permet d'enregistrer la courbe de concentration y(t) du traceur. Si toute la quantité de traceur n_0 est injectée instantanément à t = 0 ($x(t) = (n_0/Q)\ \delta(t)$), la courbe y(t) dessine la répartition des temps mis pour traverser le système. On définit dans ces conditions, la distribution des temps de séjour (DTS) par :

$$E(t) = \frac{y(t)}{\int_0^\infty y(t)dt} \qquad (1)$$

E(t)dt représente la fraction de fluide en écoulement qui séjourne dans le système pendant un temps compris entre t et t+dt.

Tout notre propos va maintenant consister à caractériser et modéliser la DTS E(t) pour un écoulement en milieu poreux.

Si E(t) n'est qu'une caractéristique globale de l'écoulement (entre les sections d'entrée et de sortie), elle n'en est pas moins une propriété intrinsèque. Le signal d'entrée x(t) étant queconque on peut poser :

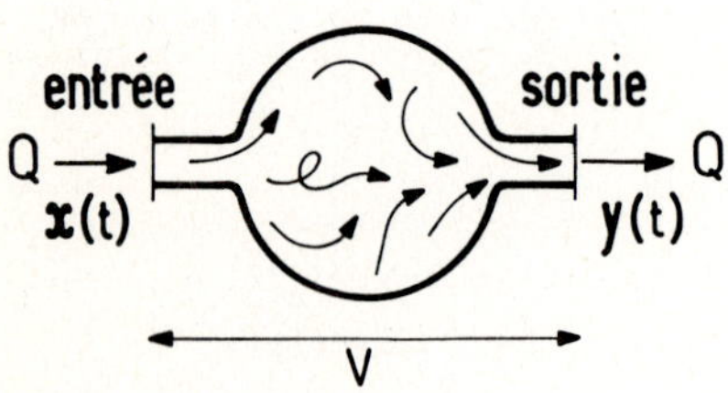

Figure 1. Schéma de principe d'une mesure de DTS.

$$x(t) = \int_0^t x(u)\, \delta(t-u)du = x(t) * \delta(t) \quad (2)$$

Le traçage d'un écoulement étant de plus un processus linéaire, chaque composante $\delta(t-u)$ de $x(t)$ concoure pour $E(t-u)$ au signal de sortie $y(t)$. On a ainsi :

$$y(t) = \int_0^t x(u)\, E(t-u)du = x(t) * E(t) \quad (3)$$

En particulier, si $x(t)$ est un échelon unité $H(t)$, la réponse $y(t)$ est :

$$F(t) = \int_0^t E(u)du \,. \quad (4)$$

Sur le plan experimental cette courbe $F(t)$ s'obtient en remplaçant rapidement l'alimemtation en fluide pur par du fluide contenant du traceur au niveau de la section d'entrée. Si aucune excitation $x(t)$ simple comme $\delta(t)$ ou $H(t)$ ne peut être envisagée expérimentalement, un couple de signaux $x(t),y(t)$, $x(t)$ étant quelconque, permet de retrouver la DTS au moyen de (3).

Pour caractériser la distribution $E(t)$ on emploie ses moments définis par :

$$\mu_n = \int_0^\infty t^n\, E(t)dt \quad (5)$$

Par construction $\mu_o = 1$. Le moment d'ordre 1 est le temps de séjour moyen $\bar{t}$. On montre que :

$$\bar{t} = \mu_1 = \frac{\text{volume accessible au fluide tracé}}{\text{Débit volumique de la phase tracée}} = \frac{V}{Q} \quad (6)$$

En outre on caractérise l'étalement de la DTS par sa variance :

$$\sigma^2 = \mu_2 - \mu_1^2 = \int_0^\infty (t-\bar{t})^2\, E(t)dt \quad (7)$$

Enfin, le produit de convolution apparaissant dans (3) et les modèles d'écoulement en termes d'équations différentielles suggèrent l'emploi de la transformation de Laplace. Pour une fonction $f(t)$, sa transformée $\bar{f}(s)$ est :

$$\bar{f}(s) = \int_0^\infty f(t)\, e^{-st}\, dt \quad (8)$$

$\bar{E}(s)$ est alors la fonction génératrice des moments de $E(t)$:

$$\mu_n = (-1)^n \lim_{s \to 0} \left(\frac{d^n\, \bar{E}(s)}{ds^n}\right) \quad (9)$$

Des compléments et précisions à cette brève introduction à la DTS peuvent être trouvés dans l'ouvrage de Villermaux [1] .

2.2. Exemples

Pour exprimer la DTS il suffit de résoudre l'équation de conservation du traceur en régime transitoire. Dans un écoulement piston parfait unidirectionnelle et à vitesse constante on a :

$$u \frac{\partial C}{\partial z} + \frac{\partial C}{\partial t} = 0 \qquad (10)$$

E(t) est la solution en z = L moyennant les conditions :

$$t = 0 \quad c = 0 \quad \forall z \quad , \quad z = 0 \quad c = \delta(t) \qquad (11)$$

On obtient ainsi $E(t) = \delta(t - L/u)$ et :

$$\bar{t} = L/u \quad , \quad \bar{E}(s) = e^{-s\bar{t}} \quad , \quad \sigma^2 = 0 \qquad (12)$$

Pour un écoulement dans un mélangeur parfait où la composition est uniforme on a :

$$Q\, C_o = QC + V \frac{dc}{dt} \qquad C_o = \delta(t) \qquad (13)$$

et par conséquent :

$$\bar{t} = V/Q \;,\; E(t) = \frac{1}{\bar{t}} e^{-t/\bar{t}} \quad , \quad \left.\begin{matrix} \bar{E}(s) = (1 + s\bar{t})^{-1} \\ \sigma^2 = \bar{t}^2 \end{matrix}\right\} \qquad (14)$$

3. Modèles d'écoulement piston-dispersion simple en milieux poreux

3.1. La traditionnelle loi de Fick

Même dans les milieux poreux les plus réguliers la DTS d'un écoulement unidirectionnel n'obéit pas au schéma simpliste modélisé par (12) : E(t) est plus ou moins étalée autour de $\bar{t}$ et σ^2 est non nul. La première approche pour tenir compte de ces observations fut de superposer au flux convectif uC, un terme dispersif répondant à la loi de Fick :

$$F = - D_e \frac{\partial C}{\partial z} \qquad (15)$$

D_e est le coefficient de dispersion axiale. Il contient l'effet de la diffusion moléculaire mais aussi l'effet de la dispersion statistique des filets de fluide qui parcourent le milieu poreux. La conservation du traceur obeit alors à :

$$D_e \frac{\partial^2 C}{\partial z^2} = u \frac{\partial C}{\partial z} + \frac{\partial C}{\partial t} \qquad (16)$$

L'analyse dimensionnelle permet de montrer que la DTS n'est fonction que de deux paramètres : $\bar{t} = L/u$ et le nombre de Péclet :

$$Pe = \frac{uL}{D_e} \qquad (17)$$

Toutefois l'expression de E(t) est fonction de la nature des conditions limites imposées à (16) : la dispersion axiale a-t-elle lieu ou non en amont et en aval des sections d'entrée et de sortie respectivement ? D'élégantes mais inutiles équations permettent de répondre à cette question (voir [1] Villermaux). En fait,dans la pratique, on ne sait que rarement si la condition limite mathématique choisie est la bonne traduction de l'expérience. En outre, ces diverses conditions conduisent à des

DTS indiscernables dès que $P_e \geq 100$. Enfin toute valeur de Péclet inférieure à 100 doit être jugée comme suspecte. Dans ces conditions normales de Péclet élevé on peut utiliser :

$$\bar{E}(s) = \exp\left[\frac{Pe}{2}\left(1 - \sqrt{1 + 4s\frac{\bar{t}}{Pe}}\right)\right]$$

$$\bar{t} = L/u \quad \text{indépendant de } D_e \tag{18}$$

$$\sigma^2 = 2\bar{t}^2/Pe \; .$$

L'expression de E(t) est suffisamment complexe pour être d'emploi malaisé. De nombreuses expériences ont été interpétées par ce modèle qui a permis de corréler le Péclet au Reynolds caractérisant l'écoulement (Fig.2). Dans les situations les plus courantes (Péclet grand) la DTS est symétrique et ressemble à une gaussienne. En outre, la corrélation de la figure 2 montre que dans une large gamme on a :

$$Pe \;\#\; 0.5\; L/dp \tag{19}$$

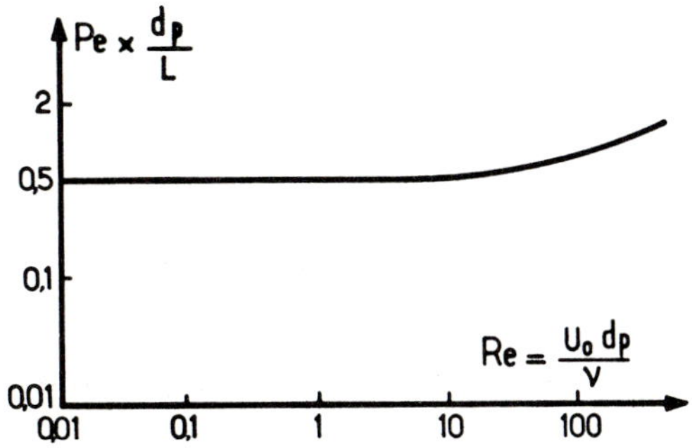

Figure 2. Corrélation entre les membres de Péclet et de Reynolds pour un écoulement de liquide en milieu poreux.

3.2. Le modèle des mélangeurs en cascade

Si les modèles précédents sont bien connus il ont l'inconvénient de fournir des DTS d'expressions complexes. Une intégration numérique de (16) permettrait de s'en affranchir, mais alors surgissent les problèmes de dispersion numérique. Une alternative simple à ces problèmes est fournie par le modèle des mélangeurs en cascade où le milieu poreux est assimilé à une cascade de J petites zones identiques de composition uniforme. Dans chaque mélangeur (13) s'applique en divisant V par J. Pour chaque zone on a :

$$\bar{E}_k(s) = \left(1 + \frac{\bar{t}}{J}s\right)^{-1} \qquad 1 \leq k \leq J \tag{20}$$

En remarquant que la sortie de la zone k est l'entrée de la zone k+1, puis en utilisant (3) dans le domaine de Laplace on obtient pour la cascade :

$$\bar{E}(s) = \left(1 + \frac{\bar{t}}{J}s\right)^{-J}, \quad E(t) = \left(\frac{J}{\bar{t}}\right)^J \frac{t^{J-1}}{(J-1)!}\exp(-Jt/\bar{t})$$
$$\bar{t} = V/Q \qquad \sigma^2 = \bar{t}^2/J \tag{21}$$

Villermaux [1] a montré que ce modèle était indiscernable des précédents si on pose :

$$Pe = 2(J-1) \quad \text{et} \quad J \geq 50 \tag{22}$$

Cette similitude repose sur deux observations :

- Le modèle des mélangeurs en cascade n'est qu'une version discrète de (16) où la dispersion numérique, traduite par J, est ajustée au moyen de (22) pour simuler la dispersion physique.

- En comparant (22) et (19) on trouve que L/J # $4d_p$. Autrement dit un mélangeur a un encombrement de l'ordre de 4 diamètres de grain. Ceci devient naturel si on assimile ces mélangeurs aux zones où les filets de fluide qui parcourent le milieu poreux, viennent se mélanger avant de se séparer à nouveau.

La figure 3 illustre les déformations typiques de E(t) en fonction de J. Les courbes sont sensiblement symétriques sauf pour les faibles valeurs de J (forte dispersion). En conséquence quelle que soit l'approche, loi de Fick ou mélangeurs en cascade, les DTS ont une asymétrie négligeable.

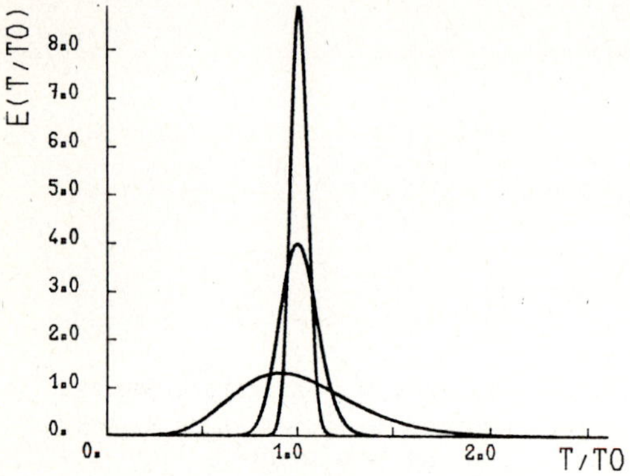

Figure 3. Rôle de J dans le modèle des mélangeurs en cascade $t_o = \bar{t}$.

De la courbe la plus étroite à la plus large : J = 500 ; 100 ; 10.

4. Ecoulement avec zone stagnante

Lorsque le milieu poreux est assez régulier et que le fluide en écoulement le sature totalement, les modèles piston dispersion précédents décrivent généralement bien l'expérience. Lorsque le milieu poreux comporte plusieurs échelles de structure (saturation incomplète par le fluide ou plusieurs fluides, répartition granulométrique à plusieurs maxima, grains de solide eux mêmes poreux) les DTS présentent une traînée et une asymétrie prononcées qu'une dispersion constante ne peut pas représenter C'est en prenant en compte de façon simple mais réaliste le rôle de la géométrie de la répartition du fluide que l'on peut représenter ces comportements.

4.1. Modèle à zone stagnante simple

Le fluide est divisé en deux zones distinctes :

- La première est un écoulement convectif et dispersif. Elle occupe la fraction ε du volume total V accessible au fluide. Nous supposerons qu'elle obéit au modèle des mélangeurs en cascade.

- La seconde zone qui occupe la fraction $1-\varepsilon$ du volume V est supposée immobile. Elle ne peut échanger avec la précédente que par diffusion moléculaire.

Le modèle d'écoulement est constitué par les équations de conservation du traceur dans chaque mélangeur. Dans le k^{e}, phases mobile et immobile réunies on a :

$$Q\,C_{k-1} = QC_k + \frac{V}{J}\left[\varepsilon \frac{dC_k}{dt} + (1-\varepsilon)\frac{dC'_k}{dt}\right] \tag{23}$$

Dans la phase immobile il est inutile de décrire toute la complexité de l'échange diffusionnel au moyen d'une loi de Fick appliquée sur une zone stagnante à géométrie inconnue. Villermaux [2] a montré que cet échange était très bien décrit par une constante de temps de transfert de matière t_m telle que :

$$\frac{dC'_k}{dt} = \frac{C_k - C'_k}{t_m} \qquad (24)$$

Cette loi de transfert entre les zones mobile et immobile est la plus simple assurant :

- un échange d'autant plus marqué que les deux zones sont à des concentrations différentes,

- l'absence d'échange à l'équilibre en régime permanent quand $C_k = C'_k$.

L'élimination de C'_k entre (24) et (23) puis la résolution dans le domaine de Laplace de manière semblable à celle du paragraphe 3.2 donne :

$$\bar{E}(s) = \left[1 + \frac{s\bar{t}}{J}\left(\varepsilon + \frac{1-\varepsilon}{1+t_m s}\right)\right]^{-J}$$

$$\bar{t} = V/Q \qquad \sigma^2 = \frac{\bar{t}^2}{J} + 2(1-\varepsilon)\,\bar{t}\,t_m \qquad (25)$$

L'expression de E(t) est ici inutilisable.

La figure 4 illustre des courbes E(t) et F(t) typiques en fonction de la valeur de t_m tout autre paramètre fixé. Les temps sont normés par $t_o = \varepsilon\bar{t}$ qui représente le temps de séjour dans la zone mobile. Pour les valeurs de t_m petites devant t_o, les courbes sont toujours symétriques et centrées autour de $\bar{t}$. La situation change radicalement pour t_m de l'ordre ou supérieur à t_o : les courbes présentent une trainée marquée, mais elles sont toujours centrées autour de $\bar{t}$. Les DTS peuvent présenter deux maxima (cas $t_m = 0.3\ t_o$) et un pic proocé en $t = t_o$. Ce pic est consitué des molécules qui sont restés dans la zone mobile du fait du transfert difficile (t_m grand) en zone immobile. La trainée qui flanque ce pic est très longue et représente les molécules qui ont pu passer en zone stagnante mais qui ont aussi de la difficulté à en sortir.

La "largeur" de ces courbes mesurée par σ^2 fait apparaître une propriété importante : les contributions à l'étalement de la DTS par dispersion (J) et par trans-

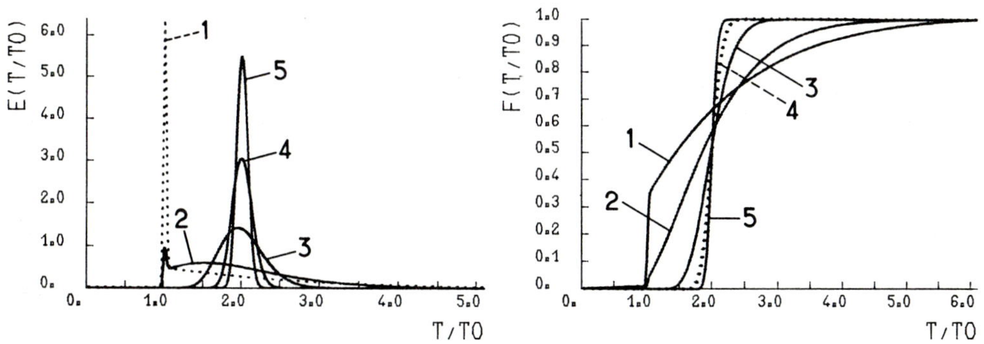

Figure 4. Rôle de t_m. Courbes E et F. $\varepsilon = 0{,}5$; J = 3000. t_m/t_o = 0,8 (1) - 0,3 (2)- 0,04 (3) - 0,008 (4) - 0,002 (5) temps normé par $t_o = \varepsilon\bar{t}$.

fert à une zone stagnante (t_m) sont additives. Dans un milieu poreux à section constante, si on double le parcours on double le volume V et par conséquent le temps de séjour moyen $\bar{t}$. Mais, en vertue de (19) et (20) on double aussi J (pour J > 50). Il en résulte que toute chose égale par ailleurs, σ^2 est proportionnel à $\bar{t}$ que l'étalement de la DTS soit dû à la dispersion axiale en zone mobile ou au transfert en zone stagnante. On a à faire à un phénomène diffusif classique où l'étalement de la DTS est proportionnel au chemin parcouru. La figure 5 illustre la déformation progressive de E(t) quand la longueur de ce chemin augmente. Les temps sont encore normés pat $t_o = \varepsilon\bar{t}$ proportionnel à la longueur du chemin. On constate qu'au bout d'une distance suffisante la DTS devient à nouveau symétrique comme pour une dispersion pure. Toutefois, les contributions de dispersion et de transfert de matière restent toujours dans le même rapport (10^{-3} pour 1 respectivement figure 5). Ceci montre que sur de longs parcours il devient très difficile de faire la part entre une dispersion axiale et un transfert de matière à une zone stagnante.

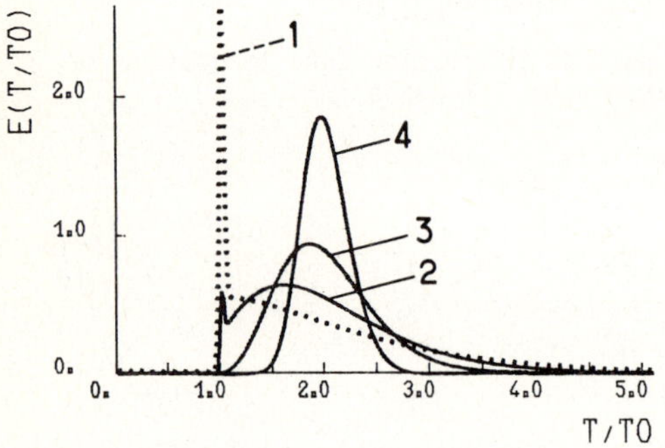

Figure 5. Rôle de la longueur du chemin entre l'entrée et la sortie du système. $\varepsilon = 0,5$; $t_m = 0,5$ unité de temps, $J/t_o = 2000\,(\text{temps})^{-1}$; $t_o = \varepsilon L/u$; u constante. t_o = 1(1) - 2(2) - 5(3) - 20(4) unité de temps.

La signification physique de t_m s'obtient de par son roigine diffusionnelle. Si L' est une dimension caractéristique des éléments de milieu poreux constituant la zone stagnante on a :

$$t_m = a \frac{L'^2}{D_m} \qquad (26)$$

où a est un facteur de forme. Par exemple pour une phase stagnante piégée dans le porosité interne de grains sphériques de diamètre dp on a L' = dp et a = 1/60. Si l'on connait donc l'origine physique précise de t_m, comme ci-dessus, on peut évaluer à priori son ordre de grandeur et par conséquent son rôle par rapport à celui de la dispersion.

Au sens de la physique statistique la relation (24) décrit une distribution de temps d'attente $\psi(t)$ dans les modèles de marche au hasard sur un réseau régulier, monodimensionnel dans notre cas. C'est ici une distribution exponentielle :

$$\psi(t) = \frac{1}{t_m} e^{-t/t_m} \qquad (\psi = C'_k\ ,\ C_k = \delta(t) \text{ dans } (24)) \quad (27)$$

Elle est analogue à celle de Montroll et Weiss [3] qui avaient montré qu'elle impliquait un comportement dispersif classique au temps de séjours longs. L'aptitude du modèle des mélangeurs en cascade à décrire la dispersion est semblable : la relation (20) peut être considérée comme la transformée de Laplace de la distribution des temps d'attente dans le mélangeur k. $E_k(t)$ est encore une exponentielle :

$$E_k(t) = \frac{J}{\bar{t}} e^{-Jt/\bar{t}} \qquad (28)$$

En d'autres termes, le modèle des mélangeurs en cascade est une marche au hasard discrète avec distribution exponentielle des temps d'attente sur chaque site (mélangeur).

Si tous ces comportements conduisent à une dispersion classique pour les longs parcours ($\bar{t}$ et J grands), nous venons de voir leurs aspects particuliers (disymétrie prononcée) pour les petits parcours.

4.2. Distribution des temps de transfert (DTT)

Dans les systèmes réels, la valeur de t_m peut ne pas être unique. Par exemple dans un milieu poreux constitué de grains de dimensions caractéristiques L' différentes, (26) donne des valeurs différentes de t_m. Dans un écoulement diphasique, les bras morts de la phase percolante ne sont pas tous identiques ce qui conduit nécessairement à une répartition des t_m. Cette distribution peut être prise en compte de la façon suivante : notons $\bar{E}_o(s)$ la distribution des temps de séjour (21) obtenue en ignorant le transfert à la zone stagnante, (25) donne alors :

$$\bar{E}(s) = \bar{E}_o(S) \qquad S = s\left[\varepsilon + \frac{1-\varepsilon}{1+t_m s}\right] \tag{29}$$

On voit que S n'est fonction que des propriétés intrinsèques du transfert à la zone stagnante. Pour tenir compte d'une loi de transfert générale (mais linéaire!) on peut poser :

$$S = s\,[\varepsilon + (1-\varepsilon)\,\bar{\psi}(s)] \tag{30}$$

Nous nous intéresserons au cas où :

$$\bar{\psi}(s) = \int_o^\infty \frac{f(t_m)}{1+t_m s}\, dt_m \tag{31}$$

qui représente une famille de sites stagnants de temps de transfert t_m en proportion $f(t_m)\, dt_m$. On obtient ainsi :

$$\bar{E}(s) = \left(1 + \frac{s\bar{t}}{J}\left[\varepsilon + (1-\varepsilon)\int_o^\infty \frac{f(t_m)}{1+t_m s}\, dt_m\right]\right)^{-J}$$

$$\bar{t} = V/Q \qquad \sigma^2 = \frac{\bar{t}^2}{J} + 2(1-\varepsilon)\,\bar{t}\,\bar{t}_m \tag{32}$$

$$\bar{t}_m = \int_o^\infty t_m\, f(t_m)\, dt_m$$

La figure 6 illustre les courbe E(t) pour trois distributions $f(t_m)$: monodispersé à $\bar{t}_m$, exponentielle et uniforme. Pour toutes ces distributions on a conservé les mêmes valeurs de J, $\bar{t}$, ε et $\bar{t}_m$. D'après (32) toutes les courbes ont donc même position moyenne et même variance σ^2 ! Cette dernière ne dépend en effet que du moment d'ordre 1 de $f(t_m)$. Toutes les DTS de la figure 6 ne diffèrent donc qu'à partir de leur moments d'ordre 3. Ceci montre clairement combien l'étude d'une distribution en se limitant à son moment d'ordre 2 est incomplète.

La figure 7 illustre le cas d'une DTT à deux sites :

$$f(t_m) = x\,\delta(t_m - t_{m_1}) + (1-x)\,\delta(t_m - t_{m_2}) \tag{33}$$

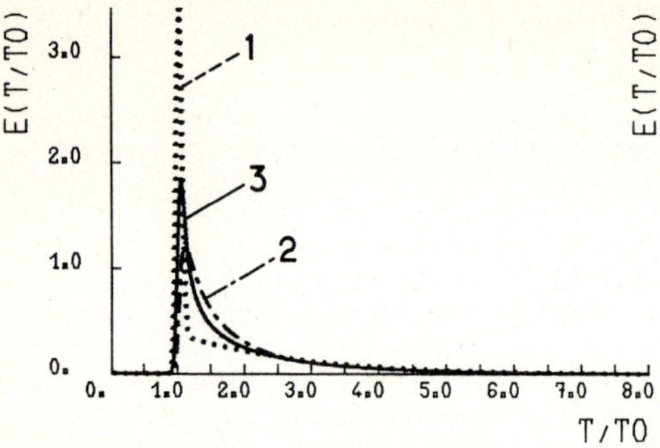

Figure 6. Rôle de $f(t_m)$ sur E(t). $\varepsilon = 0{,}5$; J = 500 ; $\bar{t}_m = 1$;
1: $f(t_m) = \delta(t_m - 1)$;
2: $f(t_m) = \exp(-t_m)$;
3: $f(t_m) = 0{,}5$; pour $0 \leq t_m \leq 2$, zero ailleurs. Temps normés par $t_o = \varepsilon\bar{t}$.

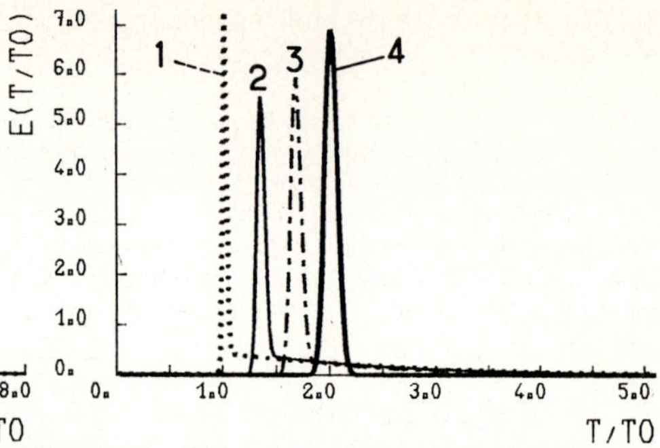

Figure 7. Rôle du poids respectif de chaque site pour une DTT à deux sites (relation (33)). $\varepsilon = 0{,}5$; J = 3000 ; $t_{m_1} = 0{,}9\ t_o$; $t_{m_2} = 0{,}01\ t_o$; x = 1(1) 0,66(2) - 0,33(3) - 0(4). Temps normés par $t_o = \varepsilon\bar{t}$.

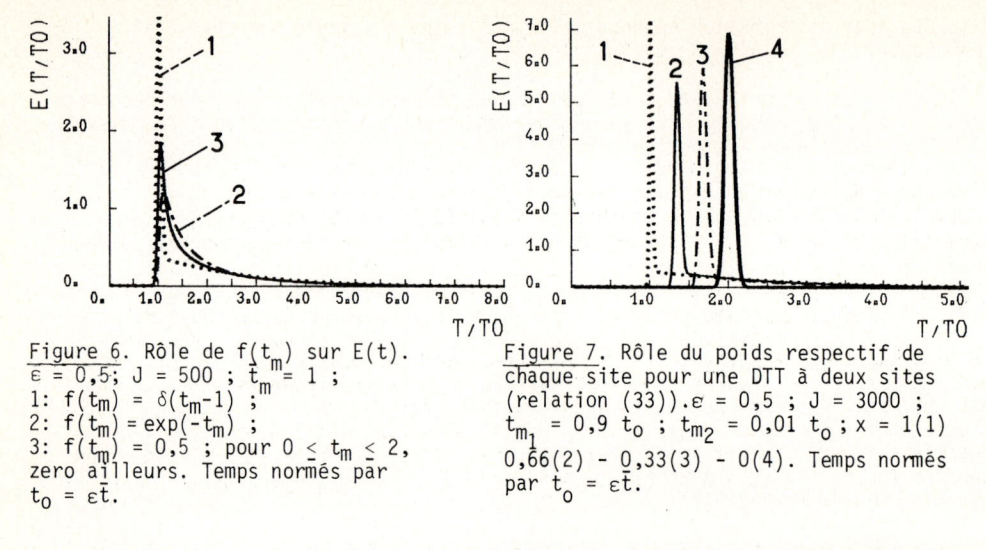

où seul x varie. Toutes les courbes ont même $\bar{t}$ et pourtant x agit fortement sur leur forme.

Comme au paragraphe précédent on peut voir $\bar{\psi}(s)$ comme la transformée de Laplace d'une distribution de temps d'attente $\psi(t)$. Tant que cette distribution à un moment d'ordre $1(\bar{t}_m)$, le comportement de E(t) pour $\bar{t}$ grand est celui d'une dispersion classique.

Il n'en est plus ainsi quand f n'a plus de moment au-delà d'un certain ordre. Si on suppose :

$$f(t_m) \sim t_m^{-1-\alpha} \qquad t_m \to \infty \quad , \quad 0 < \alpha < 1 \tag{34}$$

On peut montrer que :

$$\bar{t}_m \text{ et } \sigma^2 \text{ divergent}$$

$$\bar{\psi}(s) \sim 1 - A\,s^{\alpha} \ , \quad s \to 0,\ A = \text{constante} \tag{35}$$

$$E(t) \sim t^{-\alpha-2} \quad , \quad t \to \infty$$

La figure 8 illustre quelques exemples pour :

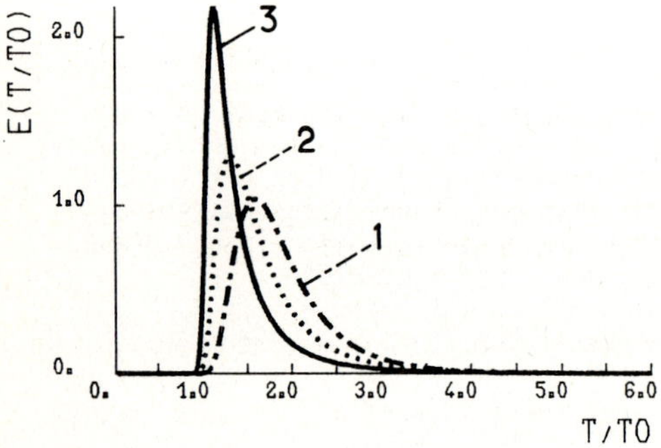

Figure 8. Rôle d'une DTT sans moment d'ordre 1. $\varepsilon = 0{,}5$; J = 500; $\theta = 0{,}1\ t_o$; α = 0,8(1) - 0,4(2) - 0,2(3). Temps normé par $t_o = \varepsilon\bar{t}$.

$$f(t_m) = \frac{\beta}{\pi} \sin(\frac{\pi}{\beta}) \frac{\theta^{\beta-1}}{t_m^\beta + \theta^\beta}$$

$$\beta = 1 + \alpha \qquad (36)$$

Les courbes présentent toujours une trainée marquée oui est à l'origine de la divergence de la variance σ^2. Elles sont nénamoins centrées autour de $\bar{t} = V/Q$. Ce comportement est analogue au mouvement "subdiffusif" de Shlesinger, Klafter et Wong [4] . On notera que l'asymétrie est d'autant plus accentuée que α s'éloigne de 1.

4.3. Quelques généralisations

La plus simple consiste à envisager des zones stagnantes de fluide placées en série et non plus en parallèle comme dans le paragraphe précédent. La zone mobile est en contact avec une première zone stagnante ou se transfert le soluté. Cette première zone stagnante isole la zone mobile d'une seconde zone stagnante où le transfert s'opère depuis la première, et ainsi de suite. Villermaux [2] a étudié ce problème et a montré que les comportements d'un tel système n'étaient pas significativement différents de ceux déjà rencontrés.

La seconde généralisation est beaucoup plus intéressante. De nombreux milieux poreux n'obéissent pas au modèle piston-dispersion monodimensionnel pour ce qui concerne la zone mobile du fluide. C'est le cas par exemple d'un puits d'injection et d'un puits de pompage d'eau ou de pétrole dans une couche géologique. On peut alors généraliser (29) et (30) de la manière suivante : si $\bar{E}_o(s)$ désigne une DTS quelconque représentant un écoulement en l'absence de zone stagnante, le transfert à cette zone est pris en compte par :

$$\bar{E}(s) = \bar{E}_o(S) \quad , S = s[\varepsilon + (1-\varepsilon)\, \bar{\psi}(s)] \qquad (37)$$

$\bar{\psi}(s)$ représentant la loi de transfert à la zone stagnante. Ce résultat démontré par Villermaux [2] constitue une généralisation de celui de Montroll et Weiss [3] concernant les marches au hasard avec noyau à mémoire.

5. Exemples d'applications

Outre les mesures classiques de dispersion, la mise en évidence et la caractérisation d'une zone stagnante a été particulièrement bien illustrée par Gaudet [5] dans l'étude de l'écoulement de l'eau dans un sol-insaturé. Les problèmes de chromatographie, traités par Villermaux [2], sont semblables quoique plus généraux.

Dans les écoulements diphasiques, la modélisation qui vient d'être présentée permet d'accéder à une description simple de la géométrie de la phase étudiée :

- Un bilan global sur la quantité de phase injectée donne la quantité globale retenue par le milieu poreux.
- La mesure de $\bar{t}$, connaissant Q, fournit la quantité exclusivement présente dans l'amas percolant ; par différence avec la précédente on accède à la quantité constituant les amas piégés et isolés.
- L'ajustement d'un modèle à zone stagnante permet d'estimer la fraction de phase effectivement mobile, puis par (26), une longueur caractéristique moyenne des bras morts de fluide stagnant.

En chromatographie par perméation de gel Costa, He, Villermaux et Schweich [6] ont pu montrer que la tortuosité de la phase gel était fonction de la taille des macromolécules à séparer, suggérant ainsi une structure fractale.

Enfin signalons que ces techniques de modélisation sont largement utilisées dans l'industrie chimique pour caractériser la nature de l'écoulement dans les réacteurs industriels.

6. Conclusions

Nous veons de voir comment une approche géométrique simple permet de caractériser un écoulement qui s'écarte du traditionnel piston-dispersion : il suffit de découper par la pensée le fluide en deux zones, l'une mobile, l'autre stagnante et d'invoquer des lois de transfert ($\psi(s)$) simples mais réalistes et ayant un support physique (relation (26)). Point n'est besoin de "torturer" la loi de Fick avec des coefficients de dispersion variables et par conséquent impossibles à extrapoler d'une situation de laboratoire à un problème appliqué. Ce n'est, en effet, pas la loi de Fick qui peut rendre compte de la structure géométrique d'un fluide en écoulement.

Il ne faut toutefois pas en conclure que l'approche présentée résoud tous les problèmes. Certains, et non des moindres, restent sans solution :

- où se situe la frontière entre fluide mobile et stagnante ?

- Comment identifier convenablement la distribution $f(t_m)$? (le seul exemple convaincant est une distribution à deux sites comme sur la figure 7. Voir Villermaux [2]).

Enfin, peut être le plus important et la plus grave : lorsque le milieu poreux est le siège d'un interaction physico-chimique non linéaire, la connaissance de $E(t)$ ne suffit plus pour caractériser convenablement l'écoulement. Ce fait a été illustré sur un exemple très simple par Schweich [7] . C'est le détail intime de la géométrie de l'écoulement qui devient important. C'est donc bien sur la relation entre la géométrie du milieu poreux et la nature de l'écoulement qu'il faut porter ses efforts.

Notations

a	facteur de forme
C	concentration dans le fluide en écoulement
C'	concentration dans le fludie stagnant
C_o	concentration d'entrée
D_e	dispersion axiale
D_m	diffusion moléculaire corrigée de l'effet de totuosité
d_p	diamètre de grain
$E(t)$	distribution des temps de séjour
$E_k(t)$	distribution des temps de séjour du $k^{ième}$ mélangeur
$f(t_m)$	distribution des temps de transfert
$F(t)$	primitive de $E(t)$
F	flux de dispersion
J	nombre de mélangeurs
L, L'	longueur du milieu poreux, longueur caractéristique
n_o	quantité de traceur
Pe	nombre de Péclet (uL/De)
Q	débit volumique
s	variable de Laplace
$t, \bar{t}, t_m, t_{m1}, t_{m2}, t_o$	temps, temps de séjour moyen, temps de transfert, temps de transfert moyen ; $\varepsilon\bar{t}$
u, u_o	vitesse réelle du fluide, vitesse en fût vide ou de Darcy
V	volume accessible du fluide

$x(t)$	signal d'entrée
x	fraction de site
$y(t)$	signal de sortie
z	distance

Lettres grecques

α	exposant de $f(t_m)$ et $\bar{\psi}(s)$ à temps long
ε	fraction de fluide mobile
θ	temps de référence, relation (36)
μ_n	moment d'ordre n de $E(t)$
$\psi(t)$	distribution des temps d'attente
σ^2	variance de $E(t)$
ν	viscosité cinamatique

Références

1. J. Villermaux : Génie de la réaction chimique, conception et fonctionnement des réacteurs. Tec.doc, Lavoisier, Paris 1982.
2. J. Villermaux : "Theory of linear chromatography" in Percolation Processes Theory and Applications, NATO adv. Stu. Inst. E.33, Sijthoff et Noordhoff (1981).
3. E.W. Montroll, G.H. Weiss : "Random walks on lattice II", J. Math. Phys. 6 167-181 (1965).
4. M.F. Schlesinger, J. Klafter, Y.M. Wong:"Random walks with infinite spatial and temporal moments", J. Stat. Phys., 27 499-512 (1982).
5. J.P. Gaudet : "Transfert d'eau et de soluté dans les sols non saturés. Mesures et simulation". Thèse d'état, Univ. Grenoble, France 1978.
6. M.R.N. Costa, D. Schweich; X.W. He, J. Villermaux "Band spreading modeling and correction in size exclusion chromatography" à paraître.
7. D. Schweich :"Influence of mixing state of carrier fluid in non-linear equilibrium echange system", Am. Inst. Chem. Eng. J. 29, 935 (1983).

Part IX **Transitions**

Wetting of a Random Surface: Statistics of the Contact Line

J.F. Joanny
Collège de France, Laboratoire de Physique de la Matière Condensée[1],
11, Place Marcelin-Berthelot, F-75231 Paris Cedex 05, France

1. Introduction

The wetting properties of a solid surface by a liquid are very sensitive to surface heterogeneities : the contact line becomes wiggly, and the final equilibrium structure of the liquid droplet or liquid film on the surface is history dependent on a heterogeneous surface (in the case of partial wetting, this phenomenon is known as contact angle hysteresis). We study here the statistics of a wiggly contact line on a heterogeneous solid, both for totally or partially wetting fluids.

The macroscopic parameter controlling the spreading of a solid by a liquid is the spreading power S related to the interfacial tensions liquid-vapor γ_{LV}, solid-liquid γ_{SL} and solid-vapor γ_{SV} :

$$S = \gamma_{SV} - \gamma - \gamma_{SL} \tag{1}$$

Whenever S is negative, the liquid wets partially the solid and at equilibrium forms a spherical cap with a contact angle θ given by Young law

$$S = \gamma \left[\cos\theta - 1\right] \simeq -\gamma \frac{\theta^2}{2} \ (\theta \text{ small}) \tag{2}$$

If S is positive the liquid wets totally the solid. For a non-volatile liquid (dry wetting in the language of ref. [1]) S is in general non-zero and the equilibrium shape of the liquid is a flat pancake, due to the balance between spreading power and disjoining pressure. The properties of this pancake are briefly reviewed in the next section.

On a random surface S is not a constant but varies randomly from point to point

$$S(x,y) = S + h(x,y) \tag{3}$$

There are two major sources of heterogeneities of a solid surface [2].

- Chemical contamination leads to a fluctuation of the spreading power h proportional to the local concentration of contaminants.

- Surface roughness leads to a fluctuation proportional to the local slope of the solid surface.

The liquid tends to cover regions of high spreading power ($h > 0$) and the contact line is distorted. Those distortions are studied in section 3 for partially wetting liquids and in section 4 for totally wetting liquids.

[1] Laboratoire rattaché au C.N.R.S. : U.A. 792.

2. Effect of long Range Forces [1][3][4]

When the liquid film spread on the solid surface become very thin, the macroscopic description becomes clearly inadequate, one must take into account the long-range character of the molecular forces. This can be done by introducing the so-called disjoining pressure $\Pi_d(\zeta)$ that one must apply on the surface of a film to maintain its thickness to a given value ζ. We will consider here only Van der Waals forces for which it is given by

$$\Pi_d = \frac{A}{6\Pi\zeta^3} \tag{4}$$

A is an effective Hamaker constant which is supposed to be positive.

The free energy functional of a film of thickness $\zeta(x)$ is

$$F = \int \left\{ -S + \frac{1}{2}\gamma \left(\frac{\partial\zeta}{\partial x}\right)^2 + P(\zeta) + \lambda\zeta \right\} dx \tag{5}$$

The first term is related to the interfacial tensions, the second represents the interfacial energy when the liquid surface is not flat (the slope $\partial\zeta/\partial x$ is small), the third term represents the Van der Waals forces

$$P(\zeta) = \int_{\zeta}^{+\infty} \Pi_d(z)dz = \gamma \frac{a^2}{2\zeta^2} \tag{6}$$

where $a = (A/6\Pi\gamma)^{1/2}$ is a molecular length.

λ is a Lagrange multiplier that must be introduced to fix the total volume of the wetting liquid $\Omega = \int \zeta dx$.

The profile of the liquid surface is obtained by minimization of the free energy functional (5).

a. In a case of partial wetting ($S < 0$), the effect of Van der Waals forces is very small, the liquid retains the shape of a spherical cap, except in a tiny region of size $x_0 \simeq a/\theta^2$ around the contact line. We will consider phenomena at length scales larger than x_0 and thus ignore the effect of long-range forces for partial wetting.

b. In a case of total wetting ($S > 0$), the liquid makes a thin pancake on the solid, resulting from the competition between the spreading power S tending to spread the liquid and the disjoining pressure Π_d tending to thicken the liquid film.

For large liquid volumes Ω, minimization of the free energy F leads to a pancake thickness

$$e(S) = a\sqrt{\frac{3\gamma}{2S}} \tag{7}$$

and to a healing length

$$\xi = \frac{e^2}{\sqrt{3}\,a} \tag{8}$$

If S and γ are of the same order of magnitude $e \sim a$ and the liquid film has a molecular thickness. However, if $S/\gamma \ll 1$, the thickness e can be much larger than a molecular thickness and reaches 100 Å for $S/\gamma = 10^{-3}$.

The energy of this equilibrium pancake can be calculated from (5). We find an energy per unit length (along the contact line) independent of the liquid volume Ω

$$T = \gamma a \,\mathrm{Log}\,\frac{e}{a} \quad (9)$$

The effect of the Van der Waals forces on the wetting properties can therefore be described in terms of a line tension T in a case of total wetting.

3. Partial Wetting of a Random Surface : Statistics of the Contact line

We study a surface with a random heterogeneity h ($< h > = 0$) partially wetted by a liquid ($S < 0$). In the absence of heterogeneity,the contact line is rectilinear ; on a random surface, the liquid covers regions of positive h and the contact line is distorted by a value $\eta(x)$ perpendicular to its unperturbed position (x is the coordinate along the pure contact line). The statistics of the distortion $\eta(x)$ is fixed by a balance between the fluctuations of spreading power h and the elasticity of the contact line.

We first fix a distance x and then wait to determine the average fluctuation $\overline{\eta}^2(x)$ of η over a distance x. The argument presented here is due to D. HUSE.

The elastic energy of an average displacement $\overline{\eta}$ of the contact line has been studied in great details in ref. [2]. It is independent of the length scale x

$$U_{el} = \frac{1}{2}\overline{\eta}^2\,|S| \quad (10)$$

The coupling energy with the heterogeneity h can be determined through an argument first proposed by IMRY and MA [7] for random fields in magnetic systems. If the field is correlated over a length d, the solid surface is paved with squares of size d^2 where the heterogeneity is $\pm\,\overline{h}$ with a probability 1/2. The contact line is distorted in such a way that it covers more squares with an heterogeneity $+\,\overline{h}$. If n squares are covered,the coupling energy is proportional to $\sqrt{n}$. Two limits should be distinguished

a. Ultraweak Heterogeneity

If the average distortion $\overline{\eta}$ is smaller than d, the number of covered squares is $n = x/d$ and the surface covered in each square is $d\overline{\eta}$

The coupling energy is then

$$U_{coup} = -\overline{h}\,\overline{\eta}\sqrt{dx} \quad (11)$$

The average distortion is obtained by a minimization of $U_{el} + U_{coup}$

$$\overline{\eta} = \frac{\overline{h}}{|S|}\sqrt{dx} \quad (12)$$

A result already derived by different arguments in ref. [2] and [6].

b. Weak Heterogeneity

If the average distortion $\overline{\eta}$ is larger than d the number of covered squares is $n = x\overline{\eta}/d^2$ and the whole squares are covered. This leads to a coupling energy

$$U_{coup} = -\overline{h}\sqrt{x\overline{\eta}}\;d \quad (13)$$

Minimization of the total energy gives the average distortion at a scale x

$$\overline{\eta} = \left|\frac{\overline{h}}{S}\right|^{2/3} d^{2/3}\, x^{1/3} \quad (14)$$

For a drop size of 1 mm and a correlation of the heterogeneity of 10 Å if $|\overline{h}/S| = 1$ the distortion is 1 μm and could be seen directly under a microscope. In practice, except for very small heterogeneities, the expected distortion is given by (14).

4. Total Wetting of a Random Surface : Statistics of the Contact line [3]

The solid surface has a random heterogeneity h but the average spreading power S is positive. The average shape of the liquid pancake on such a solid has been studied by de GENNES [9] using a geometric argument. We suppose here that this average position is known. The average contact line is a circle with a very large radius that we assimilate to a straight line. We study the fluctuations $\overline{\eta}(x)$ of this contact line at a scale x.

As in the partial wetting problem, the statistics of the contact line results from the competition between two terms : a coupling with the heterogeneity and a restoring force. The coupling with the heterogeneity is still of random field type and for not too weak heterogeneities is given by (12).

The major difference between partial wetting and total wetting in the statistics of the contact line comes from the restoring force. For partial wetting, it is an elastic force related to the increase of area of the liquid-vapor interface as the contact line is distorted. For total wetting it is a line tension force directly proportional to the length of the contact line and related to the long-range Van der Waals forces. On a length x for a fluctuation $\eta(x)$ the line energy can be written

$$U_{lt} = \frac{1}{2}\int dx\ T\left(\frac{d\eta}{dx}\right)^2 \sim \frac{T\overline{\eta}\,(x)^2}{x} \qquad (15)$$

The root mean square distortion is obtained by minimization of the total energy $U = U_{coup} + U_{lt}$

$$\overline{\eta} = \left[\frac{\overline{h}d}{T}\right]^{2/3} \quad x = \left[\frac{\overline{h}d}{\gamma a\ \mathrm{Log}\frac{e}{a}}\right]^{2/3} x \qquad (16)$$

The distortion increases thus linearly with the length scale, the contact line is rough. In fact the energy $U = U_{lt} + U_{coup}$ exactly has the same structure as the interfacial energy studied by GRINSTEIN and MA for a random field Ising system. The statistics of the contact line is thus the same as the one of an interface in a random Ising system in two dimensions.

5. Conclusion

We studied the fluctuations of the contact line on a randomly heterogeneous solid plane both for partial and total wetting, treating the heterogeneity as a random field. As far as we know, there exists no systematic experimental study of wiggly contact lines. This could probably be done by direct observation of the contact line perhaps using fluorescent liquids.

As mentioned in the introduction, heterogeneities of the solid surface are also a source of hysteresis in wetting phenomena.

For partial wetting, contact angle hysteresis can be quantitatively explained, by describing the heterogeneities in terms of defects of the solid surface [2].

For total wetting, regions of low spreading power (with a negative fluctuation h) can create holes of unwetted solid in the liquid pancake. Creating a hole of radius R we gain an energy $|h|\Pi R^2$ but we must pay the increase in the length of contact line $2\Pi RT$. A hole where the fluctuation is h is thus stable if its size R is larger than 2T/h. The position and the existence of these holes are much dependent on the way the liquid pancake has been obtained. For total wetting also we expect a strong hysteresis.

References

1. J.F. Joanny, P.G. de Gennes: C.R.A.S. 299 II , 279 (1984)
2. J.F. Joanny, P.G. de Gennes: Journal of Chemical Physics 81 , 552 (1984)
3. J.F. Joanny: Thesis University Paris VI (1985)
4. P.G. de Gennes: Reviews of Modern Physics (in press)
5. D. Huse: Private Communication
6. Y. Pomeau, J. Vannimenus: Journal of Colloid and Interface Science (in press)
7. Y. Imry, S. Ma: Physical Review Letters 35 , 1399 (1975)
8. G. Grinstein, S. Ma: Physical Review Letters 49 , 685 (1982)
9. P.G. de Gennes: C.R.A.S. (to be published)

Wettability of Solid Surfaces: A Phenomenon Where Adsorption Plays a Major Role

Jacques Chappuis
Lafarge Coppee Recherche G.I.E., B.P. 26, 2, Avenue Albert Einstein,
F-78192 Trappes Cedex, France

1 Behaviour of liquids relative to solid surfaces

A drop of a liquid, placed on the surface of a solid, can exhibit two different behaviours :

- The drop may spread completely over the solid, and after a certain length of time, form a very thin film

- The drop may become distorted, and finally take an equilibrium configuration

In the second case, at one point on the edge of the drop (also called the three-phase line), the tangential plane to the liquid surface forms with the plane of the solid surface an angle θ , different from zero and called the "contact angle" (Figs. 1 and 2).

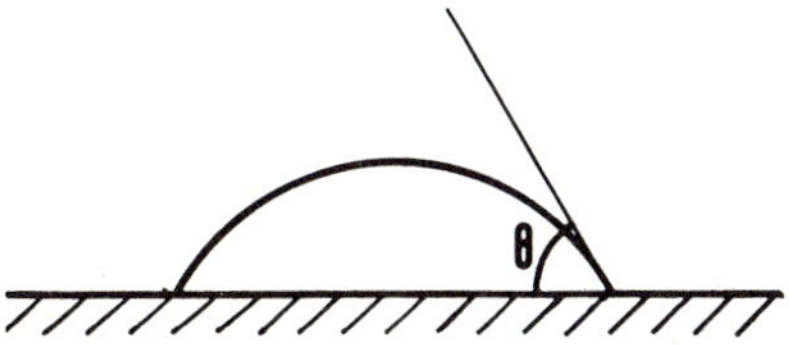

Fig. 1. A liquid drop on a solid with contact angle θ less than 90°

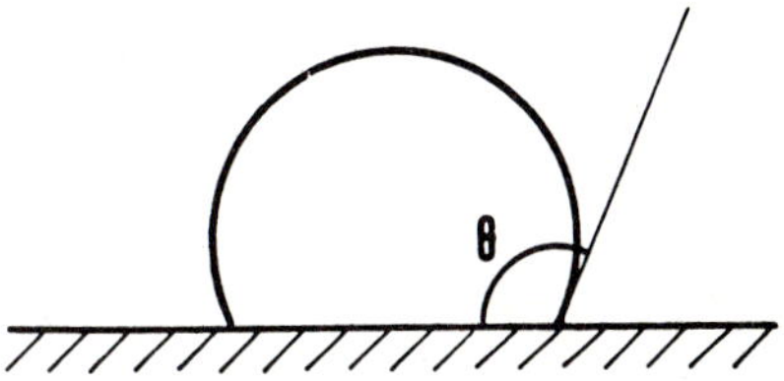

Fig. 2. A liquid drop on a solid with contact angle θ greater than 90°

When the liquid completely spreads on the solid surface (contact angle zero), the attraction of the molecules of the liquid by the atoms of the solid is larger than the mutual attraction of the molecules of the liquid ; when there is not complete spreading, the larger the contact angle is, the smaller the affinity of the liquid's molecules for the solid, in proportion to the mutual affinity of the liquid molecules. This reasoning was first expressed by YOUNG in 1805, [1] , and was proved to be very relevant, since.

2 Contact angle hysteresis

Let us consider the following experiment. A drop is placed on a plane surface. By introducing a syringe into the drop, liquid is added (Fig. 3). The volume of the drop increases without any variation of the liquid solid interfacial area. Contact angle increases to value θ_A, above which the interfacial area would grow. θ_A , the maximal possible value for the contact angle, is called "advancing angle". If we withdraw liquid instead of adding it (Fig. 4), the drop's volume diminishes, and the contact angle attains a minimal value θ_R, below which the interfacial area recedes. θ_R is called the "receding angle".

A drop placed on a solid surface, and presenting a contact angle θ which is found between θ_A and θ_R , is in equilibrium. The difference between these two values is called "contact angle hysteresis".

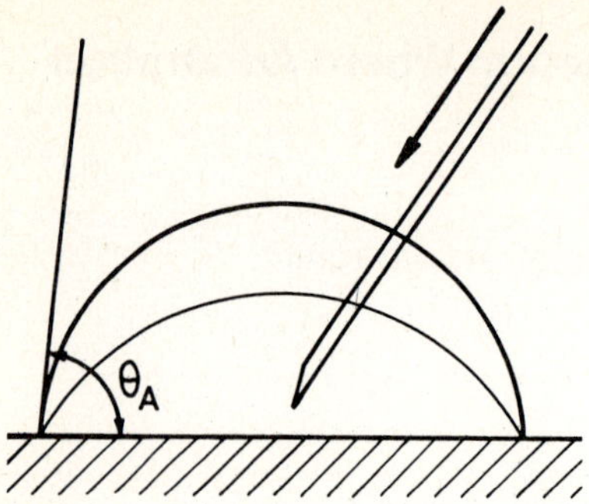

Fig. 3. Advancing contact angle

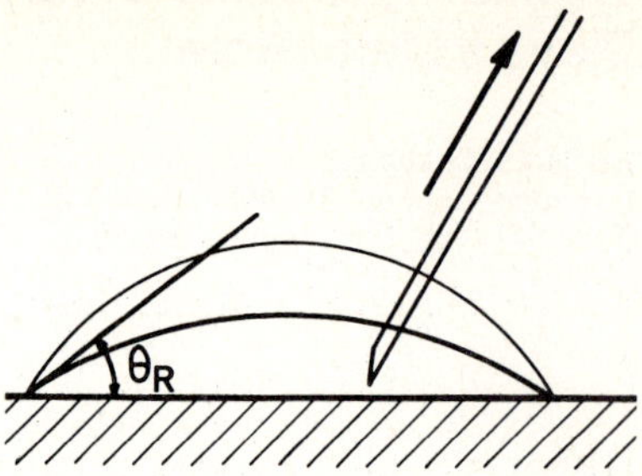

Fig. 4. Receding contact angle

Three causes are mentioned in the literature (for example [2], [3] and [4]), to explain contact angle hysteresis :

1. The heterogeneity of solid surfaces
2. The roughness of solid surfaces
3. Modification of the solid resulting from its contact with the liquid.

On a perfectly smooth and homogeneous solid surface, the contact angle would take a single value.

3 Young's equation

In 1805, YOUNG [1] proposed a relation that is now classic to explain the behaviour of a drop of liquid on an "ideal" solid surface (perfectly smooth and homogeneous). The first demonstration, based on a mechanical analysis of the problem, supposes that the three phase separating line (solid, liquid, and vapor phases) is in equilibrium because of the action of the three interfacial tensions γ_{LV}, γ_{SV} , and γ_{SL} , which are exerted respectively in the planes of the liquid vapor, solid vapor, and solid-liquid interfaces (Fig. 5). The projection of these tensions upon the plane of the solid's surface leads to the classic relation

$$\gamma_{SV} - \gamma_{SL} = \gamma_{LV} \cos \theta \qquad (1)$$

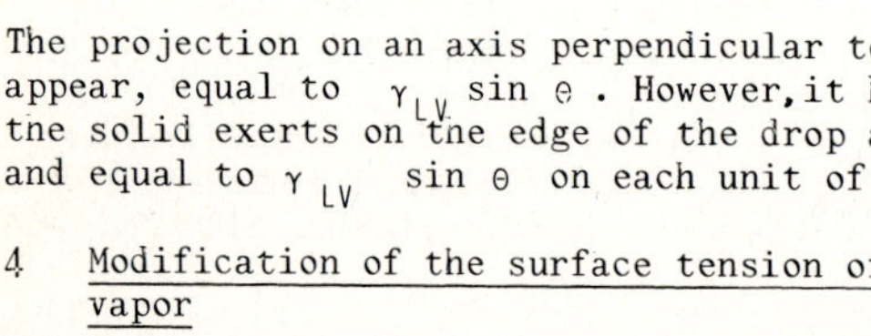

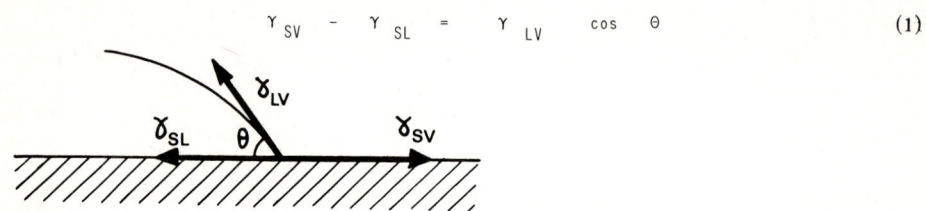

Fig. 5. A liquid drop on a solid. Each unit of length of the three phase line is subjected to three interfacial tensions

The projection on an axis perpendicular to the surface, makes a nonnul component appear, equal to $\gamma_{LV} \sin \theta$. However, it has been possible to establish, [2], that the solid exerts on the edge of the drop a vertical reaction, R_S, directed downwards and equal to $\gamma_{LV} \sin \theta$ on each unit of length of the three-phase line (see Fig. 6).

4 Modification of the surface tension of solids by adsorption of molecules from the vapor

The fixing of foreign molecules on a solid modifies its surface tension. The surface tension γ_{SV} of a solid in equilibrium with a vapor is different from the surface tension γ_{SO} of the same solid surface placed in a vacuum.

BANGHAM and RAZOUK, [5] have extended Gibbs' adsorption equation to solid gas interfaces :

$$\gamma_{SO} - \gamma_{SV} = RT \int_0^P \Gamma(P)\, d(\ln P) \tag{2}$$

in which

R is the universal gas constant

T is the temperature of the system

P is the pressure of the gas

$\Gamma(P)$ is the quantity adsorbed on the solid surface at pressure P , expressed in moles per unit of surface area

Since all the quantities located in the second member of Eq. (2) are positive, the quantity $\gamma_{SO} - \gamma_{SV}$ should also be positive. This signifies that the surface tension of a solid in equilibrium with its vapor (γ_{SV}) is weaker than the surface tension of the same solid in vacuum (γ_{SO}) when there is adsorption of the vapor. There is no vapor adsorption in the cases where such might lead to an increase in the solid's surface tension.

We note that the decrease of surface tension due to the adsorption phenomenon can be calculated when the adsorption isotherm of the gas on the surface $\Gamma = f(P)$ is known.

The quantity $\gamma_{SO} - \gamma_{SV}$ is usually called "spreading film pressure" and is then written π_e :

$$\pi_e = \gamma_{SO} - \gamma_{SV} \tag{3}$$

Hence, Young's equation can also be written as follow :

$$\gamma_{LV} \cos\Theta = \gamma_{SO} - \gamma_{SL} - \Pi e \tag{4}$$

as represented by Fig. 6

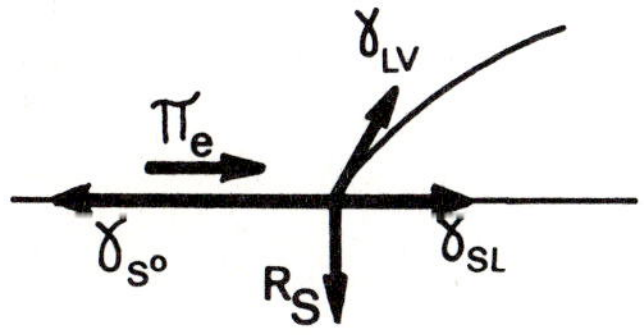

Fig. 6. Vectorial equilibrium of the edge of a drop when there is an adsorbed film at the solid vapor interface

5 Work of adhesion ; Work of cohesion

DUPRE [6] gives the name "work of adhesion W_A of a liquid to a solid" to work received (or furnished) upon contact (or separation) brought about in a reversible manner of one unit of liquid area with one unit of solid area (in the presence of the liquid's vapor), with the intention of forming one unit of solid liquid area.

$$W_A = \gamma_{SV} + \gamma_{LV} - \gamma_{SL} \tag{5}$$

If we place two units of liquid area in contact, there is a disappearance of these surfaces, and the work received, called "work of cohesion of a liquid" is equal to :

$$W_C = \gamma_{LV} + \gamma_{LV} - \gamma_{LL}$$

In that the surface formed is located in the mass of the liquid, it cannot be the seat of any tension. Thus $\gamma_{LL} = 0$, and

$$W_C = 2\gamma_{LV} \tag{6}$$

6 Contact angles of different liquids on the same solid surface

With a high energy surface (metal, oxide, ...) one usually gets no correlation between the measured contact angles and the surface tension of the liquids.

However by taking a low energy surface (polymer) and plotting cos Θ as a function of the surface tension γ_{LV} of the liquids, it is possible to see that the experimental points are assembled on a single curve : for instance, we reproduce on Fig. 7 the curve obtained by FOX and ZISMAN [7] , from the measurement of the advancing contact angles of more than 60 liquids on a smooth surface of PTFE (polytetrafluoro ethylene).

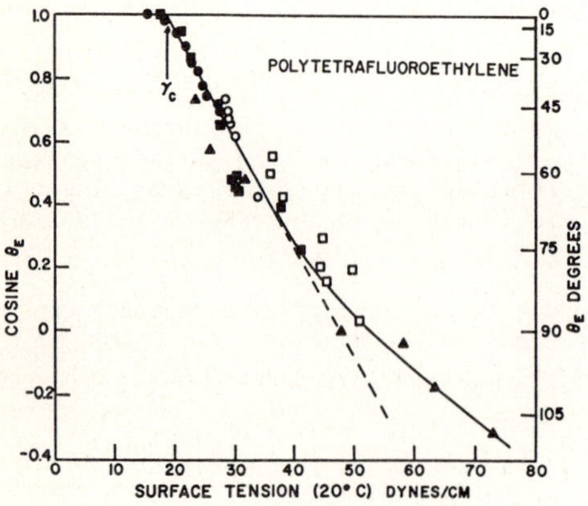

Fig. 7. Contact angles of liquids of different surface tensions on PTFE : cosine Θ versus surface tension

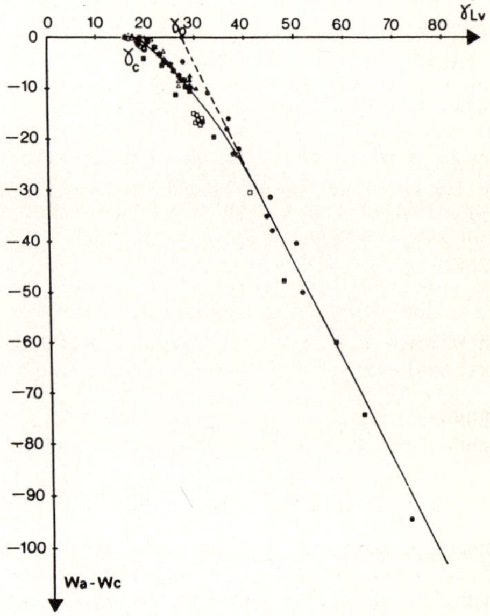

Fig. 8. Contact angles of liquids of different surface tensions on PTFE : $W_A - W_C$ versus surface tension

From this curve ZISMAN [8] has defined the "critical surface tension" of the solid γ_C ; liquids with a surface tension smaller than γ_C spread on the solid and liquids with a surface tension higher than γ_C exhibit a finite contact angle.

Another interesting representation of these experimental results is obtained by plotting the quantity $W_A - W_C$ vs the surface tension of the liquids (Fig. 8). From equations (1), (5) and (6), it is possible to write :

$$W_A - W_C = \gamma_{LV} (\cos\theta - 1) \tag{7}$$

$W_A - W_C = f(\gamma_{LV})$ is a straight line with a slope equal to − 2, for liquids having a surface tension larger than 40 dynes cm^{-1}.

7 Current theories issued from these results

All have been established by assuming that the speading film pressure Πe is negligible for all liquids tested on PTFE.

Then, the combination of Young's equation and of a mathematical expression of the experimental relation between cos θ and γ_{LV} pemits to express γ_{SL} as a function of the surface tensions of the solid γ_{SO} and of the liquid γ_{LV} . For instance, ZISMAN's treatment [8] is based on the fact that the curve $\cos\theta = f(\gamma)$ is almost a straight line of slope $-b$ for liquids of surface tension smaller than 40 dynes/cm ; he finally gets :

$$\gamma_{SL} = \gamma_{SO} - \gamma_{LV} (1 + b\gamma_c) + b\gamma_{LV}^2 \quad (8)$$

Some other authors first express γ_{SL} from theoretical considerations and then, study how such relations can fit the experimental results :

GOOD [9, 10] has extrapolated to the solid-liquid interfacial tension, the relation

$$\gamma_{12} = \gamma_1 + \gamma_2 - 2\Phi\sqrt{\gamma_1 \gamma_2} \quad (9)$$

established in the case of the interfacial tension between two nonmiscible liquids 1 and 2. In this relation, Φ is a characteristic number of the studied system. Its expression brings into action dipolar moment, polarizability, ionization energy, and molecular radius of the two phases molecules.

FOWKES [11] distinguishes, in the liquid superficial tension, between the part due to London dispersion forces (which he denotes γ^d), and the part due to other molecular forces.

FOWKES expresses the interfacial tension between two liquids as :

$$\gamma_{12} = \gamma_1 + \gamma_2 - 2\sqrt{\gamma_1^d \gamma_2^d} \quad (10)$$

In the case of the interfacial tension between a solid and a liquid, FOWKES [12] has extrapolated relation 10.

NEUMANN [13] has also proposed an equation of state of contact angles. His approaches use GOOD's interaction parameter.

8 Critical analysis of the current theories

From the point of view of the author, none of these theories is correct, because all of them are based on a wrong assumption : they all assume that the spreading film pressure is negligible for the vapor of all the tested liquids : that will be proved to be wrong.

For a sessile drop, the saturating vapor pressure p of the liquid, is, according to the curvature of the liquid's surface, greater than the saturating vapor pressure p_o of a plane surface of the same liquid. The molecules of the saturating vapor are hence able to get adsorbed on the solid surface in the vicinity of the edge of the drop. With different liquids on the same solid, we proceed from the case of nonwetting (θ finite), to the case of complete wetting ($\theta = 0°$) when the attraction of liquid molecules by the solid becomes greater than the attraction of liquid molecules among themselves (see Sec. 1). It is logical to think that when one takes liquids with lower and lower superficial tensions, and as one nears this transition, the attraction of molecules of the liquid's saturating vapor by the solid becomes great enough to form a dense layer of absorbed molecules, which has a nonnegligible spreading pressure Π_e.

Consequently, on a low energy surface (polymer), only liquids of higher surface tension do not become adsorbed. By contrast, on the same solid, liquids of low superficial tension become adsorbed and form a more or less dense film, which brings about the reduction of the solid's superficial tension. This reduction will be greatest for those liquids having a low superficial tension. Moreover, these latter are the ones for which the saturating vapor pressure is greatest.

9 Experimental evidence of the adsorption of low surface tension liquids on PTFE

In order to justify our ideas on the adsorption of the vapors of low surface tension liquids on PTFE, we have imagined and conducted the following experiment :we have measured the advancing contact angle of glycerol on PTFE in a vessel containing the saturating vapor of some alkanes. Glycerol is a liquid of high surface tension (γ_{LV} = 63.4 dynes/cm) for which we expect no adsorption, and alkanes are low surface tension liquids for which we expect a nonnegligible spreading pressure Π_e. If there were no adsorption of the alkanes on PTFE, the spreading film pressure would be null or limited, and that would not modify the contact angle of glycerol. On the other side, a nonnegligible spreading pressure of an alkane would lower the surface tension of PTFE, ($\gamma_{SV} = \gamma_{SO} - \Pi_e$), resulting, from Young's equation, in a smaller value of $\cos\theta$, that is to say, in a higher contact angle value.

Experimentally, the measurement took place in a closed vessel in which we had placed a few hours before, the sample of PTFE and a small beaker containing a few cubic centimeters of the alkane. The results are given in the following table.

Nature of the alkane	Surface tension of the alkane	Advancing contact angle of glycerol on PTFE
none	-------	96 °
Hexane	18,4	112 °
Heptane	20,3	108 °
Octane	21,8	106 °
Decane	23,9	104 °
Dodecane	25,4	103 °
Hexadecane	27,6	101 °

These results definitely indicate that the spreading film pressure of low surface tension liquids cannot be neglected on PTFE.

10 The author's approach, taking adsorption into account

The complete theoretical approcah has been published elsewhere [2] . We will limit ourselves here to the hypotheses and conclusions.
From what has been explained above, we expect on low energy surfaces :

- A null value of Π_e for liquids having the highest superficial tensions
- A nonnull value of Π_e for liquids having the lowest superficial tensions
- A larger and larger value of Π_e for liquids with lower and lower superficial tensions

An examination of the diagrams (Fig. 7 and 8) shows a discontinuity in the outline of each of them, corresponding to an abcissa of the order of 40 dyn cm-1 . Our interpretation is based on the single hypothesis :

$\Pi_e = 0$ when $\gamma_{LV} > 40$ dyn cm^{-1} and $\Pi_e \neq 0$ when $\gamma_{LV} < 40$ dyn cm^{-1}

Our conclusions are easily illustrated on Fig. 9.

On fig. 9 the experimental curve has been expressed as :

$$W_A - W_C = 2(\gamma_o - \gamma_{LV}) - \Pi_e \tag{11}$$

and for each value of the liquid surface tension, the value of Π_e can then directly be measured.
In the representation $\cos\theta = f(\gamma_{LV})$ the general equation is :

$$\cos\theta = (2\gamma_o - \gamma_{LV} - \Pi_e)/\gamma_{LV} \tag{12}$$

And for the expression of γ_{SL}, we have obtained :

$$\gamma_{SL} = \gamma_{SO} + \gamma_{LV} - 2\gamma_o \tag{13}$$

If we consider a liquid of surface tension γ_o there would be a null contact angle if

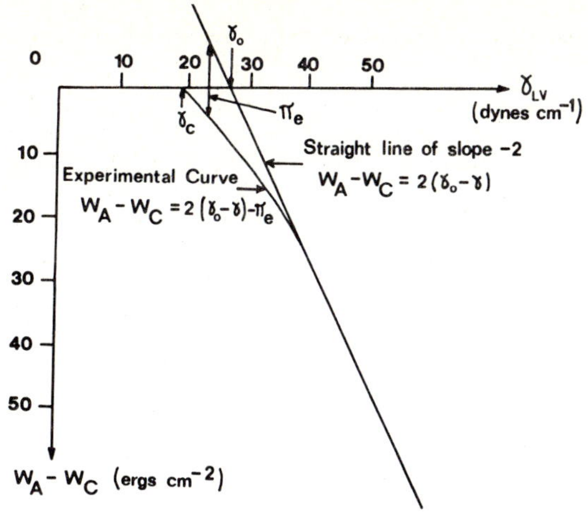

Fig. 9. Equation of the experimental curve and representation of the quantity Π_e

there were no adsorption of the vapor of the liquid. For such a liquid we then expect a null solid liquid interfacial tension :

$$\gamma_{SL} = 0 \quad \text{when} \quad \gamma_{LV} = \gamma_o$$

Combined with (13) we then obtain :

$$\gamma_{SO} = \gamma_o \tag{14}$$

And :

$$\gamma_{SL} = \gamma_{LV} - \gamma_{SO} \tag{15}$$

In conclusion we can see that the complexity of contacts angle phenomena does not arise from the expression of γ_{SL} (see eq. 15) but from adsorption, which often produces a non-negligible spreading pressure Π_e.

References

1 T. YOUNG, Miscellaneous works, Vol. 1, G. Peacock, Ed., London : J. Murray (1855)

2 J. CHAPPUIS, "Contact Angles" in "Multiphase Science and Technology", Vol. 1, G.F. Hewitt, J.M. Delhaye and N. Zuber Eds., Hemisphere Publ. Corp., pp. 387-505 (1982)

3 R.E. JOHNSON Jr. and R.H. DETTRE :"Wettability and contact angles" in"Surface and Colloîd Science", Vol. 2, E. Matijevic, Ed., New York,Wiley. Interscience pp. 85-153 (1969)

4 A.W. NEUMANN and R.J. GOOD : J. Colloîd Interface Sci. 38, 341 (1972)

5 D.H. BANGHAM and R.I. RAZOUK : Trans. Faraday Soc. 33, 1459 (1937)

6 A. DUPRE : Théorie Mécanique de la chaleur, Paris : Gauthier Villars (1869)

7 H.W. FOX and W.A. ZISMAN : J. Colloîd Sci., 5, 514 (1950)

8 W.A. ZISMAN, Adv. Chem. Ser., 43, 1 (1964)

9 R.J. GOOD, J. Am. Chem. Soc., 74, 504 (1952)

10 R.J. GOOD, J. Colloîd Interface Sci., 59, 398 (1977)

11 F.M. FOWKES, Adv. Chem. Ser., 43, 99 (1964)

12 F.M. FOWKES in"Wetting", Soc. Chem. Ind. Monograph, 25, 3 (1967)

13 A.W. NEUMANN, Adv. Colloîd Interface Sci., 4, 105 (1974)

Binary Liquid Gels

J.V. Maher*
Department of Physics and Astronomy, University of Pittsburgh,
Pittsburgh, PA 15260, USA

In a recent paper W.I. Goldburg, D.W. Pohl, M. Lanz and I reported the fabrication of several binary liquid gels, swollen gels whose solvent is a binary liquid mixture near its critical temperature [1]. One might expect novel features for such gels, since the presence of polymer strands spaced on average at something like the gel-pore size should inhibit hydrodynamic degrees of freedom in the liquid's phase separation, and the polymer strands in turn should have a preference for one of the separated liquid phases. The minimum effect one might expect is a slowing down of the gravitationally-induced late stages of spinodal decomposition by the mechanical presence of the polymer network. Much more dramatic effects would be expected if the polymer plays a more active role, and forms a ternary thermodynamic system with the solvents. Our observation is that the polymer strands interact with the liquid mixture strongly enough to frustrate the phase separation and dramatically alter the light scattering properties of the system in the critical region ; however, the interaction is sufficiently weak that light scattering does not resemble that expected of a ternary system.

The first of our binary liquid gels were agarose samples (1% by weight) whose solvents were either isobutyric acid + water (IBW) or 2,6 lutidine + water (LW). "Coexistence curves" for these gels are shown in Figure 1. The main effect on the liquid mixture of imprisonment in the gel is to broaden the phase-transition dramatically . In each of the systems of Figure 1 the dashed curve and open circles represent the coexistence curve of the free liquid, the solid curves represent the subjectively determined points at which the gel becomes : 1, slightly opalescent ; 2, strongly opalescent ; and 3, opaque. In each sample the degree of opalescence depended on temperature and not on time. As soon as the sample reached thermal equilibrium, the degree of opalescence reached its indicated value, and this opalescence relaxed only on a very long time-scale (hundreds of hours for samples of dimension ~ 1 cm and tens of hours for ~ 1 mm sized samples). I will say more about relaxation times below, but in the interim I will regard the gels as remaining cloudy indefinitely.

For our light scattering studies we switched to polyacrylamide gels (7% by weight) because the polyacrylamide, PAC, shows very much weaker light scattering than does agarose, both with pure water as a solvent and with either of the solvent mixtures (IBW or LW) in the one phase region. The PAC samples are somewhat more difficult to make than the agarose, since gelation is much looser if the liquid mixture is added before cross-linking. We were, however, able to make the samples by cross linking the polymer in water and diffusing in the second liquid later. This produced excellent IBW-PAC gels which gave extremely weak light scattering at temperatures above 30°C. Very clear LW-PAC samples were also made by diffusion, but the lutidine did appear to attack the top surface of the gel, and streaks of a syrup of dissolved gel were generally evident in the supernatant liquid. While we did not measure the "coexistence curves" as carefully as for the agarose sample, we found the IBW-PAC samples followed the shape shown on the right-hand side of Figure 1 rather closely.

*Supported by the U.S. D.O.E. under grant n° DE-F602-84ER45131.

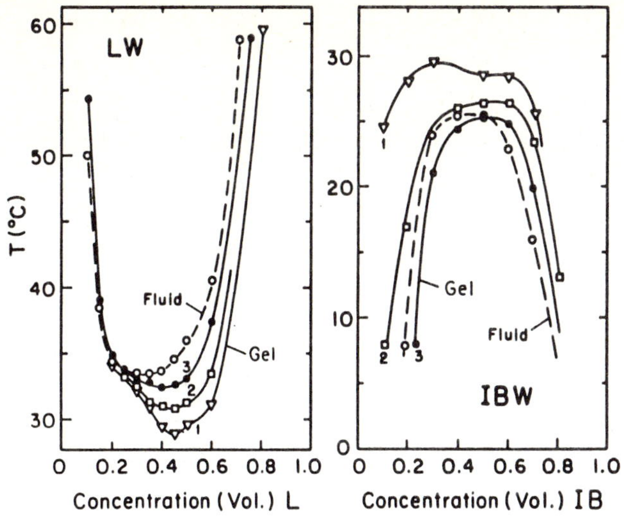

Figure 1.

Our extensive light scattering work has been performed on IBW-PAC samples of near critical (45 % isobutyric acid) composition. At this composition.the gel has deswollen to ~ 90 % of its original volume (if the gel is placed in pure water after fabrication, it eventually equilibrates at almost exactly twice its original swelling ; adding NaCl to make a 0.1 molal solution only deswells the sample to ~ 1.7 times the original swelling).

The light scattering measurements show two dramatic features :

1) temporal fluctuations are frozen out. Correlation times of order 10^{-4} and 10^{-3} sec were observed, and these are close to those reported for PAC + water [2] and for the free binary liquid mixture [3]. However, these fluctuations came from a tiny part of the scattering signal ; the main signal is essentially dc with fluctuations appearing only on times scales ~ 1 hour.

2) As shown by the "coexistence curves" the strong, approximately dc signal increases steadily over a broad range of temperatures. Figure 2 shows scattering intensity as a function of temperature of several scattering angles,and Figure 3 shows full angular distributions for several temperatures. The angular distributions at high temperature fall off as a power law $\sim q^{-3}$. As the temperature is reduced, intensity increases at all angles,but enough more rapidly at the larger angles that the power-law shape evolves into a Lorentzian (or some power of Lorentzian).

Isobutyric acid and water are closely matched in index of refraction,so the free IBW mixture shows intense equilibrium light scattering only near ($\lesssim$ 100 mK) the critical point. While quenching the free-liquid system deep into the two-phase region does give strong transient scattering while the spinodal decomposition process unfolds, this scattering disappears rapidly as the phase separation proceeds. Thus,the broad temperature-range of the gel scattering is so different from that of the free-liquid system that several possible models of the gel behavior can immediately be dismissed. For example, the gel cannot merely be offering a passive steric blocking of droplet growth,because this could not explain the observed broadening of the transition. On the other hand, if the gel structure were a full-participant in a ternary system,one would expect from results on binary-liquid-polymer solutions[4] to see a greater deformation of the coexistence curve and no broadening of the transition. Thus the gel matrix appears to be playing an active but not-too-strong role in the interaction. If this role were to be that of a chemical potential field with nonzero average over macroscopic lengths, it could take the system away from the critical point, broadening but weakening the critical scattering. However, in the binary liquid case (as opposed to magnetic systems) the trajectory should eventually reach the coexistence curve, and a strong first order phase-transition should be observed. This does not fit our case ; our observed scattering is too strong for us to be very far from the critical point, and no evidence of a first order transition is observed.

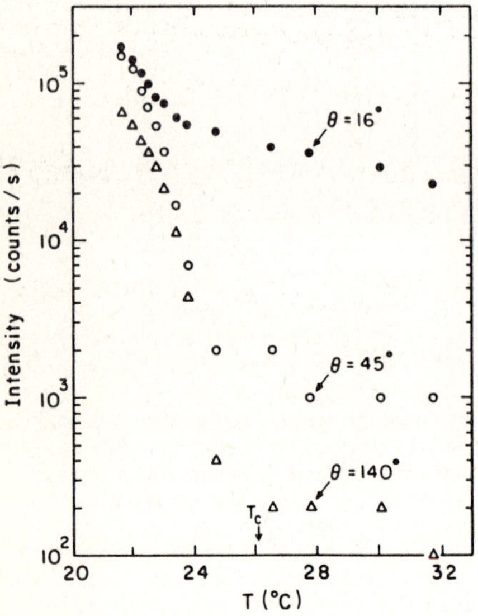

Figure 2

A remaining possibility is that the gel matrix acts like a random field whose average value is zero [5,6]. Such a random field could broaden the transition, reduce the amplitude of the temporal fluctuations and, if strong enough, destroy the transition [7] (assuming that the lower critical dimension is two rather than three). The expected intensity of scattering from this random field should follow a form [6] ;

$$I(q) = \frac{A}{q^2 + K^2} + \frac{B}{(q^2 + K^2)^2} \qquad (1)$$

where q is the wave number of the scattering and

$$K^2 = \xi^{-2} + (K^2 \ell^4)^{-1} \qquad (2)$$

where $\xi = \xi_o (T - T_c/T_c)^{-\nu}$ is the free liquid correlation length ($\nu = 0.62$) and ℓ is a length characteristic of the interaction between the liquids and the gel matrix. de Gennes [6] suggests

$$\ell = aN^{1/2}/\Delta \qquad (3)$$

where a is the monomer diameter, N is the number of monomers between cross links and Δ is related to the difference between the polymer-water interaction and the polymer-isobutyric acid interaction.

Since the Lorentzian term in equation (1) should be associated with temporal fluctuations, our lack of temporal fluctuations inclines us to say that in eqn. (1) A is very small, and the angular distributions seen on the left-hand side of Figure 3 are really Lorentzian-squared, L^2 (the data could support either an L or an L^2 interpretation). On the assumption that the angular distributions are indeed L^2 we have extracted values of K from our data ; these are shown in Figure 4. Different symbols are used for data taken during different passes through the temperature range. Temperature shifts have been applied to all but one set of data, as discussed in reference 1. Two points marked "33" show the results for a gel with larger pore size ; all the other data were taken with the same sample whose pore size should be ~ 25 Å and whose solvent is globally near critical composition. The solid curve was calculated from a fit of equation 2 to the open circles. This fit used $\xi_o = 2$ Å (appropriate for IBW in the two-phase region) and gave $T_c = 24.34°C$ (the free liquid $T_c = 26.12°C$) and $\ell = 119$ Å. Since a is of order 10 Å, $\ell \sim 120$ Å implies that $N^{1/2}/\Delta$ is of order 10. It is difficult to estimate Δ since nothing is known about the isobutyric acid-polyacrylamide interaction, but $N^{1/2}/\Delta \sim 10$ seems a reasonable number for this strongly crosslinked gel, and a not too great difference in monomer interaction with the two liquids. The results of Figure 4 are thus encouraging for the random field picture. Some of the scatter in the length scales, K, as temperature is lowered may result from slipping into metastability traps [8]. Since B/A should be $\sim 1/\ell^2 \sim 10^{12}$ cm^{-2}, the extracted value of ℓ is consistent with our lack of temporal fluctuations when $K \sim 10^5$ cm^{-1}.

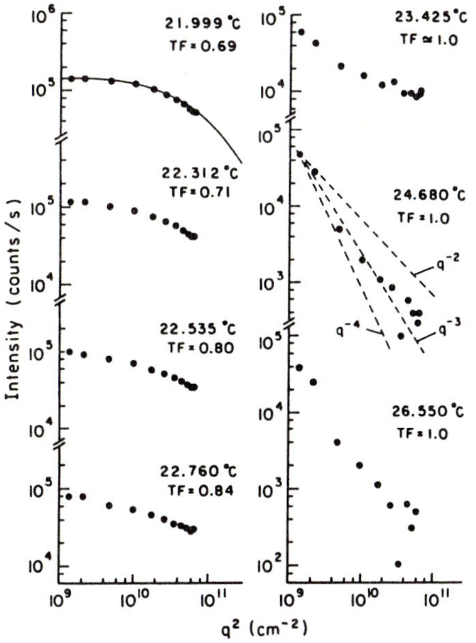

Figure 3

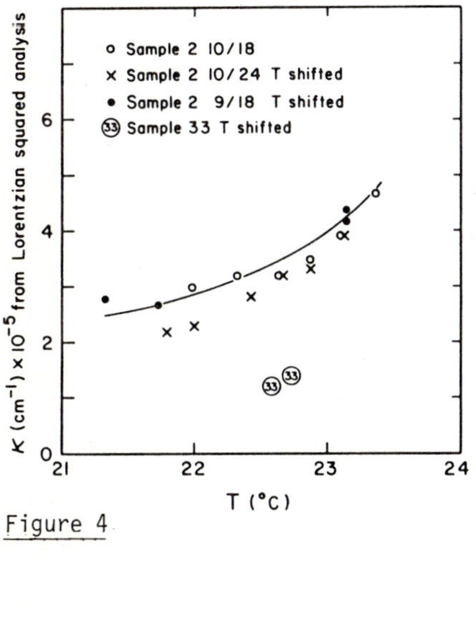

Figure 4

On the other hand, we do not understand the power law angular distributions we observe at higher temperatures. If the high temperature scattering came primarily from the gel matrix pore size, or from either of the two terms of equation 1, the angular distributions would be isotropic, since K should exceed the inverse wavelength of light.

It would be very interesting to be able to relate the very long relaxation times we observe for our gels to the long times predicted by Villain and other [8] to escape metastable traps. Unfortunately, we cannot separate such time scales from the times needed for the gel to come to equilibrium with its supernatant liquid and, as mentioned above, our long relaxation time does decrease with sample size. This, along with the fact that we have never observed a meniscus within the gel, suggests that equilibration with the supernatant liquid proceeds faster than the evolution of the mestasable domains, setting a lower limit of ~ 10 hours for the metastable time-scale.

We are currently following several lines of investigation to learn more about length scales in these gels, and thus further test the random field picture. The most complete of our recent results comes from a neutron scattering measurement which involves collaborators [9] from the University of Delaware, the Exxon Research Laboratories and the University of Pittsburgh. We chose to use PAC gels made with D_2O and 2,6 lutidine because this mixture was easy to keep in the one phase region during transport, and the gel turns cloudy while the supernatant liquid is still in one phase. Unfortunately, this system is not one whose light scattering properties or whose coexistence curve we know well, although we do know that (at 33% lutidine by volume which is very close to the critical composition of the free liquid) the gel turns slightly cloudy near 25°C and is opaque at 27°C. The supernatant liquid phase separates near 28°C. The neutron scattering results show a Lorentzian background of characteristic length ~ 20 Å which is close to the pore size and presumably comes from the polymer matrix. Superimposed on this is a Lorentzian whose length scale grows with increasing temperature for $T \leq 23.2$°C. Above T = 24.4°C the neutron scattering distributions are Lorentzian squared (q^{-4}) with too long a characteristic length for the neutrons to measure. At these same temperatures, preliminary light scattering results [10] indicate values of K in the range $1\text{-}5 \times 10^5$ cm^{-1}. The neutron scattering measurements were made in relatively large temperature steps, and the crossover from Lorentzian to Lorentzian-squared is sufficiently abrupt that we never see both in one angular distribution. If the characteristic wavenumber K seen for the Lorentzian below 24°C is to comply with equation 2, the critical temperature would have to be shifted down by ~ 15 K from the free liquid value. This is to be compared with the ~ 2 K shift for the IBW system discussed above. However, as noted above, the lutidine attacks the gel and might be expected to produce somewhat bizarre behavior. Despite these problems, the LW-PAC gels do seem to have much in common with the IBW-PAC gels. Both show a very broadened transition from being very weak light scatterers to giving strong scattering in a Lorentzian-squared pattern.

From the discussion above, one can see that the observed properties of our binary liquid gels suggest that the gel matrix acts in some important respects like a random field, but there is not detailed agreement with the simplest picture provided by equation 1-3. We are currently working to measure light scattering from the LW gels to coordinate with the neutron scattering (this light scattering work is currently hindered by problems of quantitative but not qualitative irreproducibility, presumably of the sort we learned to overcome in working with IBW gels, but exacerbated by problems of working with the gels at very low temperature when we must reach equilibrium deep in the one phase region). We are also measuring gels of much larger pore size, and looking at light scattering from entangled ($c \gtrsim 100^*$) solutions.

REFERENCES

1. J.V. Maher, W.I. Goldburg, D.W. Pohl and M. Lanz, Phys. Rev. Letters 53, 60 (1984).
2. T. Tanaka, L. Hocker and G. Benedek, J. Chem. Phys. 59, 5151 (1973).
3. B. Chu et al., Phys. Rev. A 7, 353 (1973).
4. See, for example, Y. Izumi et al., Macromol. Chem. 180, 2483 (1979).
5. see, for example, A. Aharony, J. Magn. Mater. 31-34, Pt. 3, 1434 (1983); and other references in this volume.
6. F. Brochard and P.G. de Gennes, J. Phys. (Paris) Lett. 44L, 785 (1983); P.G. de Gennes, J. Phys. Chem. 88, 6469 (1985).
7. M. Hagen et al, Phys. Rev. B 28, 2602 (1983); C. De Dominicis, Phys. Rev. B 18, 4913 (1978) ; Y.Y. Goldschmidt and B. Schaub, to be published.
8. J. Villain, Phys. Rev. Lett. 52, 1543 (1984) ; D. Andelman and J.F. Joanny, to be published.
9. The neutron scattering measurements have been performed in collaboration with W.I. Goldburg, J.S. Huang, S. Satiga and S. Sinha.
10. These measurements are progressing in collaboration with L. Rendon and W.I. Goldburg.

Phase Separation and Metastability in Binary Liquid Mixtures in Gels and Porous Media

D. Andelman[1] and J.F. Joanny
Physique de la Matière Condensée[2], Collège de France,
F-75231 Paris Cedex 05, France

1. Introduction

Recently, experiments were performed [1] with gels immersed in binary liquid mixtures such as water-lutidine or water-isobutyric acid chosen for their well studied demixing curves and consolute points conveniently located at room temperatures. The gels used are either agarose or polyacrylamide.

One of the most striking phenomena that is observed as the sample approaches the demixing curve from the one-phase region is a gradual appearance of opalescence in a range of a few degrees. This opalescence should be distinguished from critical opalescence seen in the pure A/B mixtures only in the vicinity of the consolute point (in a range of temperatures not wider than mK°). On the contrary, the opalescence here persists even in the two-phase region of the pure A/B mixture, and one does not observe a complete phase separation inside the gel. Nevertheless, since the onset of opalescence resembles qualitatively the demixing curve of the pure A/B mixture, it was suggested [1,2] that a phenomenon of droplets pinning by the random gel structure inhibits a complete phase separation. This effect is quite similar to domain states that are seen [3] in diluted antiferromagnets in presence of a magnetic field. In those systems,as one cools down in presence of the field, neutron scattering experiments show that even at low temperatures the system is composed of domains that are pinned by the random field.

The gel + A/B system has two characteristic lengths : one is the mesh size of the gel (L) and the other is the bare fluid-fluid correlation length (ξ). The behavior is determined by a comparison between these two lengths. In section 2 we show that for $\xi > L$, an analogy can be made between the gel + A/B mixture and random field in magnetic systems After discussing the equilibrium state in section 3, we generalize a theory proposed by VILLAIN [4] and apply it to the gel case. An explanation of the observed opalescence via droplets pinning by the random structure is given in section 4 and our conclusions in section 5.

Finally we would like to note that although reported experiments exist only for the gel + A/B systems, similar behavior (at least in the regime $\xi > L$) is expected for A/B mixtures in porous solids.

2. Connection with Random Fields

Pure A/B liquid mixtures are usually described by lattice gas models; we would like to extend these models to binary liquid mixtures in a gel. The gel always has an heterogeneous chemical structure. We focus here on rigid gels,which thus act on the liquid mixture as a source of quenched disorder (the same assumption is correct for porous solids). In the opposite limit of flexible gels, thermal fluctuations become important,and one should consider the binary liquid mixture in the gel as a ternary system.

---

[1]Also : Corporate Research Science Laboratory, Exxon Research and Engineering Co., Annandale, New Jersey 08801, U.S.A.
[2]Laboratoire rattaché au C.N.R.S. : U.A. 792.

The main effect of the gel on the A/B mixture is a preferential adsorption [2] of one of the liquid components (in our case the water-component A).

We will describe the system in the regime $L < \xi$ (tight gels or temperatures close to the consolute point). Since ξ is large, the system can be divided into blocks of size L (coarse graining) ; in each block there is an effective random field H(r) proportional to the average gel concentration in the block. The average of H(r), $\langle H(r) \rangle$, expresses the total preference of the gel plus a contribution from the concentration difference between the liquid inside the gel and in the A/B reservoir into which the gel is immersed. We first discuss the case $\langle H \rangle = 0$ and then include the effect of a non-zero $\langle H \rangle = h$. The order parameter of the system (analogous to the magnetization in magnetic systems) may be defined as $M = (C_A - C_B)/(C_A + C_B)$ where C_A (C_B) is the number density of A (B). The strength of the random field is characterized by the r.m.s. deviation H.

3. Equilibrium Behavior

For convenience,we study a system with an upper critical point for which phase separation occurs at low temperatures. As the system is cooled down from the one-phase region, at thermodynamical equilibrium, a phase separation should be seen as well as a consolute point that depends on the strength of the random field, $T_c(H) < T_c(0)$. This statement is based on the fact that the lower critical dimension (d_1) of the random field Ising model is believed to be two, $d_1 = 2$ [5]. Thus demixing occurs in d = 3 for small enough random fields,and the shift of the critical temperature is expected to scale as $T_c(H) - T_c(0) \sim H^{-2/\gamma}$ where γ is the compressibility exponent. In the gel, the strength of the random field (i.e. the fluctuations of the gel structure) cannot be varied in a controlled way, thus such a scaling law cannot be verified. Moreover, good estimates for H are difficult to obtain.

In actual experiments, the gel is in contact with an A/B reservoir and the demixing curve is shifted towards higher concentrations of B since the concentration inside the gel and in the reservoir are not the same. Rather in order to balance the chemical potential difference due to the gel preference, the system reaches its equilibrium by creating a difference in concentrations ; C_B (in the gel) $< C_B$ (in the reservoir).

To summarize, we expect to see for $\xi > L$ and rigid gels, phase separation in the gel + binary mixture at thermodynamical equilibrium for small enough randomness. Additionally, a shift towards higher C_B values and towards lower temperatures is expected for the demixing curve as a function of the *total* (gel + reservoir) concentration of B. These conclusions coming from equilibrium considerations are not very satisfactory,since they do not provide an explanation to the observed opalescence. In the next section such an explanation based on metastability is proposed.

4. Metastability and Droplet States

In this section we follow some ideas of VILLAIN [4] and propose a theory that explains how the system can be trapped in metastable domain states although phase separation is energetically more preferable. The theory is constructed by assuming a rapid quench from high temperatures to temperatures below T_c. Our starting point is to write the energy barrier between two metastable states. We concentrate on a single domain of "spins down" (a droplet rich in B) with a well-defined radius R embedded in a region of "spins up" (rich in A). At equilibrium,this domain disappears, since phase separation is expected in three dimensions ($d_1 = 2$). However, due to effects of random field pinning [4,6] changes in the domain interface are carried out by jumps over energy barriers that exist between "successive" metastable states. Each one of these states is characterized by an average displacement w (which should be distinguished from the width of the interface (the correlation length ξ). As long as $R > \xi$ the domain is well defined. An expression of the energy barrier would include the following three terms [6] :

$$\Delta F(w,b) = -g\left[b^{d-1} \left(\frac{w}{b}\right)^2 + b^{d-1} \left(\frac{w}{R}\right)\right] + HM \left(b^{d-1}w\right)^{1/2}, \tag{1}$$

where g is the interfacial tension between A and B, H is the strength of the random field, M is the magnetization and b is the length scale ($b < R$) for which we consider the interface roughness (w). For each b, maximizing ΔF with respect to w gives the maximal energy barrier :

$$\Delta F_{max}(b) = g\,(HMb/g)^{4/3} \quad \text{when } w/R > (w/b)^2 \text{ and } d = 3 \tag{2}$$

and

$$\Delta F_{max}(b) = Rg\,(HM/g)^2 \quad \text{when } w/R < (w/b)^2. \tag{3}$$

ΔF_{max} is thus an increasing function of b. The minimal domain size is obtained by comparing ΔF_{max} in (3) with the available thermal fluctuation kT :

$$R_{min}(H) = gkT\,(HM)^{-2} \tag{4}$$

For $R < R_{min}$, $kT > \Delta F_{max}$ for all b and the domain will disappear due to thermal fluctuations, whereas for $R > R_{min}$, $kT > \Delta F_{max}$ only for the smaller lengths thus pinning of domains is possible for some larger length scales ($b < R$). Equation (4) expresses the temperature-dependence of R_{min} as the temperature is quenched rapidly from high T. For slow cooling, a freezing of domains is rather expected at T close to T_c since at lower temperatures there is even less thermal fluctuation. This could be the source of opalescence in the gel + A/B experiments.

We consider now a random field with non-zero average h [6] proportional to the distance, in terms of concentration, to the coexistence curve. For the energy barrier (1), a constant field will add a term of the form $hM\,(wb^{d-1})$ which has the same dependence on w and b as the second term in (1), namely it will just renormalize the radius R_{min}. In the presence of a constant field h, $R_{min}(H,h)$ is related to $R_{min}(H,0)$

$$R_{min}^{-1}(H,h) = R_{min}^{-1}(H,0) - hM/g \tag{5}$$

or using (4)

$$R_{min}^{-1}(H,h) = (HM)^2/gkT - (hM)/g < R_{min}^{-1}(H,0). \tag{6}$$

For a fixed H, there is a line in the T - h plane for which $R_{min}(H,h) \to \infty$. This "freezing line"

$$kT_g = (HM)^2/(hM) \tag{7}$$

is the separation between equilibrium (reversible) behavior at high T and h and between metastable (frozen) behavior at low T and h.

For the gel + A/B this means that freezing of metastable states will occur all around the consolute point ($h \neq 0$). We believe that this is the reason why the onset of opalescence follows the demixing curve. Hysteresis, which usually is associated with metastability, can be seen if one cools in constant field $h > 0$ or one cools with $h = 0$ and then applies a constant field. For the two procedures different curves for the onset of opalescence is predicted [7].

5. Conclusions

The behavior of binary liquid mixtures immersed in a gel cannot be explained only from equilibrium considerations, and we rather propose an explanation related to random field effects. For $\xi > L$, the opalescence seen in these systems is explained as a phenomenon of domains freezing due to the random structure of the gel. In the other limit $\xi < L$, the problem resembles biphasic flows in porous media [2]. Experimentally, this situation can be achieved by cooling a rich-A gel ($M \simeq M_0$) to low temperatures and then increasing the concentration of B in the reservoir ($M \simeq -M_0$). In this regime, $\xi < L$, interesting quantities such as the threshold of macroscopic invasion of the other phase into the gel depends on the details of the gel structure [8].

As for additional experiments, it will be interesting to perform experiments where the liquids concentration is also a varying parameter as well as the temperature. Moreover, a method that measures directly the concentration inside the gel (rather than the total gel + reservoir) will facilitate making more quantitative predictions.

References

1. J.V. Maher, W.I. Goldburg, D.W. Pohl, and M. Lanz: Phys. Rev. Lett. 53, 60 (1984)
2. P.G. de Gennes: J. Phys. Chem. 88, 6469 (1985)
3. M. Hagen, A. Cowley, S.K. Satija, H. Yoshizawa, G. Shirane, R.J. Birgeneau, and H.J. Guggenheim: Phys. Rev. B 28, 2602 (1983); D.P. Belanger, A.R. King, and V. Jaccarino: Phys. Rev. Lett. 48, 1050 (1982)
4. J. Villain: Phys. Rev. Lett. 52, 1543 (1984)
5. G. Grinstein and S.-k. Ma: Phys. Rev. B 28, 2588 (1983); J.Z. Imbrie: Phys. Rev. Lett. 53, 1747 (1984)
6. D. Andelman and J.F. Joanny: Collège de France preprint (1985)
7. D. Andelman and J.F. Joanny (to be published)
8. F. Brochard and P.G. de Gennes: J. Phys. Lett. (Paris) 44, 785 (1983)

Index of Contributors